structure of
Hilbert Space
Operators

Chunlan Jiang
Hebei Normal University, China
Zongyao Wang
East China University of Science and Technology, China

structure of
Hilbert Space
Operators

 **World Scientific**

NEW JERSEY • LONDON • SINGAPORE • BEIJING • SHANGHAI • HONG KONG • TAIPEI • CHENNAI

Published by

World Scientific Publishing Co. Pte. Ltd.

5 Toh Tuck Link, Singapore 596224

USA office: 27 Warren Street, Suite 401-402, Hackensack, NJ 07601

UK office: 57 Shelton Street, Covent Garden, London WC2H 9HE

British Library Cataloguing-in-Publication Data
A catalogue record for this book is available from the British Library.

STRUCTURE OF HILBERT SPACE OPERATORS

ISBN-13 978-981-256-616-4
ISBN-10 981-256-616-3

Printed in Singapore

Preface

In the matrix theory of finite dimensional space, the famous Jordan Standard Theorem sufficiently reveals the internal structure of matrices. Jordan Theorem indicates that the eigenvalues and the generalized eigenspace of matrix determine the complete similarity invariants of a matrix. It is obvious that the Jordan block in matrix theory plays a fundamental and important role. When we consider a complex, separable, infinite dimensional Hilbert space $\mathcal{H}$ and use $\mathcal{L}(\mathcal{H})$ to denote the class of linear bounded operators on $\mathcal{H}$, we face one of the most fundamental problems in operator theory, that is how to build up a theorem in $\mathcal{L}(\mathcal{H})$ which is similar to the Jordan Standard Theorem in matrix theory, or how to determine the complete similarity invariants of the operators. Two operators A and B in $\mathcal{L}(\mathcal{H})$ are said to be similar if there is an invertible operator X, XA is equal to BX. The complexity of infinite dimensional space makes it impossible to find generally similarity invariants. The main difficulty behind this is that it is impossible for people to find a fundamental element in $\mathcal{L}(\mathcal{H})$, similar to Jordan's block, so as to construct a perfect representation theorem. We appreciate such a mathematical point of view as people's being not powerful enough to deal with a complicated mathematical problem then that reflects their lack of sufficient knowledge and understanding of some fundamental mathematical problems. It is because of the sufficient study of the *-cyclic self-adjoint (normal) operators that people have set up the perfect spectral representation theorem for self-adjoint (normal) operators and commutative C^*-algebra. It is also because of the introduction of the concept of irreducible operators by Halmos, P.R. in 1968 that Voiculescu, D. obtained the well-known Non-commutative-Weyl-von Neumann Theorem for general C^*-algebra. But irreducibility is only a unitary invariant and can not reveal the general internal structure of operator algebra and non-self-adjoint op-

erators. Since the 1970s, some mathematicians have showed their concern
for the problem on Hilbert space operator structure in two aspects. In one
aspect, the mathematicians, such as Foias, C., Ringrose, J.R., Arveson,
W.B., Davidson, K.R. etc. have made great efforts to study the structures of different classes of operators or operator algebras, such as Toeplitz
operator, weighted shift operator, quasinilpotent operator, triangular and
quasitriangular operators, triangular and quasitriangular algebras etc. In
the other aspect, they have set up the approximate similarity invariants
for general operators by introducing the index theory and fine spectral picture as tools. One of the most typical achievements, made by Apostol, C.,
Filkow, L.A., Herrero, D.A. and Voiculescu, D. is the theorem of similarity
orbit of operators. This theorem suggests that the fine spectral picture is
the complete similarity invariant as far as the closure of similarity orbit of
operators are concerned. Besides, in the 1970s, Gilfeather, F. and Jiang,
Z.J. proposed the notion of strongly irreducible operator ((SI) operator)
respectively. And Jiang, Z.J. first thought that the (SI) operators could be
viewed as the suitable replacement of Jordan block in $\mathcal{L}(\mathcal{H})$. An operator
will be considered strongly irreducible if its commutant contains no nontrivial idempotent. In the theory of matrix, strongly irreducible operator
is Jordan block up to similarity. Through more than 20 years' research,
the authors and their cooperators have founded the theorems concerning
the unique strongly irreducible decomposition of operators in the sense of
similarity, the spectral picture and compact perturbation of strongly irreducible operators, and have formed a theoretical system of (SI) operators
preliminarily. But, with the research going deeper and deeper, the authors
and their cooperators are badly in need of new ideas and new tools to be
introduced so as to further their research. In the 1980s, Elloitt, G. classified
AF-algebra successfully by using of K-theory language, which stimulated
us to apply the K-theory to the exploration of the internal structure of operators, which features this book. In 1978, Cowen, M.J. and Douglas, R.G.
defined a class of geometrical operators, Cowen-Douglas operator, in terms
of the notion of holomorphic vector bundle. They, for the first time, applied
the complex geometry into the research of operator theory. Cowen, M.J.
and Douglas, R.G. have proved the Clabi Rigidity Theorem on the Grass-
man manifold, defined a new curvature function and indicated that this
curvature is a complete unitary invariant of Cowen-Douglas operators. It is
these perfect results that have inspired us since 1997, to combine K-theory
with complex geometry in order to seek the complete similarity invariants
of Cowen-Douglas operators and their internal structures. Cowen-Douglas

operator is a class of operators with richer contents and contains plenty of triangular operators, weighted shift operators, the duals of subnormal operators and hypernormal operators. Its natural geometrical properties support a very exquisite mathematical structure. Based on some of our successful research on Cowen-Douglas operators, we have made headway in the study of other operator classes.

This monograph covers almost all of our own and our cooperators' research findings accumulated since 1998. The book consists of six chapters. Chapter 1 provides the prerequisites for this book. Chapter 2 explains the Jordan Standard Theorem again in K_0-group language, and gives readers a new point of view to understand the complete similarity invariants in the theory of matrix. And this chapter also helps readers get well-prepared for the study of operator structure in terms of K-theory in later chapters. Chapter 3 mainly discusses how to set up the theorem on the approximate (SI) decomposition of operators by using the (SI) operators as the basic elements. Meanwhile, to meet the needs of the study of the structure, it also reports the relationship between (SI) operators and the compact perturbation of operators, and proves that each operator is a sum of two (SI) operators. Chapter 4 describes the unitary invariants and similarity invariants of operators in K_0-group language by observing the commutants. This chapter contains the following four aspects: (1) Gives a complete description of the unitary invariants of operators using K_0-group and lists some properties of lattices of reducing subspaces of operators. (2) Illustrates the establishment of the relationship between the unique (SI) decomposition of operator up to similarity and the K_0-group of its commutant, and at the same time, carefully states the complete unitary invariants and complete similarity invariants, and the uniqueness of (SI) decomposition of the operator weighted shift and analytic Toeplitz operators using the results of (1) and (2). (3) Makes a concrete description of the commutant of (SI) Cowen-Douglas operators by using complex geometry. (4)Discusses Sobolev disk algebra, the internal structure of the multiplication operators on it and their commutants by using Sobolev space theory, complex analysis and the results in (3). Chapter 5 focuses the discussion mainly on the complete similarity invariants of Cowen-Douglas operator and proves that the K_0-group is the complete similarity invariant of it. In addition, our discussion is extended to the other classes of operators which are related to Cowen-Douglas operators. Chapter 6 concerns some applications of operator structure theorem, including the determination of K_0-group of some Banach algebras, the distribution of zeros of analytic functions in the

Structure of Hilbert Space Operators

unit disk and a sufficient condition for a nilpotent similar to an irreducible operator.

We would hereby like to give sincere thanks to all the following professors: Davidson, K.R., Douglas, R.G., Elloitt, G., Gong, G.H., Lin, H.X., Yu, G.L., Zheng, D.C., Ge, L.M. etc. For their many years' encouragement and support. We would like to give special thanks to Gong, G.H., Yu, G.L. and Ge, L.M., for, since 2000, both of them have enthusiastically lectured on K-theory and geometry in our seminar, which have enabled us to make greater progress with our research. We are also grateful to Academician Gongqing Zhang and professor Zhongqin Xu at Beijing university and professor Yifeng Sun at Jilin university. They have given us enormous concern and encouragement since the early days of our research. It is their encouragement and support that have encouraged us to unshakably finish the course of research. We also wish to thank Mr. Xianzhou Guo for the technical expertise with which he typed the manuscript of this monograph.

C.L. Jiang
Z.Y. Wang

Contents

Chapter 1

Background

In this chapter, we review briefly some of the facts about operator algebra and operator theory which will be needed to read this book. Most of the material can be found in books or papers such as [Admas (1975)], [Apostal, C., Bercobici, H., Foias, C. and Pearcy, C. (1985)], [Blanckdar, B. (1986)], [Conway, J.B. (1978)], [Cowen, M.J. and Douglas, R. (1977)], [Douglas, R.G. (1972)], [Herrero, D.A. (1990)], [Herrero, D.A. (1987)], [Jiang, C.L. and Wang, Z.Y. (1998)] and [Rudin, W. (1974)].

1.1 Banach Algebra

A Banach algebra is a Banach space $\mathcal{A}$ over $\mathbf{C}$ which is also an (associative) algebra over $\mathbf{C}$ such that $\|ab\| \leq \|a\| \|b\|$ for all a, b in $\mathcal{A}$. When $\mathcal{A}$ has a unit e, the spectrum of a $\sigma(a)$ (or $\sigma_{\mathcal{A}}(a)$ if $\mathcal{A}$ needs to be clarified) is the set $\{\lambda \in \mathbf{C} : \lambda e - a$ is not invertible in $\mathcal{A}$ $\}$. The left spectrum of a $\sigma_l(a)$ is the set $\{\lambda \in \mathbf{C} : \lambda e - a$ is not left invertible in $\mathcal{A}$ $\}$; the right spectrum of a $\sigma_r(a)$ is the set $\{\lambda \in \mathbf{C} : \lambda e - a$ is not right invertible in $\mathcal{A}$ $\}$. The resolvent set of a $\rho(a) := \mathbf{C} \backslash \sigma(a)$. The left and right resolvent set of a are $\rho_l(a) := \mathbf{C} \backslash \sigma_l(a)$ and $\rho_r(a) := \mathbf{C} \backslash \sigma_r(a)$ respectively. $\sigma(a)$ is a non-empty compact subset of $\mathbf{C}$ and $\sigma(a) = \sigma_l(a) \cup \sigma_r(a)$. Let f be holomorphic in a neighborhood Ω of $\sigma(a)$ and let c be a finite union of Jordan curves such that $ind_c(\lambda) = 1$ for every λ in $\sigma(a)$. Define

$$f(a) = \frac{1}{2\pi i} \int_c f(z)(ze - a)^{-1}dz.$$

Let $Hol(\sigma(a))$ denote the set of all functions which are holomorphic in a neighborhood of $\sigma(a)$. We have the following theorems.

Riesz Functional Calculus   *Let a be an element of a Banach algebra $\mathcal{A}$ with identity, then for every $f \in Hol(\sigma(a))$, $f(a)$ is well defined independent of the curve c. The mapping $f \mapsto f(a)$ is an algebra homomorphism and maps each polynomial $p(z) = \sum_{k=0}^{n} c_k z^k$ to $c_0 e + \sum_{k=1}^{n} c_k a^k$.*

Spectral Mapping Theorem   *For $f \in Hol(\sigma(a))$, $\sigma(f(a)) = f(\sigma(a))$.*

Upper Semi-continuity of the Spectrum   *Let a be an element of a Banach algebra $\mathcal{A}$ with identity. Given a bounded open set Ω, $\Omega \supset \sigma(a)$, there exists $\delta > 0$ such that $\sigma(b) \subset \Omega$, provided $\|a - b\| < \delta$ and $b \in \mathcal{A}$.*

Let $\mathcal{H}$ be a complex, separable, infinite dimensional Hilbert space and let $\mathcal{L}(\mathcal{H})$ denote the algebra of linear bounded operators on $\mathcal{H}$. For each $T \in \mathcal{L}(\mathcal{H})$, $\sigma(T), \sigma_l(T)$, $\sigma_r(T), \rho(T), \rho_l(T), \rho_r(T)$ and $f(T)$ are defined as above, where $f \in Hol(\sigma(T))$.

Riesz Decomposition Theorem   *Assume that $\sigma(T) = \sigma_1 \cup \sigma_2$, $\sigma_1 \cap \sigma_2 = \emptyset$, where σ_1, σ_2 are non-empty compact sets, then $\mathcal{H}$ is the direct sum of two invariant subspaces $\mathcal{H}_1$ and $\mathcal{H}_2$ of T, such that $\sigma(T|_{\mathcal{H}_i}) = \sigma_i$ and $\mathcal{H}_i$ is the range of Riesz idempotent corresponding to $\sigma_i (i = 1, 2)$, where $T|_{\mathcal{H}_i}$ is the restriction of T on $\mathcal{H}_i$.*

Let $\mathcal{A}$ be an abelian Banach algebra with identity. A multiplicative linear functional ϕ is an algebra homomorphism $\phi : \mathcal{A} \longrightarrow \mathbf{C}$ with $\|\phi\| = 1$. The collection $\sum$ of all maximal ideals of $\mathcal{A}$ is a compact Hausdorff space in the sense of weak-$*$ topology. The Gelfand transform $\hat{a}$ of a is the function $\hat{a} : \sum \longrightarrow \mathbf{C}$ defined by $\hat{a}(\phi) = \phi(a)$.

Gelfand's Theorem   *If $\mathcal{A}$ is an abelian Banach algebra with identity, $a \in \mathcal{A}$ and $C(\sum)$ is the space of continuous functions on $\sum$, then the Gelfand transform $\hat{a}$ of a belongs to $C(\sum)$ and $\sigma(a) = \{\hat{a}(\phi) : \phi \in \sum\}$. The mapping $a \longrightarrow \hat{a}$ is a continuous homomorphism of $\mathcal{A}$ into $C(\sum)$.*

Let $\mathcal{A}$ be a Banach algebra with identity. A two-sided ideal $rad\mathcal{A}$ of $\mathcal{A}$ is the Jacobson Radical if it is the intersection of all maximal left (right) ideals of $\mathcal{A}$. Equivalently, $rad\mathcal{A} = \{a : \sigma(ab) = \sigma(ba) = \{0\}$ for all $b \in \mathcal{A}\}$.

A C^*-algebra $\mathcal{C}$ is a Banach algebra with a conjugation operator $*$ such that $(a^*)^* = a, (ab)^* = b^* a^*, (\alpha a + \beta b)^* = \overline{\alpha} a^* + \overline{\beta} b^*$ and $\|a^* a\| = \|a\|^2$ for all $a, b \in \mathcal{C}$ and $\alpha, \beta \in \mathbf{C}$. A $*$-homomorphism ρ of a C^*-algebra $\mathcal{C}$ is a $*$-homomorphism from $\mathcal{C}$ into $\mathcal{L}(\mathcal{H}_\rho)$, where $\mathcal{H}_\rho$ is a Hilbert space. If $\mathcal{C}$ has an identity e and $\rho(e) = I$, then ρ is unital; if $ker\rho = \{0\}$, ρ is faithful. It is obvious that ρ is faithful if and only if ρ is a $*$-isometric isomorphism from $\mathcal{C}$ onto $\rho(\mathcal{C})$.

Gelfand-Naimark-Segal Theorem *Every abstract C^*-algebra $\mathcal{C}$ with identity admits a faithful unital $*$-representation ρ in $\mathcal{L}(\mathcal{H}_\rho)$ for a suitable Hilbert space $\mathcal{H}_\rho$, i.e., $\mathcal{C}$ is isometrically $*$-isomorphic to a C^*-algebra of operators. Furthermore, if $\mathcal{C}$ is separable, then $\mathcal{H}_\rho$ can be chosen separable.*

von-Neumann Double Commutant Theorem *Let $\mathcal{A} \subset \mathcal{L}(\mathcal{H})$ be a unital C^*-algebra. Then the closure of $\mathcal{A}$ in any of weak operator, strong operator and weak-$*$ topologies is the double commutant $\mathcal{A}''$, where*

$$\mathcal{A}'' = \{\mathcal{A}'\}'$$

and

$$\mathcal{A}'(T) = \{T \in \mathcal{L}(\mathcal{H}) : AT = TA \quad \text{for all} \;\; A \in \mathcal{A}\}.$$

We call $\mathcal{A}'(T)$ commutant of T.

1.2 K-Theory of Banach Algebra

K_0-**group.** Let $\mathcal{A}$ be a Banach algebra with identity, and let e and f be idempotents in $\mathcal{A}$. e and f are said to be algebraic equivalent, denoted by $e \sim_a f$, if there are $x, y \in \mathcal{A}$ such that $xy = e$ and $yx = f$; e and f are said to be similar, denoted by $e \sim f$, if there is an invertible $z \in \mathcal{A}$ such that $zez^{-1} = f$. It is obvious that $e \sim_a f$ and $e \sim f$ are equivalent relations. Let $M_\infty(\mathcal{A})$ be the set of all finite matrices over $\mathcal{A}$, $Proj(\mathcal{A})$ be the set of algebraic equivalent classes of idempotents in $\mathcal{A}$. Set $\bigvee(\mathcal{A}) = Proj(M_\infty(\mathcal{A}))$, then $\bigvee(M_n(\mathcal{A}))$ is isomorphic to $\bigvee(\mathcal{A})$. If p, q are idempotents in $Proj(\mathcal{A})$, $p \sim_s q$ if and only if $p \oplus r \sim_a q \oplus r$ for some $r \in Proj(\mathcal{A})$, then "$\sim_s$" is called stable equivalence. $K_0(\mathcal{A})$ is the Grothendieck group generated by $\bigvee(\mathcal{A})$ [B. Blackadar [1]]. The pair (G, G^+) is said to be an ordered group if G is an abelian group and G^+ is a subset of G satisfying

   i. $G^+ + G^+ \subseteq G^+$;
   ii. $G^+ \cap (-G^+) = \{0\}$;
   iii. $G^+ - G^+ = G$.

   An ordered relation "$\leq$" can be defined in G by $x \leq y$ if $y - x \in G^+$ for $x, y \in G$.

Isomorphism Theorem *Let $\mathcal{A}$, $\mathcal{A}_1$ and $\mathcal{A}_2$ be Banach algebras and*

$$\mathcal{A} = \mathcal{A}_1 \oplus \mathcal{A}_2,$$

then

$$\bigvee(\mathcal{A})\simeq\bigvee(\mathcal{A}_1)\oplus\bigvee(\mathcal{A}_2),\ \ K_0(\mathcal{A})\simeq K_0(\mathcal{A}_1)\oplus K_0(\mathcal{A}_2),$$

$$\bigvee(M_n(\mathcal{A}))\simeq\bigvee(\mathcal{A})\ \ and\ \ K_0(M_n(\mathcal{A}))\simeq K_0(\mathcal{A}),$$

where "$\simeq$" means isomorphism.

Let $\mathcal{A}$ and $\mathcal{B}$ be two Banach algebras and let α be a homomorphism from $\mathcal{A}$ into $\mathcal{B}$. Then there is a homomorphism α_* induced by α from $K_0(\mathcal{A})$ into $K_0(\mathcal{B})$.

Six-Term Exact Sequence   *Let $\mathcal{A}$ be a unital Banach algebra and let $\mathcal{J}$ be its ideal, then we have the following standard exact sequence*

$$0\longrightarrow\mathcal{J}\overset{i}{\longrightarrow}\mathcal{A}\overset{\pi}{\longrightarrow}\mathcal{A}/\mathcal{J}\longrightarrow0$$

and the following exact cyclic sequence

$$\begin{array}{ccc} K_0(\mathcal{J}) & \overset{i_*}{\longrightarrow}\ K_0(\mathcal{A})\ \overset{\pi_*}{\longrightarrow} & K_0(\mathcal{A}/\mathcal{J}) \\ \partial\uparrow & & \partial\downarrow \\ K_1(\mathcal{A}/\mathcal{J}) & \longleftarrow\ K_1(\mathcal{A})\ \longleftarrow & K_1(\mathcal{J}), \end{array}$$

where $K_1(\mathcal{B})$ is the K_1-group of Banach algebra $\mathcal{B}$.

1.3   The Basic of Complex Geometry

Let Λ be a manifold with a complex structure and let n be a positive integer. (E,π) is called a holomorphic vector bundle of rank n over Λ if π is a holomorphic map from E onto Λ such that each fibre $E_x=\pi^{-1}(x)$ is isomorphic to $\mathbf{C}^n(x\in\Lambda)$ and such that for each $z_0\in\Lambda$, there exist a neighborhood Δ of z_0 and holomorphic functions $e_1(z),e_2(z),\cdots,e_n(z)$ from Δ to E such that $e_1(z),e_2(z),\cdots,e_n(z)$ form a basis of $E_z=\pi^{-1}(z)$ for each $z\in\Delta$. The functions $e_1,e_2,\cdots,e_n$ are said to be a holomorphic frame for E on Δ. The bundle is said to be trivial if Δ can be assigned to Λ.

Let E and F be two holomorphic bundles over a complex manifold Λ. A map φ from E to F is a bundle map if φ is holomorphic and $\varphi:E_\lambda\longrightarrow F_\lambda$ is a linear transformation for every $\lambda\in\Lambda$.

A Hermitian holomorphic vector bundle E over Λ is a holomorphic bundle such that each fibre E_λ is an inner product space. Two Hermitian holomorphic vector bundles E and F over Λ are said to be equivalent if there exists an isometric holomorphic bundle map from E onto F.

Let $\mathcal{H}$ be a separable complex Hilbert space and let n be a positive integer. Denote $G_r(n, \mathcal{H})$, the Grassmann manifold, the set of all n- dimensional subspaces of $\mathcal{H}$. For an open connected subset Λ of $\mathbf{C}^k$, a map $f : \Lambda \longrightarrow G_r(n, \mathcal{H})$ is said to be holomorphic if at each $\lambda_0 \in \Lambda$ there is a neighborhood Δ of λ_0 and n holomorphic $\mathcal{H}$-valued functions $e_1(z), e_2(z), \cdots, e_n(z)$ such that $f(z) = \bigvee\limits_{j=1}^{n} \{e_j(z)\}$. If $f : \Lambda \longrightarrow G_r(n, \mathcal{H})$ is a holomorphic map, then an n-dimensional Hermitian holomorphic vector bundle E_f over Λ and a map ϕ can be induced by f, i.e.,

$$E_f := \{(x, z) \in \mathcal{H} \times \Lambda : x \in f(z)\}$$

and

$$\phi : E_f \longrightarrow \Lambda, \phi(x, z) = z \in \Lambda.$$

Given two holomorphic maps f and $g : \Lambda \longrightarrow G_r(n, \mathcal{H})$, we have two vector bundles E_f and E_g over Λ. If there exists a unitary operator U on $\mathcal{H}$ such that $g = Uf$, then f and g are said to be unitarily equivalent. If there is an open subset Δ of Λ such that $E_f|_\Delta$ is unitarily equivalent to $E_g|_\Delta$, then E_f and E_g are said to be locally unitarily equivalent.

Rigidity Theorem   *Let Λ be an open connected subset of $\mathbf{C}^k$ and let f and g be holomorphic maps from Λ to $G_r(n, \mathcal{H})$ such that*

$$\bigvee_{z \in \Lambda} f(z) = \bigvee_{z \in \Lambda} g(z) = \mathcal{H}.$$

Then f and g are unitarily equivalent if and only if E_f and E_g are locally unitarily equivalent.

1.4   Some Results on Cowen-Douglas Operators

Let Ω be a connected open subset of $\mathbf{C}$, n is a positive integer, the set $\mathcal{B}_n(\Omega)$ of Cowen-Douglas Operators of index n is the set of operators $T \in \mathcal{L}(\mathcal{H})$ satisfying

(i) $\Omega \subset \sigma(T)$;

(ii) $\operatorname{ran}(z - T) := \{(z - T)x : x \in \mathcal{H}\} = \mathcal{H}$ for each $z \in \Omega$;

(iii) $\bigvee\limits_{z \in \Omega} \ker(z - T) = \mathcal{H}$;

(iv) $\dim \ker(z - T) = n$ for each $z \in \Omega$.

It can be proved that if Ω_0 is a nonempty open subset of Ω, then $\mathcal{B}_n(\Omega) \subset \mathcal{B}_n(\Omega_0)$. For an operator $T \in \mathcal{B}_n(\Omega)$, the mapping $z \mapsto ker(z - T)$ defines a Hermitian holomorphic vector bundle of rank n. Let (E_T, π) denote the subbundle of trivial bundle $\Omega \times \mathcal{H}$ given by

$$E_T := \{(z, x) \in \Omega \times \mathcal{H} : x \in ker(z - T) \text{ and } \pi(z, x) = z\}.$$

Let $\mathcal{A}'(T)$ be the commutant of T, i.e., $\mathcal{A}'(T) := \{A \in \mathcal{L}(\mathcal{H}) : TA = AT\}$, then for $T \in \mathcal{B}_n(\Omega)$, there is a monomorphism Γ_T from $\mathcal{A}'(T)$ into $H^\infty_{\mathcal{L}(E_T)}(\Omega)$ satisfying $\Gamma_T X = X|_{ker(z-T)}$ for $X \in \mathcal{A}'(T)$ and $z \in \Omega$, or $\Gamma_T X(z) = X|_{ker(z-T)} := X(z)$, where $H^\infty_{\mathcal{L}(E_T)}(\Omega)$ is the set of all bounded bundle endomorphisms from E_T to E_T.

To summarize the above and Section 1.3, we can find a holomorphic frame $(e_1(z), \cdots, e_n(z))$ such that

$$ker(z - T) = \bigvee_{k=1}^{n} e_k(z), \; z \in \Omega \text{ for } T \in \mathcal{B}_n(\Omega).$$

Fix a $z_0 \in \Omega$, denote

$$\begin{aligned}
\mathcal{H}_1 &= ker(z_0 - T), \\
\mathcal{H}_2 &= ker(z_0 - T)^2 \ominus ker(z_0 - T), \\
&\;\;\vdots \\
\mathcal{H}_m &= ker(z_0 - T)^m \ominus ker(z_0 - T)^{m-1}.
\end{aligned}$$

We have:

Theorem CD1 [Cowen, M.J. and Douglas, R. (1977)]

(i) $\displaystyle\sum_{k=1}^{m} \oplus \mathcal{H}_k = \bigvee\{e_j^{(k)}(z_0) : 1 \leq j \leq n, 0 \leq k \leq m - 1\}$;

(ii) $\displaystyle\sum_{k=1}^{\infty} \oplus \mathcal{H}_k = \mathcal{H}$;

(iii) $\{e_j^{(k)}(z_0) : 1 \leq j \leq n, 0 \leq k \leq m - 1\}$ *is a basis of* $ker(z_0 - T)^m, m = 1, 2, \cdots$, *where* $e_j^{(k)}(z_0)$ *denote the k-th derivative of* $e_j(z)$ *at* $z = z_0$.

The following theorems will often be used in the chapters hereafter.

Theorem H [Herrero, D.A. (1990)] *If* $T \in \mathcal{B}_n(\Omega)$, *then* $\sigma_p(T^*) = \emptyset$, *where* T^* *is the adjoint of* T *and* $\sigma_p(T^*)$ *is the point spectrum of* T^*.

Theorem JW1 [Jiang, C.L. and Wang, Z.Y. (1998)] *Let* $T \in \mathcal{B}_n(\Omega)$ *and let* P_z *be the orthogonal projection from* $\mathcal{H}$ *onto* $ker(z - T)$ *for* $z \in \Omega$, *then* $(I - P_z)|_{ker(z-T)^\perp}$ *is similar to* T.

Theorem JW2 [Jiang, C.L. and Wang, Z.Y. (1998)] *Let $T \in \mathcal{L}(\mathcal{H})$. For given number $\varepsilon > 0$, there exist a positive integer n and Cowen-Douglas operators $\{A_i\}_{i=1}^{l}$, $\{b_j\}_{j=l+1}^{n}$ such that*

$$\|T - (\bigoplus_{j=1}^{l} A_j) \oplus (\bigoplus_{i=l+1}^{n} B_i^*)\| < \varepsilon.$$

Theorem JW3 [Jiang, C.L. and Wang, Z.Y. (1998)] *Given $T \in \mathcal{B}_1(\Omega)$, there exist compact operators $K_1, K_2, \cdots, K_n, \cdots$ with $\|K_i\| < \frac{\varepsilon}{2^i}$ such that $T + K_i \in \mathcal{B}_1(\Omega)$ and $ker\tau_{T+K_i, T+K_j} = \{0\}, i \neq j$, where $\tau_{A,B}$ is the Rosenblum operator from $\mathcal{L}(\mathcal{H})$ to $\mathcal{L}(\mathcal{H})$ given by $\tau_{A,B}(X) = AX - XB$ for $X \in \mathcal{L}(\mathcal{H})$.*

1.5 Strongly Irreducible Operators

Operator T is strongly irreducible if there is no nontrivial idempotent in $\mathcal{A}'(T)$ ([Gilfeather, F. (1972)], [Jiang, Z.J. (1979)], [Jiang, Z.J. (1981)]). Operator T is irreducible if there is no nontrivial orthogonal projection in $\mathcal{A}'(T)$ ([Halmos, P.R. (1968)]). It is obvious that strongly irreducibility is invariant under similarity while irreducibility is just unitarily invariant. Denote (SI) and (IR) the set of all strongly irreducible operators and irreducible operators, respectively, on $\mathcal{H}$.

Let $\mathcal{K}(\mathcal{H})$ be the ideal of compact operators on $\mathcal{H}$ and let

$$\pi : \mathcal{L}(\mathcal{H}) \longrightarrow \mathcal{A}(\mathcal{H}) := \mathcal{L}(\mathcal{H})/\mathcal{K}(\mathcal{H})$$

be the canonical quotient mapping, $\mathcal{A}(\mathcal{H})$ is called the Calkin algebra. The essential spectrum of operator T is $\sigma_e(T) = \{\lambda \in \mathbf{C} : \lambda - \pi(T)$ is not invertible in $\mathcal{A}(\mathcal{H})\}$ and the Fredholm domain of T is $\rho_F(T) = \mathbf{C} \backslash \sigma_e(T)$. It is well known that

$$\sigma_e(T) = \sigma_{le}(T) \cup \sigma_{re}(T),$$

where

$$\sigma_{le}(T) := \sigma_l(\pi(T))$$

and

$$\sigma_{re}(T) := \sigma_r(\pi(T)).$$

Operator T is a Fredholm operator if $0 \in \rho_F(T)$. T is a semi-Fredholm operator if the range of T, $ranT$, is closed and either

$$nulT := dimkerT$$

or

$$nulT^* := dimkerT^*$$

is finite. In this case the index $indT$ of T is defined by

$$indT := nulT - nulT^*.$$

The Wolf spectrum $\sigma_{lre}(T)$ of T is given by

$$\sigma_{lre}(T) := \sigma_{re}(T) \cap \sigma_{le}(T)$$

and $\rho_{s-F}(T) := \mathbf{C} \backslash \sigma_{lre}(T)$ is the semi-Fredholm domain of T. The spectral picture $\Lambda(T)$ of T consists of the compact set $\sigma_{lre}(T)$ and the index $ind(T - \lambda)$ on the bounded connected components of $\rho_{s-F}(T)$.

Spectral picture theorem of strongly irreducible operators [Jiang, C.L. and Wang, Z.Y. (1996b)] *Let T be in $\mathcal{L}(\mathcal{H})$ with connected spectrum $\sigma(T)$. Then there exists a strongly irreducible operator L satisfying*
   (i) $\Lambda(L) = \Lambda(T)$;
   (ii) $T \in \mathcal{S}(L)^-$;
   (iii) If there is another strongly irreducible operator L_1 with $\Lambda(L_1) = \Lambda(T)$, then $L_1 \in \mathcal{S}(L)^-$, where $\mathcal{S}(L)$ is the similarity orbit of L, i.e.,

$$S(L) := \{XTX^{-1} : X \in \mathcal{L}(\mathcal{H}) \text{ is invertible}\}$$

and $\mathcal{S}(L)^-$ is the norm closure of $\mathcal{S}(L)$.

Spectral picture theorem of Cowen-Douglas operators [Jiang, C.L. and Wang, Z.Y. (1998)] *Let T be in $\mathcal{L}(\mathcal{H})$ with connected $\sigma(T)$ and $\sigma(T) \backslash \rho^0_{s-F}(T)$. If $\rho_F(T) \neq \emptyset$, then there exists a Cowen-Douglas operator $A \in (SI)$ such that*

$$\Lambda(T) = \Lambda(A)$$

and if there is another operator $B \in \mathcal{S}(T)^-$, then $B \in \mathcal{S}(A)^-$, where

$$\rho^0_{s-F}(T) := \{\lambda \in \rho_{s-F}(T) \cap \sigma(T) : ind(\lambda - T) = 0\}.$$

Commutant theorem of strongly irreducible operators [Fang, J.S. and Jiang, C.L. (1999)] *Operator T is strongly irreducible if and only if $\sigma(A)$ is connected for each A in $\mathcal{A}'(T)$.*

The following theorem will be used frequently in this book.

Theorem CD2 [Cowen, M.J. and Douglas, R. (1977)] *Each operator in $\mathcal{B}_1(\Omega)$ is strongly irreducible.*

1.6 Compact Perturbation of Operators

We introduce only two famous theorems on compact perturbation of operator here.

Brown-Douglas-Fillmore theorem *If T_1 and T_2 are essentially normal operators on $\mathcal{H}$, then a necessary and sufficient condition that T_1 be unitarily equivalent to some compact perturbation of T_2 is that*

$$\sigma_e(T_1) = \sigma_e(T_2)$$

and

$$ind(\lambda - T_1) = ind(\lambda - T_2)$$

for all $\lambda \notin \sigma_e(T_1)$.

An operator T is essentially normal if $T^*T - TT^*$ is compact.

Voiculescu's theorem *Let $T \in \mathcal{L}(\mathcal{H})$ and ρ be a unital faithful $*$-representation of a separable C^*-subalgebra of the Calkin algebra $\mathcal{A}(\mathcal{H})$ containing the canonical images $\pi(T)$ and $\pi(I)$ on a separable space $\mathcal{H}_\rho$. Let $A = \rho(\pi(T))$ and k be a positive integer. Given $\varepsilon > 0$, there exists $K \in \mathcal{K}(\mathcal{H})$, with $\|K\| < \varepsilon$, such that*

$$T - K \cong T \oplus A^{(\infty)} \cong T \oplus A^{(k)},$$

where "$\cong$" means unitarily equivalent.

1.7 Similarity Orbit Theorem

Complex number λ is a normal eigenvalue of T if λ is an isolated point of $\sigma(T)$ and the dimension of $H(\lambda, T)$, the range of the Riesz idempotent corresponding to λ, is finite. Denote the set of all normal eigenvalues of T by $\sigma_0(T)$. The minimal index of $\lambda - T, \lambda \in \rho_{s-F}(T)$, is defined by

$$min \cdot ind(\lambda - T) := min\{nul(\lambda - T), nul(\lambda - T)^*\}.$$

Similarity orbit theorem   *Given $T, A \in \mathcal{L}(\mathcal{H})$ satisfying*
   (i) $\sigma_0(A) \subset \sigma_0(T), dimH(\lambda, A) = dimH(\lambda, T)$ for all $\lambda \in \sigma_0(A)$;
   (ii) Each component of $\sigma_{lre}(A)$ meets $\sigma_e(T)$;
   (iii) $\rho_{s-F}(A) \subset \rho_{s-F}(T)$,
       $ind(\lambda - A) = ind(\lambda - T)$,
       $min \cdot ind(\lambda - A)^k \geq min \cdot ind(\lambda - T)^k$ for all $\lambda \in \rho_{s-F}(A)$ and $k \geq 1$;
   (iv) There is no isolated point in $\sigma_e(T)$;
   then $A \in \mathcal{S}(T)^-$.

Note that this is only a sufficient condition of the similarity orbit theorem. The reader is referred to [Apostal, C., Fialkow, L.A. Herrero, D.A. and Voiculescu, D. (1984)] for the general form of the theorem.

1.8   Toeplitz Operator and Sobolev Space

Let D and C be the open unit disk and unit circle in the complex plane respectively, and let μ be the Lebesgue measure on C, normalized so that $\mu(C) = 1$. If $e_n(z) = z^n$ for $z \in C$ and $n = 0, \pm 1, \pm 2, \cdots$, then $\{e_n, -\infty < n < +\infty\}$ is an orthonormal basis (ONB) for $L^2(C, \mu)$. Let $H^2 := span\{e_n : n \geq 0\}$ and $H^\infty = L^\infty(C, \mu) \cap H^2$. If P is the orthogonal projection from $L^2(C, \mu)$ onto H^2 and if $\phi \in L^\infty(C, \mu)$, then the Toeplitz operator T_ϕ with symbol ϕ is defined by $T_\phi f = P(\phi f)$ for all $f \in H^2$. If $\phi \in H^\infty$, T_ϕ is called an analytic Toeplitz operator.

A function $m \in H^2$ is inner if $|m(z)| = 1$ a.e. on C. If ϕ is a positive measurable function on C such that $log\phi \in L^1(C, \mu)$ and if $Q(z) = c \cdot exp\{\int_C \frac{\omega+z}{\omega-z} log\phi(\omega)d\mu(\omega)\}$ for $z \in C$, then Q is called an outer function, where c is a constant and $|c| = 1$.

The Blaschke product is a class of inner functions with the form

$$B(z) = cz^k \prod_{j=1}^{\infty} \frac{\overline{\lambda_j}}{|\lambda_j|} \frac{\lambda_j - z}{1 - \overline{\lambda_j}z},$$

where $k \geq 0, c \in \mathbf{C}$, with $|c| = 1$ and $\{\lambda_j\}$ is a sequence of nonzero numbers in D satisfying $\sum_{j=1}^{\infty} (1 - |\lambda_j|) < +\infty$.

Factorization theorem *If $f{\in}H^2$ and $f \neq 0$, then f is the product of an inner function m and an outer function Q, i.e., $f = mQ$.*

Beurling's theorem *A subspace H_1 of H^2 is invariant under the operator T_z if and only if $H_1 = \phi H^2 := \{\phi f : f{\in}H^2\}$ for some inner function ϕ.*

Let Ω be an analytic Cauchy domain in the complex plane and let $W^{22}(\Omega)$ be the Sobolev space

$$W^{22}(\Omega) := \left\{ f \in L^2(\Omega, dm) : \begin{array}{l} \text{the distributional derivative of first and} \\ \text{second order of } f \text{ belong to } L^2(D, dm) \end{array} \right\}$$

where dm denotes the planar Lebesque measure.

For $f, g{\in}W^{11}(\Omega)$, we define $(f,g) = \sum_{|\alpha|\leq 2} \int D^\alpha f \overline{D^\alpha g} dm$, then $W^{22}(\Omega)$ is a Hilbert space and a Banach algebra with identity under an equivalent norm. By Sobolev embedding theorem, $f{\in}W^{22}(\Omega)$ implies that $f{\in}C(\overline{\Omega})$ and

$$\|f\|_{C(\overline{\Omega})}{\leq}M\|f\|_{W^{22}(\Omega)}$$

for some $M > 0$. Thus a sequence of functions $\{f_n\}_{n=1}^\infty$ converges to f in $W^{22}(\Omega)$ implies that f_n converges to f uniformly on $\widetilde{\Omega}$. For $f{\in}W^{22}(\Omega)$, the multiplication operator M_f on $W^{22}(\Omega)$ is defined as follows

$$M_f g = fg, \quad g{\in}W^{22}(\Omega).$$

Let $W(\Omega) := \{M_f : f{\in}W^{22}(\Omega)\}$, then $W(\Omega)$ is a strictly cyclic operator algebra with strictly cyclic vector $e(s,t) \equiv 1$. An operator algebra $\mathcal{A}$ on a Hilbert space $\mathcal{H}$ is said to be strictly cyclic if there exists a separating vector e such that

$$\mathcal{A}e := \{Ae : A{\in}\mathcal{A}\} = \mathcal{H}.$$

Theorem JW4 [Jiang, C.L. and Wang, Z.Y. (1996a)]
  (i) $\sigma(M_z) = \sigma_{lre}(M_z) = \overline{\Omega}$;
  (ii) $\mathcal{A}'(M_z) = W(\Omega)$;
  (iii) $\mathcal{A}^\alpha(M_z) = R(\Omega)$, where $\mathcal{A}^\alpha(M_z)$ is the algebra generated by the rational function of M_z with poles outside $\overline{\Omega}$ and $R(\Omega)$ is the closure in $W^{22}(\Omega)$ of all rational functions with poles outside $\overline{\Omega}$.

Chapter 2

Jordan Standard Theorem and K_0-Group

2.1 Generalized Eigenspace and Minimal Idempotents

Recall that a $k \times k$ Jordan block

$$J_k(\lambda) = \begin{bmatrix} \lambda & & & 0 \\ 1 & \lambda & & \\ & \ddots & \ddots & \\ 0 & & 1 & \lambda \end{bmatrix}$$

has the following properties

    (i) $J_k(\lambda)$ is strongly irreducible on $\mathbf{C}^k$, i.e., there is no nontrivial idempotent in $\mathcal{A}'(J_k(\lambda))$;

    (ii) $nul(\lambda - J_k(\lambda)) = 1, ker(\lambda - J_k(\lambda))^j \subsetneqq ker(\lambda - J_k(\lambda))^{j+1}, 1 \leq j \leq k-1$ and $ker(\lambda - J_k(\lambda))^j = ker(\lambda - J_k(\lambda))^k = \mathbf{C}^k$ for all $j \geq k$;

    (iii) $\mathcal{A}'(J_k(\lambda))$ consists of all $k \times k$ lower triangular matrices B

$$B = \begin{bmatrix} a_1 & & & 0 \\ a_2 & a_1 & & \\ \vdots & & \ddots & \\ a_k & \cdots & a_2 & a_1 \end{bmatrix}, \quad \text{where } a_i \in \mathbf{C}, i = 1, 2, \cdots, k;$$

    It follows from (iii) directly that

    (iv) $\mathcal{A}'(J_k(\lambda))/rad\mathcal{A}'(J_k(\lambda)) \simeq \mathbf{C}$.

    For $A \in M_n(\mathbf{C})$ and $\lambda \in \sigma(A), ker(\lambda - A)$ is called the eigenspace of A related to λ. If there is a positive integer $m, m < n$, such that

$$ker(\lambda - A)^m = ker(\lambda - A)^{m+1},$$

then $ker(\lambda - A)^m$ is called the generalized eigenspace of A related to λ. By elementary matrix theory, we have the following proposition.

Proposition 2.1.1 *Let $A \in M_n(\mathbf{C})$ and $\lambda \in \sigma(A)$ with $nul(\lambda - A) = 1$ and let $\mathcal{N}$ be the generalized eigenspace of A related to λ, then $A|_{\mathcal{N}} \sim J_k(\lambda)$ for some $k, k < n$.*

A nonzero idempotent $P \in \mathcal{A}'(T)$ is minimal, if for each idempotent $Q \in \mathcal{A}'(T)$, $ranQ \subset ranP$ implies $Q = 0$.

Proposition 2.1.2 *Let $A \in M_n(\mathbf{C})$ and let P be a minimal idempotent in $\mathcal{A}'(A)$, then there exists a number $\lambda \in \mathbf{C}$, such that $A|_{ranP} \sim J_k(\lambda)$, where $k = dimranP$.*

Proof Since $P \in \mathcal{A}'(T)$, $ranP$ and $ran(I - P)$ are invariant subspaces of A.

{Claim 1} $B := A|_{ranP}$ has a unique eigenvalue λ_B. Otherwise, it follows from the theory of linear algebra that P is not minimal.

{Claim 2} B is strongly irreducible and $nul(\lambda_B - B) = 1$. Since P is minimal, $B \in (SI)$. Using the basic theory of linear algebra and the fact that P is minimal again, we have $nul(\lambda_B - B) = 1$. Thus the proposition is a conclusion of Proposition 2.1.1.

Let $\{\lambda_k\}_{k=1}^n$ be all of the eigenvalues of $A \in M_n(\mathbf{C})$, the multiplicity is included, by Proposition 2.1.2 and the theory of matrix, we have the next proposition.

Proposition 2.1.3 *Let $A \in M_n(\mathbf{C})$, then there exist minimal idempotents $\{P_{\lambda_i}\}_{i=1}^n$ such that*

(i) $\quad \sum\limits_{k=1}^{n} P_{\lambda_k} = I_{\mathbf{C}^n}$ *and* $P_{\lambda_k} P_{\lambda_j} = 0$ *for* $k \neq j$;

(ii) $\quad A|_{ranP_{\lambda_k}} \sim J_{m_k}(\lambda_k)$, *where* $m_k = dimranP_{\lambda_k}$;

(iii) $\quad A = \sum\limits_{k=1}^{n} \dot{+} A|_{ranP_{\lambda_k}}$.

2.2 Similarity Invariant of Matrix

Jordan standard theorem *Let $A \in M_n(\mathbf{C})$, then A is similar to a direct sum of finitely many Jordan blocks, i.e.,*

$$A \sim \bigoplus_{k=1}^{l} J_{m_k}(\lambda_k).$$

By Propositions 2.1.2, 2.1.3 and Jordan standard theorem, the similarity of two $n{\times}n$ matrices A and B depends completely on their eigenvalues and generalized eigenspaces. A quantity (or quantities) or property (or properties) $\mathcal{P}$ is similarity invariant if operator A has $\mathcal{P}$ and $A{\sim}B$ imply B has $\mathcal{P}$. For a subset $\mathcal{R}$ of $\mathcal{L}(\mathcal{H})$, similarity invariant (or invariants) $\mathcal{P}$ is completely similarity invariant if $A{\in}\mathcal{R}$, then $A{\sim}B$ if and only if $B{\in}\mathcal{R}$ and A and B have same $\mathcal{P}$. One of our aims is to find or determine complete similarity invariants of operators. But if $\mathcal{H}$ is an infinite dimensional separable Hilbert space, it is very difficult to obtain complete similarity invariant of operators in $\mathcal{L}(\mathcal{H})$. As a matter of fact, some operators in $\mathcal{L}(\mathcal{H})$ even have no eigenvalues. To provide a new idea for finding of the similarity invariants of Hilbert space operators, we interpret the Jordan standard theorem in the view point of K_0-theory.

The propositions below are from [Blanckdar, B. (1986)] and [Aupetit, B. (1991)].

Proposition 2.2.1 $K_0(M_n(\mathbf{C})){\cong}\mathbf{Z}$ *and* $\bigvee(M_n(\mathbf{C})){\cong}\mathbf{N}$, *where* $\mathbf{N} = \{0,1,2,\cdots\}$ *and* $\mathbf{Z} = \{0,\pm1,\pm2,\cdots\}$.

Proposition 2.2.2   *Let* $\mathcal{A}$ *be a unital Banach algebra and let P be an idempotent in $\mathcal{A}$ and $R{\in}rad\mathcal{A}$. If $P+R$ is still an idempotent in $\mathcal{A}$, then there exists an invertible element $X{\in}\mathcal{A}$ such that $X(P+R)X^{-1} = P$.*

We know that $\mathcal{A}'(J_k(\lambda))/rad\mathcal{A}'(J_k(\lambda)){\cong}\mathbf{C}$ for each Jordan block $J_k(\lambda)$. If P is an idempotent in $\mathcal{A}'(J_k(\lambda))$, it follows from the structure of $\mathcal{A}'(J_k(\lambda))$ that $P = I_{\mathbf{C}^k} + R$, where R is a lower triangular idempotent. Thus $P{\sim}_{\mathcal{A}'(J_k(\lambda))}I_{\mathbf{C}^k}$ by Proposition 2.2.2.

From Proposition 2.2.1 and the definition of K_0-group, we have the following proposition.

Proposition 2.2.3 $\bigvee(\mathcal{A}'(J_k(\lambda))){\cong}\mathbf{N}$ *and* $K_0(\mathcal{A}'(J_k(\lambda))){\cong}\mathbf{Z}$.

Lemma 2.2.4   *Let* $A_1, A_2{\in}(SI){\cap}\mathcal{L}(\mathcal{H})$ *satisfying*

$$\mathcal{A}'(A_i)/rad\mathcal{A}'(A_i){\cong}\mathbf{C}, \quad i = 1,2,$$

then at least one of the following is true
   (i) $A_1{\sim}A_2$;
   (ii)   *If* $X,Y{\in}\mathcal{L}(\mathcal{H})$ *with* $A_1X = XA_2$ *and* $YA_1 = A_2Y$, *then* $XY{\in}rad\mathcal{A}'(A_1)$ *and* $YX{\in}rad\mathcal{A}'(A_2)$.

Proof   If A_1 is not similar to A_2 and $A_1X = XA_2$, $YA_1 = A_2Y$. Thus

$$A_1XY = XA_2Y = XYA_1$$

and

$$XY \in \mathcal{A}'(A_1).$$

If $XY \notin rad\mathcal{A}'(A_1)$, then $XY = \lambda + R$ for some $\lambda \in \mathbf{C}, \lambda \neq 0$ and $R \in rad\mathcal{A}'(A_1)$, since $\mathcal{A}'(A_1)/rad\mathcal{A}'(A_1) \cong \mathbf{C}$. Therefore XY is invertible. Similarly, YX is invertible. This implies that X and Y are both invertible, i.e., $A_1 \sim A_2$. The contradiction indicates that $XY \in rad\mathcal{A}'(A_1)$. Similarly, $YX \in rad\mathcal{A}'(A_2)$.

Lemma 2.2.5   Let $A \in (M_n(\mathbf{C}))$, then $\mathcal{A}'(A)/rad\mathcal{A}'(A) \cong \sum_{i=1}^{l} M_{k_i}(\mathbf{C})$, where

$$k_1 + k_2 + \cdots + k_l = n.$$

Proof   By Proposition 2.1.3 and Proposition 2.1.2, we have

$$A \sim \sum_{i=1}^{m} \oplus J_{k_i}(\lambda_i).$$

For simplicity we prove the lemma for $m = 2$ and consider the following two cases.

{**Case 1**} $k_1 = k_2$ and $\lambda_1 = \lambda_2$, therefore $J_{k_1}(\lambda_1) = J_{k_2}(\lambda_2)$. A simple computation indicates that

$$\mathcal{A}'(A) = \left\{ \begin{bmatrix} A_{11} & A_{12} \\ A_{21} & A_{22} \end{bmatrix} : \ A_{ij} = \begin{bmatrix} \alpha_1 & & & 0 \\ \alpha_2 & \alpha_1 & & \\ \vdots & \ddots & \ddots & \\ \alpha_k & \cdots & \alpha_2 & \alpha_1 \end{bmatrix} \right\}$$

from which we have

$$rad\mathcal{A}'(A) = \left\{ \begin{bmatrix} R_{11} & R_{12} \\ R_{21} & R_{22} \end{bmatrix} : \ R_{ij} = \begin{bmatrix} 0 & & & 0 \\ \alpha & 0 & & \\ \vdots & \ddots & \ddots & \\ \beta & \cdots & \alpha & 0 \end{bmatrix} \right\}.$$

Thus $\mathcal{A}'(A)/rad\mathcal{A}'(A) \cong M_2(\mathbf{C})$.

{**Case 2**} $k_1 \neq k_2$ or $\lambda_1 \neq \lambda_2$. If $k_1 \neq k_2$, without loss of generality we assume that $k_1 > k_2$. By Lemma 2.2.4, if $X \in ker\tau_{J_{k_1}(\lambda_1), J_{k_2}(\lambda_2)}$ and $Y \in ker\tau_{J_{k_2}(\lambda_2), J_{k_1}(\lambda_1)}$, then $XY \in rad\mathcal{A}'(J_{k_1}(\lambda_1)), YX \in rad\mathcal{A}'(J_{k_2}(\lambda_2))$.

Set

$$\mathcal{J} = \left\{ \begin{bmatrix} X_{11} & X_{12} \\ X_{21} & X_{22} \end{bmatrix} : \begin{array}{l} X_{11} \in rad\mathcal{A}'(J_{k_1}(\lambda_1)), \ X_{12}X_{21} \in rad\mathcal{A}'(J_{k_1}(\lambda_1)) \\ X_{22} \in rad\mathcal{A}'(J_{k_2}(\lambda_2)), \ X_{21}X_{12} \in rad\mathcal{A}'(J_{k_2}(\lambda_2)) \end{array} \right\}.$$

It is easily seen that $\mathcal{J}$ is a two sided ideal of $\mathcal{A}'(A)$. We claim that

$$\begin{bmatrix} X_{11} - \lambda & X_{12} \\ X_{21} & X_{22} - \lambda \end{bmatrix}$$

is invertible for each $\lambda \neq 0$ and

$$\begin{bmatrix} X_{11} & X_{12} \\ X_{21} & X_{22} \end{bmatrix} \in \mathcal{J}.$$

In fact, observing that

$$\begin{bmatrix} 1 & 0 \\ -(X_{11} - \lambda)^{-1} & 1 \end{bmatrix} \begin{bmatrix} X_{11} - \lambda & X_{12} \\ X_{21} & X_{22} - \lambda \end{bmatrix}$$

$$= \begin{bmatrix} X_{11} - \lambda & X_{12} \\ 0 & (X_{22} - \lambda) - X_{12}(X_{11} - \lambda)^{-1}X_{21} \end{bmatrix},$$

where $X_{12}(X_{11} - \lambda)^{-1}X_{21} \in rad\mathcal{A}'(J_{k_2}(\lambda_2))$. Thus

$$(X_{22} - \lambda) - X_{12}(X_{11} - \lambda)^{-1}X_{21}$$

is invertible. Therefore $\begin{bmatrix} X_{11} - \lambda & X_{12} \\ X_{21} & X_{22} - \lambda \end{bmatrix}$ is invertible.

The claim above indicates that $\sigma(X) = \{0\}$ for each $X \in \mathcal{J}$. Thus

$$\mathcal{J} = rad\mathcal{A}'(A)$$

and

$$\mathcal{A}'(A)/rad\mathcal{A}'(A) \cong \mathbf{C} \oplus \mathbf{C}.$$

This complete the proof.

Theorem 2.2.6 *Let* $A \in M_n(\mathbf{C})$ *and* $A \sim \sum_{i=1}^{l} \oplus J_{k_i}(\lambda_i)^{(n_i)}$, *then* $\bigvee(\mathcal{A}'(A)) \cong \mathbf{N}^{(l)}$ *and* $K_0(\mathcal{A}'(A)) \cong \mathbf{Z}^{(l)}$, *where* $J_{k_i}(\lambda_i)^{(n_i)}$ *denotes the orthogonal direct sum of n_i copies of $J_{k_i}(\lambda_i)$.*

Proof By Lemma 2.2.5, $\mathcal{A}'(A)/rad\mathcal{A}'(A)\cong\sum\limits_{i=1}^{l}\oplus M_{n_i}(\mathbf{C})$. By Proposition 2.2.2 and Proposition 2.2.3 we have

$$\bigvee(\mathcal{A}'(A))\cong\bigvee\{\mathcal{A}'(A)/rad\mathcal{A}'(A)\}\cong\bigvee\{\sum\limits_{i=1}^{l}\oplus M_{n_i}(\mathbf{C})\}\cong\mathbf{N}^{(l)}$$

and

$$K_0(\mathcal{A}'(A))\cong\mathbf{Z}^{(l)}.$$

Theorem 2.2.6 is another form of Jordan standard theorem. The following result gives the complete similarity invariant for matrices in terms of K_0-group.

Theorem 2.2.7 *Let* $A, B\in M_n(\mathbf{C})$ *and* $A=\sum\limits_{i=1}^{m}\oplus A_{k_i}^{(n_i)}$ *with* $A_{k_i}\in(SI)$ *and* A_{k_i} *is not similar to* A_{k_j} *for* $i\neq j$. *Then* $A\sim B$ *if and only if there exists an isomorphism* h *such that*

$$h(K_0(\mathcal{A}'(A\oplus B)))\cong\mathbf{Z}^{(m)}$$

and $h[I_{\mathcal{A}'(A\oplus B)}]=2n_1e_1+2n_2e_2+\cdots+2n_ke_k$, *where* $0\neq n_i\in\mathbf{N}, i=1,2,\cdots,k$, $\{e_i\}_{i=1}^{k}$ *are the generators of the semigroup* $\mathbf{N}^{(m)}$ *of* $\mathbf{Z}^{(m)}$ *and* $I_{\mathcal{A}'(A\oplus B)}$ *is the identity of* $\mathcal{A}'(A\oplus B)$.

Proof If $A\sim B$, then $A\oplus B\sim\sum\limits_{i=1}^{m}\oplus A_{k_i}^{(2n_i)}$. The "necessary" part follows from Theorem 2.2.6.

Proof of "Sufficient" part. Since $B\in M_n(\mathbf{C})$, $B\sim\sum\limits_{j=1}^{l}\oplus B_{k_j}^{(m_j)}$, where $B_{k_j}\in(SI)$, $B_{k_j}\nsim B_{k_{j'}}$ if $j\nsim j'$. It follows from $h(K_0(\mathcal{A}'(A\oplus B)))\cong\mathbf{Z}^{(m)}$ and

$$h[I_{\mathcal{A}'(A\oplus B)}]=2n_1e_1+2n_2e_2+\cdots+2n_ke_k$$

that $l=m$. By Lemma 2.2.4, for each B_{k_j} there exists an A_{k_i} such that $B_{k_j}\sim A_{k_i}$ and $m_j=n_i$. This implies that $A\sim B$.

2.3 Remark

Theorem 2.2.6 and Theorem 2.2.7 are different forms of Jordan standard theorem and can be obtained in different ways. Lemma 2.2.4 is due to [Cao, Y., Fang, J.S. and Jiang, C.L.(2002)].

Chapter 3

Approximate Jordan Theorem of Operators

In the operator theory of finite dimensional space, or in matrix theory, Jordan canonical theorem is one of the core contents. The Jordan theorem gives the complete similarity invariant of matrices. But in the infinite dimensional Hilbert space case, it is very difficult to find complete similarity invariant for operators. We can only give an approximate Jordan theorem, or obtain complete similarity invariant for some special class of operators. In this chapter, we give some different kinds of approximate Jordan theorems considering strongly irreducible operators as the replacement of Jordan blocks in matrix theory.

3.1 Sum of Strongly Irreducible Operators

[Radjaval, H. and Rosenthal, P. (1973)] proved that every operator in $\mathcal{L}(\mathcal{H})$ is a sum of two irreducible operators. In this section, we will prove the following result.

Theorem 3.1.1 *Every bounded linear operator on $\mathcal{H}$ is a sum of two strongly irreducible operators.*

In order to prove the theorem, we need the following lemmas.

Lemma 3.1.2 *Assume that $T \in \mathcal{L}(\mathcal{H})$ with $indT = -1$ and $min\cdot indT = 0$. If B_1 and B_2 are two left inverses of T and e_0 is a unit vector in $(ranT)^{\perp}$, then there exists an $f \in \mathcal{H}$ such that $B_1 = B_2 + f \otimes e_0$.*
Proof   Set $f = (B_1 - B_2)e_0$ and $A = B_1 - B_2 - f \otimes e_0$. Since $indT = -1$ and since $min\cdot indT = 0$, $ranT$ is closed and $dim(ranT)^{\perp} = 1$. Thus for each $x \in \mathcal{H}$, there is a number $\alpha \in \mathbf{C}$ and $x_1 \in \mathcal{H}$ such that $x = \alpha e_0 + Tx_1$.

Since $B_1T = B_2T = I$,

$$Ax = \alpha(B_1 - B_2)e_0 + B_1Tx_1 - B_2Tx_1 - \alpha(f\otimes e_0)e_0 - (f\otimes e_0)Tx_1$$
$$= \alpha f + x_1 - x_1 - \alpha f - <Tx_1, e_0> f$$
$$= 0.$$

Thus, $A = 0$ and $B_1 = B_2 + f\otimes e_0$.

Lemma 3.1.3 *Let $T\in\mathcal{L}(\mathcal{H})$ with $indT = -1$ and $min\cdot indT = 0$. Let e_0 be a unit vector in $(ranT)^\perp$ and B be a left inverse of T with $Be_0 = 0$. Then $TB = I - e_0\otimes e_0$.*
Proof   Set $A = TB + e_0\otimes e_0 - I$. Since for each $x\in\mathcal{H}$ there exist $\alpha \in \mathbf{C}$ and $y\in\mathcal{H}$ such that $x = \alpha e_0 + Ty$,

$$Ax = \alpha TB e_0 + TBTy + \alpha(e_0\otimes e_0)e_0 + (e_0\otimes e_0)Ty - \alpha e_0 - Ty$$
$$= 0 + Ty + \alpha e_0 + 0 - \alpha e_0 - Ty$$
$$= 0.$$

Thus $A = 0$ and $TB = I - e_0\otimes e_0$.

Proposition 3.1.4   *Given T, e_0 and B as that in Lemma 3.1.3.   If $B\in(SI)$, then $T\in(SI)$.*
Proof   If there is a nontrivial idempotent $P\in\mathcal{A}'(T)$, $kerT^*$ is an invariant subspace of P^*. Since $indT = -1$ and $min\cdot indT = 0$, $dimkerT^* = 1$. Therefore, $P^*e_0 = \lambda e_0$ and $\lambda = 0$ or $\lambda = 1$. Assume that $\lambda = 0$ (otherwise consider $I - P$), then

$$P = \begin{bmatrix} 0 & Q \\ 0 & I_{ranP^*} \end{bmatrix} \begin{matrix} kerP \\ ranP^* \end{matrix} \quad \text{and} \quad T = \begin{bmatrix} T_1 & T_{12} \\ 0 & T_2 \end{bmatrix} \begin{matrix} kerP \\ ranP^*. \end{matrix}$$

Set

$$X = \begin{bmatrix} I_{kerP} & Q \\ 0 & I_{ranP^*} \end{bmatrix} \begin{matrix} kerP \\ ranP^* \end{matrix}, \quad \overline{P} = X^{-1}PX = \begin{bmatrix} 0 & 0 \\ 0 & I_{ranP^*} \end{bmatrix} \begin{matrix} kerP \\ ranP^* \end{matrix}$$

and

$$\tilde{T} = X^{-1}TX = \begin{bmatrix} T_1 & 0 \\ 0 & T_2 \end{bmatrix} \begin{matrix} kerP \\ ranP^*. \end{matrix}$$

Note that $\overline{P}^* = \overline{P}$, $\overline{P}\in\mathcal{A}'(\tilde{T})$, $ind\tilde{T} = -1$ and $min\cdot ind\tilde{T} = 0$. Thus T_1 and T_2 are injective with closed ranges and $indT_1 + indT_2 = ind\tilde{T} = -1$. Assume that $e_0 = e_1 + e_2$, where $e_1\in kerP$ and $e_2\in ranP^*$. Since $P^*e_0 = 0$ and $e_0 \neq 0$, $e_1 \neq 0$ and $Q^*e_1 = -e_2$. Since $T_1^*e_1 + T_{12}^*e_1 + T_2^*e_2 = 0$, $T_1^*e_1 = 0$. Thus $indT_1 = -1$ and $indT_2 = 0$. This implies that T_2 is invertible and T_1 is left invertible. Suppose that B_2 is the inverse of T_2 and B_1 is a left inverse of T_1 satisfying $\ker B_1 = (\operatorname{ran} T_1)^\perp$. Thus $B_1 e_1 = 0$. Set

$$\tilde{B} = \begin{bmatrix} B_1 & 0 \\ 0 & B_2 \end{bmatrix} \begin{matrix} \ker P \\ \operatorname{ran} P^* \end{matrix},$$

then $\tilde{B}$ is a left inverse of $\tilde{T}$ and $\tilde{B}\overline{P} = \overline{P}\tilde{B}$. Set $\hat{B} = X^{-1}BX$, then $\hat{B}$ is also a left inverse of $\tilde{T}$. It follows from $\tilde{T}^* e_1 = T_1^* e_1 = 0$ that $\frac{e_1}{\|e_1\|}$ is a unit vector in $(\operatorname{ran} T)^\perp$. By Lemma 3.1.2, there exists a vector f such that $\hat{B} = \tilde{B} + f \otimes \frac{e_1}{\|e_1\|}$. From $\hat{B}X^{-1}e_0 = X^{-1}Be_0 = 0$, $Xe_0 = e_0 + Qe_2 + e_2$ and $\hat{B}Xe_0 = 0$, we have:

$$0 = \hat{B}(e_1 + Qe_2 + e_2)$$

$$= B_1 e_1 + B_1 Qe_2 + B_2 e_2 + < e_1, \tfrac{e_1}{\|e_1\|} > f + < Qe_2, \tfrac{e_1}{\|e_1\|} > f$$

$$= B_1 Qe_2 + B_2 e_2 + \|e_1\| f - \tfrac{\|e_2\|^2}{\|e_1\|} f.$$

If $e_2 \neq 0$, it follows from $B_1 Qe_2 \perp B_2 e_2$ and $B_2 e_2 \neq 0$ that $B_1 Qe_2 + B_2 e_2 \neq 0$. Thus

$$f = \frac{\|e_1\|}{\|e_2\|^2 - \|e_1\|^2} B_1 Qe_2 + \frac{\|e_1\|}{\|e_2\|^2 - \|e_1\|^2} B_2 e_2$$

and

$$\hat{B} = \begin{bmatrix} B_1 + \frac{B_1 Qe_2 \otimes e_1}{\|e_2\|^2 - \|e_1\|^2} & 0 \\ \frac{B_2 e_2 \otimes e_1}{\|e_2\|^2 - \|e_1\|^2} & B_2 \end{bmatrix} \begin{matrix} \ker P \\ \operatorname{ran} P^* \end{matrix}.$$

Moreover,

$$\tilde{T}\hat{B} = \begin{bmatrix} I_{\ker P} - \frac{e_1 \otimes e_1}{\|e_1\|^2} + Qe_2 \otimes e_1 - \frac{<Qe_2,e_1>}{\|e_1\|^2} e_1 \otimes e_1 & 0 \\ \frac{e_2 \otimes e_1}{\|e_2\|^2 - \|e_1\|^2} & I_{\operatorname{ran} P^*} \end{bmatrix}. \tag{3.1.1}$$

By Lemma 3.1.3,

$$TB = I - e_0 \otimes e_0 = I - e_1 \otimes e_1 - e_1 \otimes e_2 - e_2 \otimes e_1 - e_2 \otimes e_2.$$

Therefore,

$$\tilde{T}\hat{B} = \begin{bmatrix} * & * \\ -e_2 \otimes e_1 & * \end{bmatrix} \begin{matrix} \ker P \\ \operatorname{ran} P^* \end{matrix}. \tag{3.1.2}$$

Compare (3.1.1) with (3.1.2) we have

$$\|e_1\|^2 - \|e_2\|^2 = 1.$$

Since $\|e_1\|^2 + \|e_2\|^2 = \|e_0\|^2 = 1, \|e_2\| = 0$ and $e_2 = 0$. Thus, $Pe_0 = Pe_1 = 0$. Therefore,

$$T(PB - BP) = P(TB) - (TB)P$$

$$= P(I - e_0 \otimes e_0) - (I - e_0 \otimes e_0)P$$

$$= -(Pe_0) \otimes e_0 + e_0 \otimes P^* e_0$$

$$= 0.$$

But T is injective, thus $PB - BP = 0$, i.e., $PB = BP$. This contradicts $B \in (SI)$. Therefore $T \in (SI)$ and the proof of the proposition is now complete.

Lemma 3.1.5 *Let $A, S \in \mathcal{L}(\mathcal{H}), \|A\| < \frac{1}{2}$ and let S be the froward unilateral shift. Then for each $\lambda \in \mathbf{C}$ with $|\lambda| < \frac{1}{2}$, $S + A - \lambda$ is a Fredholm operator and*

$$ind(S + A - \lambda) = -1, \ min\cdot ind(S + A - \lambda) = 0.$$

Proof Note that for each $x \in \mathcal{H}$,

$$\|(S + A - \lambda)x\| \geq \|Sx\| - \|Ax\| - |\lambda|\|x\| \geq (\frac{1}{2} - \|A\|)\|x\|.$$

Thus $S + A - \lambda$ is bounded below, $ran(S + A - \lambda)$ is closed and

$$dimker(S + A - \lambda) = 0, min\cdot ind(S + A - \lambda) = 0.$$

On the other hand, it follows from $\|(A - \lambda)S^*\| < 1$ that $I + (A - \lambda)S^*$ is invertible. Therefore,

$$ind(S + A - \lambda) = ind[S + (A - \lambda)S^* S]$$

$$= ind[(I + (A - \lambda)S^*)S]$$

$$= ind[I + (A - \lambda)S^*] + indS$$

$$= 0 + (-1) = -1.$$

Thus $S + A - \lambda$ is a Fredholm operator.

Lemma 3.1.6 *Assume that $T \in \mathcal{L}(\mathcal{H})$, $\lambda_0 \in \mathbf{C}$, $ind(T - \lambda_0) = -1$ and*

$$min \cdot ind(T - \lambda_0) = 0.$$

Then

$$T \sim T|_{ran(T-\lambda_0)}.$$

Proof Define $X : \mathcal{H} \longrightarrow ran(T - \lambda_0)$ by $Xx = (T - \lambda_0)x$ for $x \in \mathcal{H}$. Since $ran(T - \lambda_0)$ is closed and $T - \lambda_0$ is injective, X is invertible. Note that

$$T|_{ran(T-\lambda_0)}Xx = T(T - \lambda_0)x = (T - \lambda_0)Tx = XTx,$$

thus

$$T|_{ran(T-\lambda_0)}X = XT,$$

i.e.,

$$T|_{ran(T-\lambda_0)} \sim T.$$

Lemma 3.1.7 *Given $A \in \mathcal{L}(\mathcal{H})$ that admits the following lower triangular matrix representation*

$$A = \begin{bmatrix} \lambda_0 & & & 0 \\ a_{21} & 0 & & \\ & a_{32} & 0 & \\ & & \ddots & \ddots \\ \star & & & \end{bmatrix} \begin{matrix} e_0 \\ e_1 \\ e_2 \\ \vdots \end{matrix} \qquad (3.1.3)$$

with respect to the ONB $\{e_k\}_{k=0}^{\infty}$. If $\|A\| < \frac{1}{2}$, then $(S + A)^* \in \mathcal{B}_1(\Omega)$, where S is the unilateral shift.

Proof Since $\|A\| < \frac{1}{2}$ and $|\lambda_0| < \frac{1}{2}$, $ind(S + A - \lambda) = -1$ and

$$min \cdot ind(S + A - \lambda) = 0 \text{ for } |\lambda| < \frac{1}{2}.$$

Thus $(S + A - \lambda)^*$ is surjective and $nul(S + A - \lambda)^* = 1$. In particular,

$$ker(S + A - \lambda_0)^* = \{\alpha e_0 : \alpha \in \mathbf{C}\}.$$

Therefore $ran(S + A - \lambda_0) = \bigvee\{e_k : k \geq 1\} := \mathcal{H}_1$. Set $T = (S + A)|_{\mathcal{H}_1}$. By Lemma 3.1.6, $S + A \sim T$. Thus

$$ran(T - \lambda)^* = \mathcal{H}_1$$

and

$$nul(T - \lambda)^* = 1$$

for all $\lambda \in \Omega := \{\lambda : |\lambda| < \frac{1}{2}\}$. Clearly, T^* admits a strict upper triangular matrix representation with respect to $\{e_k\}_{k=1}^{\infty}$. Thus

$$\bigvee \{ker T^{*k}, k \geq 1\} = \mathcal{H}_1.$$

This implies that $T^* \in \mathcal{B}_1(\Omega)$. Thus $(S + A)^* \in \mathcal{B}_1(\Omega)$.

Lemma 3.1.8   *Given $B \in \mathcal{L}(\mathcal{H})$ that admits the following upper triangular matrix representation*

$$B = \begin{bmatrix} 0 & 0 & 0 & \cdots \\ & b_{11} & b_{12} & \star \\ & & b_{22} \\ 0 & & & \ddots \end{bmatrix} \begin{matrix} e_0 \\ e_1 \\ e_2 \\ \vdots \end{matrix}$$

with respect the ONB $\{e_k\}_{k=0}^{\infty}$. If $\|B\| < 1$, then $S + B \in (SI)$, where S is the unilateral shift.

Proof   Since $\|BS^*\| < 1, (I + BS^*)$ is invertible. Set

$$A = S^*(1 + BS^*)^{-1}.$$

Thus $A - \lambda = S^*(I + BS^*)^{-1}(I - \lambda(I + BS^*)S)$ for $|\lambda| < \|I + BS^*\|^{-1}$. Note that $(I - \lambda(I + BS^*)S)$ is invertible. Therefore, $A - \lambda$ is surjective and $nul(A - \lambda) = 1$. Since A admits a strict upper triangular matrix representation with respect to the ONB $\{e_k\}_{k=0}^{\infty}$, $\bigvee \{ker A^k, k \geq 1\} = \mathcal{H}$. Denote $\tilde{\Omega} = \{\lambda \in \mathbf{C} : |\lambda| < (1 + \|B\|)^{-1}\}$, then $A \in \mathcal{B}_1(\tilde{\Omega})$ and $A \in (SI)$. Because of $A(S + B) = A(I + BS^*)S = I, A$ is a left inverse of $S + B$ and

$$Ae_0 = A(e_0 + BS^*e_0)$$

$$= A(I + BS^*)e_0$$

$$= S^*(I + BS^*)^{-1}(I + BS^*)e_0$$

$$= S^*e_0 = 0.$$

It is easy to see that $e_0 \in [ran(S + B)]^{\perp}$. By Proposition 3.1.4, $S + B \in (SI)$.

Lemma 3.1.9 *Let $B \in \mathcal{L}(\mathcal{H})$ admit the following upper triangular matrix representation*

$$B = \begin{bmatrix} 0 & b_{12} & b_{13} & \cdots \\ & b_{22} & b_{23} & \star \\ & & \ddots & \ddots \\ 0 & & & \end{bmatrix} \begin{matrix} e_0 \\ e_1 \\ \vdots \end{matrix}$$

with respect the ONB $\{e_k\}_{k=0}^{\infty}$. If $\|B\| < \frac{1}{10}$, then there exist $\tilde{B} \in \mathcal{L}(\mathcal{H})$ and $\alpha \in \mathbf{C}$ satisfying

(i) $|\alpha| < \frac{1}{9}$;

(ii) $\|\tilde{B}\| < \frac{2}{9}$ *and $\tilde{B}$ admits the following matrix representation*

$$\tilde{B} = \begin{bmatrix} 0 & 0 & 0 & \cdots \\ & \tilde{b}_{22} & \tilde{b}_{23} & \cdots \\ & & \tilde{b}_{33} & \\ 0 & & & \ddots \end{bmatrix} \begin{matrix} e_0 \\ e_1 \\ e_2 \\ \vdots \end{matrix};$$

(iii) $S + B + \alpha e_0 \otimes e_0 \sim S + \tilde{B}$ *and* $S + B + \alpha e_0 \otimes e_0 \in (SI)$*, where S is the unilateral shift.*

Proof Denote $\mathcal{H}_0 = \{\lambda e_0 : \lambda \in \mathbf{C}\}, \mathcal{H}_1 = \bigvee \{e_n : n \geq 1\}$. Then

$$S = \begin{bmatrix} 0 & 0 \\ e_1 \otimes e_0 & S_1 \end{bmatrix} \begin{matrix} \mathcal{H}_0 \\ \mathcal{H}_1 \end{matrix} \quad \text{and} \quad B = \begin{bmatrix} 0 & e_0 \otimes f \\ 0 & B_1 \end{bmatrix} \begin{matrix} \mathcal{H}_0 \\ \mathcal{H}_1 \end{matrix}$$

under the decomposition of the space $\mathcal{H} = \mathcal{H}_0 \oplus \mathcal{H}_1$. Note that S_1 is unitarily equivalent to S, i.e., $S_1 \cong S$. Set

$$g = S_1(I_{\mathcal{H}_1} + S_1 B_1^*)^{-1} f = \sum_{n=0}^{\infty} (-1)^n S_1 (S_1 B_1^*)^n f.$$

Then $g \perp e_1$ and

$$\|g\| \leq \sum_{n=0}^{\infty} \|B^*\|^n \|f\| = \frac{\|f\|}{1 - \|B^*\|} < \frac{10}{9} \|f\|.$$

Since $f = B^* e_0, \|f\| \leq \|B^*\| < \frac{1}{10}$. Therefore $\|g\| < \frac{1}{9}$. Let

$$\alpha = - < e_1, g >,$$

then $|\alpha| \leq \|f\| < \frac{1}{9}$, i.e., α satisfies (i).

Set

$$\tilde{B} = \begin{bmatrix} 0 & 0 \\ 0 & B_1 + e_1 \otimes g \end{bmatrix} \begin{matrix} \mathcal{H}_0 \\ \mathcal{H}_1 \end{matrix},$$

i.e., $\tilde{B} = B - e_0 \otimes f + e_1 \otimes g$.

Then

$$\|\tilde{B}\| = \|B_1 + e_1 \otimes g\| \leq \|B\| + \|g\| < \frac{2}{9},$$

and

$$\tilde{B} = \begin{bmatrix} 0 & 0 & 0 & \cdots \\ & \tilde{b}_{22} & \tilde{b}_{23} & \cdots \\ 0 & & \ddots & \ddots \end{bmatrix} \begin{matrix} e_0 \\ e_1 \\ \vdots \end{matrix},$$

i.e., $\tilde{B}$ satisfies (ii).

Set

$$X = \begin{bmatrix} 1 & e_0 \otimes g \\ 0 & I_{\mathcal{H}_1} \end{bmatrix},$$

i.e., $X = I + e_0 \otimes g$. It is obvious that X is invertible and $X^{-1} = I - e_0 \otimes g$.

Note that

$$X^{-1}(S + B + \alpha e_0 \otimes e_0)X$$

$$= \begin{bmatrix} e_0 \otimes e_0 & -e_0 \otimes g \\ 0 & I_{\mathcal{H}_1} \end{bmatrix} \begin{bmatrix} \alpha e_0 \otimes e_0 & e_0 \otimes f \\ e_1 \otimes e_0 & S_1 + B_1 \end{bmatrix} \begin{bmatrix} e_0 \otimes e_0 & e_0 \otimes g \\ 0 & I_{\mathcal{H}_1} \end{bmatrix}$$

$$= \begin{bmatrix} \alpha e_0 \otimes e_0 - (e_0 \otimes g)(e_1 \otimes e_0) & e_0 \otimes f - (e_0 \otimes g)(S_1 + B_1) \\ e_1 \otimes e_0 & S_1 + B_1 \end{bmatrix} \begin{bmatrix} e_0 \otimes e_0 & e_0 \otimes g \\ 0 & I_{\mathcal{H}_1} \end{bmatrix}$$

$$= \begin{bmatrix} \alpha e_0 \otimes e_0 - <e_1, g> (e_0 \otimes e_0) & e_0 \otimes f - e_0 \otimes (S_1 + B_1)^* g \\ e_1 \otimes e_0 & S_1 + B_1 \end{bmatrix} \begin{bmatrix} e_0 \otimes e_0 & e_0 \otimes g \\ 0 & I_{\mathcal{H}_1} \end{bmatrix}$$

$$= \begin{bmatrix} 0 & 0 \\ e_1 \otimes e_0 & S_1 + B_1 \end{bmatrix} \begin{bmatrix} e_0 \otimes e_0 & e_0 \otimes g \\ 0 & I_{\mathcal{H}_1} \end{bmatrix}$$

$$= \begin{bmatrix} 0 & 0 \\ e_1 \otimes e_0 & e_1 \otimes g + S_1 + B_1 \end{bmatrix}$$

$$= S + \tilde{B},$$

i.e., $S + B + \alpha e_0 \otimes e_0 \sim S + \tilde{B}$. By Lemma 3.1.8, $S + \tilde{B} \in (SI)$. Therefore,

$$S + B + \alpha e_0 \otimes e_0 \in (SI).$$

The following lemma is due to [Jiang, C.L. and Wu, P.Y. (1998)].

Lemma 3.1.10 *For given $T \in \mathcal{L}(\mathcal{H})$, there exist ONB $\{e_k\}_{k=0}^{\infty}$ and $A, B \in \mathcal{L}(\mathcal{H})$ satisfying*

(i) $T = A + B$;

(ii) A admits a strict lower triangular matrix representation with respect to $\{e_k\}_{k=0}^{\infty}$;

(iii) B admits an upper triangular matrix representation with respect to $\{e_k\}_{k=0}^{\infty}$.

We are now in a position to prove Theorem 3.1.1.

Proof of Theorem 3.1.1 By Lemma 3.1.9, we can find ONB $\{e_k\}_{k=0}^{\infty}$ and $A_1, B_1 \in \mathcal{L}(\mathcal{H})$ such that $T = A_1 + B_1$, A_1 is strict lower triangular and B_1 is upper triangular with respect to $\{e_k\}_{k=1}^{\infty}$. Denote $b = <B_1 e_0, e_0>$, then b is the upper left entry of the upper triangular matrix of B_1. Set $A_2 = A_1 + be_0 \otimes e_0$, $B_2 = B_1 - be_0 \otimes e_0$ and $r = \frac{1}{10(\|A_2\|+\|B_2\|+1)}$, $B_3 = rB_2$. Then $\| - B_3\| < \frac{1}{10}$ and $-B_3$ admits the following matrix representation

$$-B_3 = \begin{bmatrix} 0 & * & * & \cdots \\ & * & & \\ & & * & \\ 0 & & & \ddots \end{bmatrix} \begin{matrix} e_0 \\ e_1 \\ e_2 \\ \vdots \end{matrix}$$

with respect to $\{e_k\}_{k=0}^{\infty}$. By Lemma 3.1.9, there is an $\alpha \in \mathbf{C}$ with $|\alpha| < \frac{1}{9}$ such that $S - B_3 + \alpha e_0 \otimes e_0 \in (SI)$, where S is the unilateral shift. Denote

$$T_1 = -\frac{1}{r}(S - B_1 + \alpha e_0 \otimes e_0),$$

then

$$T_1 \in (SI).$$

Let $A_3 = rA_2 + \alpha e_0 \otimes e_0$ and $\lambda_0 = rb + \alpha$, then

$$\|A_1\| \leq r\|A_2\| + |\alpha| < \frac{1}{4}$$

and A_3 admits the matrix representation (3.1.3) with respect to ONB $\{e_k\}_{k=0}^{\infty}$.

Since

$$|b| \leq \|A_3\|,$$

$$|\lambda_0| \leq |rb| + |\alpha| < \frac{1}{10} + \frac{1}{9} < \frac{1}{4}.$$

By Lemma 3.1.7, $S + A_3 \in (SI)$. Define $T_2 = \frac{1}{r}(S + A_3)$, then $T_2 \in (SI)$ and

$$T_1 + T_2$$

$$= \tfrac{1}{r}(B_3 - S - \alpha e_0 \otimes e_0 + S + A_3)$$

$$= \tfrac{1}{r}(rB_2 - \alpha e_0 \otimes e_0 + rA_2 + \alpha e_0 \otimes e_0)$$

$$= B_2 + A_2$$

$$= B_1 - be_0 \otimes e_0 + A_1 + be_0 \otimes e_0$$

$$= A_1 + B_1$$

$$= T.$$

Theorem 3.1.1 indicates that every bounded linear operator acting on an infinite dimensional, separable Hilbert space can be expressed as a sum of two strongly irreducible operators. In the proof, we used the strongly irreducibility of Cowen-Douglas operators of index 1 frequently. As a matter of fact, in finite dimensional Hilbert space we have the following theorem.

Theorem 3.1.11   *Every operator on a finite dimensional Hilbert space is a sum of two strongly irreducible operators.*
Proof   Let $T \in \mathcal{L}(\mathbf{C}^n)$ and $A = T - (\frac{1}{n})\mathrm{tr}T$, where $\mathrm{tr}T$ is the trace of T, i.e., the sum of diagonal elements in the matrix representation with respect to some ONB. If $rankA \leq 1$,

$$A \cong \begin{bmatrix} 0 & a & 0 & \cdots & 0 \\ & 0 & a0 & \cdots & 0 \\ & & \ddots & \ddots & \vdots \\ & & & 0 & 0 \\ 0 & & & & 0 \end{bmatrix}.$$

Let $b = (\frac{1}{2n})\mathrm{tr}T, c \neq 0$ and $c \neq a$, then

$$T \cong \begin{bmatrix} b & c & 0 & \cdots & 0 \\ & b & c & \cdots & 0 \\ & & \ddots & \ddots & \vdots \\ & & & b & c \\ 0 & & & & b \end{bmatrix} + \begin{bmatrix} b & a-c & 0 & \cdots & 0 \\ & b & a-c & \cdots & 0 \\ & & \ddots & \ddots & \vdots \\ & & & b & a-c \\ 0 & & & & b \end{bmatrix}.$$

Since $T_1 \sim T_2 \sim J_n(b) \in (SI)$, T is the sum of two (SI) operators.

If $rank A > 1$, by the matrix theory, we have

$$A \sim \begin{bmatrix} 0 & a_{12} & \cdots & a_{1\,n-1} & a_{1n} \\ a_{21} & 0 & & & a_{2n} \\ \vdots & & \ddots & & \vdots \\ a_{n-1\,1} & & & 0 & a_{n-1\,n} \\ a_{n1} & a_{n2} & \cdots & a_{n\,n-1} & 0 \end{bmatrix}$$

$$= \begin{bmatrix} 0 & a_{12} & a_{13} & \cdots & a_{1n} \\ & 0 & a_{23} & \cdots & a_{2n} \\ & & \ddots & \ddots & \vdots \\ & & & 0 & a_{n-1\,n} \\ & & & & 0 \end{bmatrix} + \begin{bmatrix} 0 & & & & 0 \\ a_{21} & 0 & & & \\ \vdots & & \ddots & \ddots & \\ a_{n-1\,1} & a_{n-1\,2} & \cdots & 0 & \\ a_{n1} & a_{n2} & \cdots & a_{n\,n-1} & 0 \end{bmatrix},$$

where $a_{ij} \neq 0 (i \neq j)$ [Choi, M.D., Lausee, C. and Radjavi, H. (1981)]. It is easily seen that T is the sum of two (SI) operators.

3.2 Approximate Jordan Decomposition Theorem

If $\sigma(T)$ is disconnected for $T \in \mathcal{L}(\mathcal{H})$, by Riesz decomposition theorem $T \notin (SI)$, and by upper semi-continuity of spectrum, there exists an $\varepsilon > 0$ such that $T + A \notin (SI)$ for all $A \in \mathcal{L}(\mathcal{H})$ with $\|A\| < \varepsilon$. If $\sigma(T)$ is connected, we have the following theorem stated in section 1.5.

Spectral picture theorem of (SI) operators *Given $T \in \mathcal{L}(\mathcal{H})$ with connected spectrum $\sigma(T)$, there exists a (SI) operator L satisfying*
   (i) $\Lambda(L) = \Lambda(T)$;
   (ii) $T \in \mathcal{S}(L)^-$;
   (iii) If there is another $L_1 \in (SI)$ such that $\Lambda(L_1) = \Lambda(T)$, then $L_1 \in \mathcal{S}(L)^-$.

By the above spectral picture theorem, we get the following theorem.

First approximate Jordan decomposition theorem[Jiang, C.L. and Wang, Z.Y. (1996b)]   *Let $T \in \mathcal{L}(\mathcal{H})$, then there exists $A \in \mathcal{L}(\mathcal{H})$ such that A is the direct sum of at most infinitely many (SI) operators and satisfying*
   (i) $T \in \mathcal{S}(A)^-$;
   (ii) If $B \in \mathcal{L}(\mathcal{H})$ and $B \in \mathcal{S}(T)^-$, then $B \in \mathcal{S}(A)^-$.

The interested reader is referred to [Jiang, C.L. and Wang, Z.Y. (1998)] for the proof of the theorem. In general, we call this theorem the unique decomposition theorem with respect to similarity orbit.

In 1988, D.A. Herrero raised the following conjecture in personal communication: Every operator with connected spectrum is a small compact perturbation of (SI) operator. Or precisely, given $T \in \mathcal{L}(\mathcal{H})$ with connected $\sigma(T)$ and given an $\varepsilon > 0$, there exists a compact operator K with $\|K\| < \varepsilon$ such that $T + K \in (SI)$. Following theorem confirms Herrero's conjecture.

Theorem 3.2.1   *Let $T \in \mathcal{L}(\mathcal{H})$ (or $T \in \mathcal{B}_n(\Omega)$) with connected $\sigma(T)$. Then given $\varepsilon > 0$, there is a compact operator K, $\|K\| < \varepsilon$, such that*

$$T + K \in (SI) \quad (or \ T + K \in \mathcal{B}_n(\Omega) \cap (SI)).$$

We will prove the theorem only in the $\mathcal{B}_n(\Omega)$ operator case. Reader is referred to [Ji, Y.Q. and Jiang, C.L. (2002)], for the general case, which is a little more complex. Before we begin our proof, we need the following lemmas.

Lemma 3.2.2[Fialkow, L.A. (1981)]   *Let $A, B \in \mathcal{L}(\mathcal{H})$, then the following are equivalent:*
   (i) $\sigma_r(A) \cap \sigma_l(B) = \emptyset$;
   (ii) $\tau_{A,B}$ is surjective;
   (iii) $ran\tau_{A,B}$ contains $\mathcal{K}(\mathcal{H})$.

Lemma 3.2.3[Herrero, D.A. and Jiang, C.L. (1990)]   *Let $A, B \in \mathcal{L}(\mathcal{H})$. Assume that $\mathcal{H} = \bigvee\{ker(\lambda - B)^k : \lambda \in \Gamma, k \geq 1\}$ for a certain subset Γ of the point spectrum $\sigma_p(B)$ of B, and $\sigma_p(A) \cap \Gamma = \emptyset$, then $\tau_{A,B}$ is injective.*

Let Ω be a non-empty open connected subset of $\mathbf{C}$ such that $(\overline{\Omega})^0 = \Omega$, where $(\overline{\Omega})^0$ denotes the interior of the closure $\overline{\Omega}$ of Ω. Given $z_0 \in \Omega$, there exists a probability measure μ supported by $\Gamma := \partial\Omega$, the boundary of Ω, satisfying $f(z_0) = \int_\Gamma f d\mu$ for every function f analytic on $\overline{\Omega}$ [Herrero,

D.A. (1990)]. Let $M(\Gamma) =$ "multiplication by λ" on $L^2(\Gamma, \mu)$. The subspace $H^2(\Gamma)$ of $L^2(\Gamma, \mu)$, spanned by the functions analytic on $\overline{\Omega}$, is obviously invariant under $M(\Gamma)$, i.e.,

$$M(\Gamma) = \begin{bmatrix} M_+(\Gamma) & Z \\ 0 & M_-(\Gamma) \end{bmatrix} \begin{matrix} H^2(\Gamma) \\ L^2(\Gamma, \mu) \ominus H^2(\Gamma) \end{matrix}.$$

Lemma 3.2.4[Herrero, D.A. (1990)] *Let $M(\Gamma), M_+(\Gamma)$ and $M_-(\Gamma)$ be as above, then*

(i) $M(\Gamma)$ is normal and both $M_+(\Gamma)$ and $M_-(\Gamma)$ are essentially normal;

(ii) $\sigma(M(\Gamma)) = \sigma_e(M(\Gamma)) = \sigma_e(M_+(\Gamma)) = \sigma_e(M_-(\Gamma)) = \Gamma$; $\sigma(M_+(\Gamma)) = \sigma(M_-(\Gamma)) = \overline{\Omega}$; $ind(\lambda - M_+(\Gamma)) = -nul(\lambda - M_+(\Gamma)) = -nul(\lambda - M_+(\Gamma))^* = -1$ for all $\lambda \in \Omega$;

(iii) If Ω is simply connected, then $\|Z\| \leq \frac{m(\Omega)}{\pi}$, where m denotes the planar Lebesgue measure.

It is obvious that $M_+(\Gamma)^* \in \mathcal{B}_1(\Omega^*)$ and $M_-(\Gamma) \in \mathcal{B}_1(\Omega)$, where $\Omega^* := \{\overline{\lambda} : \lambda \in \Omega\}$.

An operator is quasitriangular if $ind(\lambda - T) \geq 0$ for all $\lambda \in \rho_{s-F}(T)$. The Weyle spectrum $\sigma_W(T)$ of $T \in \mathcal{L}(\mathcal{H})$ is defined by $\sigma_W(T) := \bigcap \{\sigma(T + K) : K \in \mathcal{K}(\mathcal{H})\}$. It is well known that $\sigma_W(T) := \{\lambda \in \mathbf{C} : \lambda - T$ is not a Fredholm operator of index $0\}$.

Lemma 3.2.5[Herrero, D.A. (1990)] *Suppose that $T \in \mathcal{L}(\mathcal{H})$ is quasitriangular and $\sigma(T) = \sigma_W(T)$. Let $\Gamma = \{\lambda_n\}_{n=1}^{\infty} \subset \sigma(T)$ satisfy that each clopen subset of $\sigma(T)$ intersects the closure of Γ. Given $\varepsilon > 0$, there exist a compact operator $K, \|K\| < \varepsilon$, and an ONB $\{e_k\}_{k=1}^{\infty}$ such that*

$$T + K = \begin{bmatrix} a_1 & & \star \\ & a_2 & \\ 0 & & \ddots \end{bmatrix} \begin{matrix} e_1 \\ e_2 \\ \vdots \end{matrix},$$

where the set $\{a_i\}_{i=1}^{\infty} = \Gamma$ and $card\{j : a_j = a_i\} = \infty$ for each i.

Lemma 3.2.6[Herrero, D.A. (1990)] *Given $T \in \mathcal{L}(\mathcal{H})$, nonempty subset $\Gamma \subset \sigma_{lre}(T)$ and $\varepsilon > 0$, there exists a compact operator $K, \|K\| < \varepsilon$, such that*

$$T + K = \begin{bmatrix} N & \star \\ 0 & A \end{bmatrix} \begin{matrix} \mathcal{H}_1 \\ \mathcal{H}_2 \end{matrix},$$

where N is a diagonal operator of uniformly infinite multiplicity with

$$\sigma(N) = \overline{\Gamma}, \sigma(A) = \sigma(T), \sigma_{lre}(A) = \sigma_{lre}(T)$$

and

$$ind(A - \lambda) = ind(T - \lambda)$$

for each $\lambda \in \rho_{s-F}(T)$.

Lemma 3.2.7   *Suppose that*

$$T = \begin{bmatrix} T_1 & T_{12} \\ 0 & T_2 \end{bmatrix} \begin{matrix} \mathcal{H}_1 \\ \mathcal{H}_2 \end{matrix}$$

satisfies the following conditions:
  (i) $T_2 \in \mathcal{B}_n(\Omega) \cap (SI)$;
  (ii)

$$\sigma_0(T) = \sigma(T_1) \cap \Omega = \{\lambda_k\}_{k=1}^{\infty},$$

$$nul(T_1 - \lambda_k) = n, k = 1, 2, \cdots$$

and

$$\bigvee \{ker(\lambda_k - T_1) : k = 1, 2, \cdots\} = \mathcal{H}_1;$$

  (iii) $B_k := P_{ker(T_1 - \lambda_k)^*} T_{12}|_{ker(T - \lambda_k)}$ is injective, where $P_{ker(T_1 - \lambda_k)^*}$ is the orthogonal projection from $\mathcal{H}$ onto $ker(T_1 - \lambda_k)^*$.
  Then $T \in \mathcal{B}_n(\Omega) \cap (SI)$.
Proof   **{Claim}** $ker(T - \lambda_k) = ker(T_1 - \lambda_k)$.
  Assume that

$$\begin{bmatrix} T_1 - \lambda_k & T_{12} \\ 0 & T_2 - \lambda_k \end{bmatrix} \begin{bmatrix} x \\ y \end{bmatrix} = 0$$

for $x \in \mathcal{H}_1$ and $y \in \mathcal{H}_2$, then $(T_1 - \lambda_k)x + T_{12}y = 0$ and $(T_2 - \lambda_k)y = 0$. Since

$$P_{ker(T_1 - \lambda_k)^*}(T_1 - \lambda_k) = 0,$$

$$P_{ker(T_1 - \lambda_k)^*} T_{12} y = 0.$$

Since $P_{ker(T_1 - \lambda_k)^*} T_{12}$ is injective, $y = 0$. This implies that

$$ker(T - \lambda_k) = ker(T_1 - \lambda_k)$$

and

$$\bigvee\{ker(T - \lambda_k) : k \geq 1\} = \mathcal{H}_1.$$

Since $\sigma_0(T_1) \cap \Omega = \{\lambda_k\}_{k=1}^{\infty}$, there exists connected open subset $\Omega_1 \subset \Omega \cap \rho(T_1)$. Since $T_2 \in \mathcal{B}_n(\Omega), \bigvee\{ker(T_2 - \lambda) : \lambda \in \Omega_1\} = \mathcal{H}_2$. Furthermore,

$$\bigvee\{ker(T - \lambda) : \lambda \in \{\lambda_k\}_{k=1}^{\infty} \cup \Omega_1\} = \mathcal{H}_1 \oplus \mathcal{H}_2.$$

Thus $T \in \mathcal{B}_n(\Omega)$.

Now we will prove that $T \in (SI)$. Assume that $P \in \mathcal{A}'(T)$ is an idempotent. Since $ker(T - \lambda_k) = ker(T_1 - \lambda_k), k = 1, 2, \cdots, \mathcal{H}_1$ is an invariant subspace of P, i.e.,

$$P = \begin{bmatrix} P_1 & P_{12} \\ 0 & P_2 \end{bmatrix} \begin{matrix} \mathcal{H}_1 \\ \mathcal{H}_2 \end{matrix}.$$

Since $T_2 \in (SI)$, $P_2 = \delta I_{\mathcal{H}_2}, \delta = 0$ or 1. Without loss of generality, we can assume that $\delta = 0$, i.e., $P_2 = 0$. Thus $ranP \subset \mathcal{H}_1$. Note that $P \in \mathcal{A}'(T)$ and $T \in \mathcal{B}_n(\Omega)$. If $P \neq 0$, then $T|_{ranP} \in \mathcal{B}_m(\Omega)$ for some $m, 1 \leq m \leq n$. Since

$$T|_{ranP} = T_1|_{ranP},$$

$$T_{11} = T_1|_{ranP} \in \mathcal{B}_m(\Omega).$$

It is easy to see that $ranP$ is an invariant subspace of T_1. Therefore,

$$T_1 = \begin{bmatrix} T_{11} & E \\ 0 & T_{22}' \end{bmatrix} \begin{matrix} ranP \\ \mathcal{H}_1 \ominus ranP \end{matrix}.$$

By the condition (ii) of the lemma, $T_1 - \lambda$ is invertible for $\lambda \in \Omega \backslash \{\lambda_k\}_{k=1}^{\infty}$. Thus we can find an invertible $X \in \mathcal{L}(\mathcal{H})$,

$$X = \begin{bmatrix} X_{11} & X_{12} \\ X_{21} & X_{22} \end{bmatrix} \begin{matrix} ranP \\ \mathcal{H}_1 \ominus ranP \end{matrix}$$

such that

$$\begin{bmatrix} X_{11} & X_{12} \\ X_{21} & X_{22} \end{bmatrix} \begin{bmatrix} T_{11} - \lambda & E \\ 0 & T_{22}' - \lambda \end{bmatrix} = \begin{bmatrix} I & 0 \\ 0 & I \end{bmatrix}.$$

Simple computations indicate that $X_{11}(T_{11} - \lambda) = I$ Thus $T_{11} - \lambda$ is invertible. This contradicts $T_{11} \in \mathcal{B}_m(\Omega)$. Therefore $P_1 = P_2 = 0$ and $T \in (SI)$.

Lemma 3.2.8[Ji, Y.Q., Jiang, C.L. and Wang, Z.Y. (1996)] *Suppose that B is essentially normal and $B \in \mathcal{B}_n(\Omega)$. Then for given $\varepsilon > 0$, there exists a compact operator K with $\|K\| < \varepsilon$, such that $T + K \in \mathcal{B}_n(\Omega) \cap (SI)$.*

Lemma 3.2.9[Herrero, D.A. (1987)] *Suppose that $R \in \mathcal{L}(\mathcal{H})$ and satisfies the following conditions:*

(i) $\sigma(R)$ *and* $\sigma_W(R)$ *are connected and contain a connected open set* Ω;

(ii) $ind(\lambda - R) \geq 0$ *for all* $\lambda \in \rho_{s-F}(R)$;

(iii) $\rho_{s-F}(R) \supset \Omega$ *and* $ind(\lambda - R) = n$ *for all* $\lambda \in \Omega$.

Then given $\varepsilon > 0$, there exists a compact operator $K, \|K\| < \varepsilon$, such that $R - K \in \mathcal{B}_n(\Omega)$.

Lemma 3.2.10[Herrero, D.A. (1984)] *Let $T \in \mathcal{L}(\mathcal{H})$ be quasitriangular operator and $\lambda \in \sigma_e(T)$. If $\sigma_W(T)$ is connected, then given $\varepsilon > 0$ and natural number m, there is a compact operator $K, \|K\| < \varepsilon$, such that $nul T_\varepsilon^k = km, k \geq 1$, and $\bigvee \{ker T_\varepsilon^k : k \geq 1\} = \mathcal{H}$, where $T_\varepsilon = T + K - \lambda$.*

Lemma 3.2.11 *Let $T \in \mathcal{L}(\mathcal{H})$ be quasitriangular operator and let Ω be a connected component of $\rho_{s-F}^{(0)}(T)$ such that $\sigma_W(T) \cup \Omega$ is connected. Given $\varepsilon > 0$ and natural number n, there exist a sequence $\{\lambda_k\}_{k=1}^\infty$ of complex numbers in Ω and a compact operator K with $\|K\| < \varepsilon$, such that $\{\lambda_k\}_{k=1}^\infty \subset \sigma_0(T + K)$ and*

$$
T + K \cong \begin{bmatrix} \lambda_1 & & \star \\ & \lambda_2 & \\ 0 & & \ddots \end{bmatrix} \begin{matrix} \mathbf{C}^n \\ \mathbf{C}^n \\ \vdots \end{matrix},
$$

where $\rho_{s-F}^{(0)}(T) := \{\lambda \in \rho_{s-F}(T) : ind(\lambda - T) = 0\}$.

Proof Note that $\partial \Omega \subset \sigma_{lre}(T)$. We choose a dense subset $\{\mu_k\}_{k=1}^\infty$ of $\partial \Omega$. Then each clopen subset of $\sigma_W(T)$ intersects $\{\mu_k\}_{k=1}^\infty$. By Lemma 3.2.5, there exists a compact $C_1, \|C_1\| < \frac{\varepsilon}{8}$, such that

$$
\bigvee \{ker(T + C_1 - \mu_k)^j : j \geq 1, k \geq 1\} = \mathcal{H}
$$

and

$$
dim \bigvee \{ker(T + C_1 - \mu_k)^j : j \geq 1\} = \infty
$$

for all $k \geq 1$. Without loss of generality, we can assume that $\sigma_p(T^* + C_1^*) = \emptyset$. Since

$$
\Omega \subset \rho_{s-F}^{(0)}(T) = \rho_{s-F}^{(0)}(T + C_1),
$$

$$\Omega \cap \sigma(T + C_1) = \emptyset.$$

Denote

$$\mathcal{M}_1 = \bigvee \{ker(T + C_1 - \mu_1)^j : j \geq 1\},$$

then $\mathcal{M}$ is an invariant subspace of $T + C_1$ and $dim\mathcal{M}_1 = \infty$. Let

$$T_1 = (T + C_1)|_{\mathcal{M}_1},$$

then $\mu_1 \in \sigma_e(T_1)$ and $\sigma(T_1) = \sigma_W(T_1)$ is connected. If

$$dim \bigvee \{ker(T + C_1 - \mu_i)^j : i = 1, 2, \cdots, j \geq 1\} \ominus \mathcal{M}_1 < \infty,$$

then

$$ker(T + C_1 - \mu_i)^j \subset \mathcal{M}_1$$

for all $j \geq 1$. Denote

$$\mathcal{M}_k = \bigvee \{ker(T + C_1 - \mu_i)^j : 1 \leq i \leq k\} \ominus \bigvee \{ker(T + C_1 - \mu_i)^j : 1 \leq i \leq k - 1\},$$

where $k \geq 1, j \geq 1$.

Without loss of generality, we can assume that $dim\mathcal{M}_k = \infty$. Thus

$$T + C_1 = \begin{bmatrix} T_1 & & & \star \\ & T_2 & & \\ & & T_3 & \\ 0 & & & \ddots \end{bmatrix}.$$

Since $\bigvee \{ker(T_k - \mu_k)^j : j \geq 1\} = \mathcal{M}_k$, $\sigma_W(T_k) = \sigma(T_k)$ is connected and $\sigma_p(T_k^*) \subset \{\overline{\mu_k}\}$. Thus T_k is a triangular, $\mu_k \in \sigma_e(T_k)$ and $\Omega \cap \sigma(T_k) = \emptyset$ for each $k \geq 1$. By Lemma 3.2.10, there exists a compact operator $K_k, \|K_k\| < \frac{\varepsilon}{8^k}$, such that

$$T_k + K_k = \begin{bmatrix} \mu_k & & & \star \\ & \mu_k & & \\ & & \mu_k & \\ 0 & & & \ddots \end{bmatrix} \begin{matrix} \mathbf{C}^n \\ \mathbf{C}^n \\ \mathbf{C}^n \\ \vdots \end{matrix}, \quad k = 1, 2, \cdots.$$

Since $\{\mu_k\}_{k=1}^{\infty} \subset \partial\Omega$, we can choose pairwise distinct numbers $\{\lambda_j^{(k)} : k \geq 1, j \geq 1\}$ in Ω such that $|\mu_k - \lambda_j^{(k)}| < \frac{\varepsilon}{j8^k}$ for all j and k. Thus there is a compact operator $\overline{K}_k$ with $\|\overline{K}_k\| < \frac{\varepsilon}{8^k}$ such that

$$
A_k = T_k + K_k + \overline{K}_k \cong
\begin{bmatrix}
\lambda_k^{(1)} & & & \star \\
& \lambda_k^{(2)} & & \\
& & \lambda_k^{(3)} & \\
0 & & & \ddots
\end{bmatrix}
\begin{matrix}
\mathbf{C}^n \\
\mathbf{C}^n \\
\mathbf{C}^n \\
\vdots
\end{matrix}, \quad k = 1, 2, \cdots.
$$

Furthermore, there is a compact operator $C_2, \|C_2\| < \frac{\varepsilon}{2}$ satisfying

$$
T + C_1 + C_2 =
\begin{bmatrix}
A_1 & & & \star \\
& A_2 & & \\
& & A_3 & \\
0 & & & \ddots
\end{bmatrix}
\begin{matrix}
\mathcal{M}_1 \\
\mathcal{M}_2 \\
\mathcal{M}_3 \\
\vdots
\end{matrix},
$$

where $C_2 = \sum\limits_{k=1}^{\infty} \oplus (K_k + \overline{K}_k)$. Set $K = C_1 + C_2$, then K is compact and $\|K\| < \varepsilon$. Rearrange $\{\lambda_j^{(k)} : j \geq 1, k \geq 1\}$ as $\{\lambda_k\}_{k=1}^{\infty}$. It is not difficult to see that K and $\{\lambda_k\}_{k=1}^{\infty}$ satisfy all the requirements of the lemma.

Now we are in a position to prove Theorem 3.2.1.

Proof of Theorem 3.2.1   For $T \in \mathcal{B}_n(\Omega)$, assume that $\partial\Omega \subset \sigma_{lre}(T)$ (otherwise, replace Ω with the component of $\rho_{s-F}(\Omega)$ containing Ω). Denote $\Phi = (\overline{\Omega})^0$ and $\Gamma = \partial\Phi$. Then $\Gamma \subset \sigma_{lre}(T)$. By Lemma 3.2.6 and Lemma 3.2.9, we can find a compact operator $K_1, \|K_1\| < \frac{\varepsilon}{4}$ such that

$$
T + K_1 \cong
\begin{bmatrix}
T_1 & \star \\
0 & N
\end{bmatrix}
\begin{matrix}
\mathcal{H}_1 \\
\mathcal{H}_2
\end{matrix},
$$

where $T_1 \in \mathcal{B}_n(\Omega)$, N is a diagonal operator of uniformly infinite multiplicity and $\sigma(N) = \Gamma$. Let $M(\Gamma)$ be given as in Lemma 3.2.4, then $M(\Gamma)^{(n)} = \bigoplus\limits_{k=1}^{n} M(\Gamma)$ is normal and

$$
M(\Gamma)^{(n)} =
\begin{bmatrix}
\bigoplus\limits_{k=1}^{n} M_+(\Gamma) & \bigoplus\limits_{k=1}^{n} Z \\
0 & \bigoplus\limits_{k=1}^{n} M_-(\Gamma)
\end{bmatrix}.
$$

It is clear that $\sigma(M(\Gamma)^{(n)}) = \Gamma$.

By Voiculescu theorem [Voiculescu, D. (1976)] there exists a compact operator F_1 such that $\|F_1\| < \frac{\varepsilon}{8}$ and $N + F_1 \cong M(\Gamma)^{(n)}$. Thus

$$T + K_1 + K_2 \cong \begin{bmatrix} T_1 & \star & \star \\ 0 & \bigoplus\limits_{k=1}^{n} M_+(\Gamma) & \bigoplus\limits_{k=1}^{n} Z \\ 0 & 0 & \bigoplus\limits_{k=1}^{n} M_-(\Gamma) \end{bmatrix},$$

where K_2 is compact, $K_2 = 0 \oplus F_2$ and $\|K_2\| < \frac{\varepsilon}{8}$. Set

$$B_1 = \begin{bmatrix} T_1 & \star \\ 0 & \bigoplus\limits_{k=1}^{n} M_+(\Gamma) \end{bmatrix}.$$

It is obvious that $\sigma(B_1) = \sigma(T)$, $\Omega \subset \rho_F^{(0)}(B_1)$ and $ind(B_1 - \lambda) > 0$ for all λ in $\sigma(B_1) \cap \rho_F(B_1) \backslash \Omega$. Therefore B_1 is quasitriangular, Ω is a connected component of $\rho_F^{(0)}(B_1)$ and $\Omega \cap \sigma_W(B_1) = \sigma(T)$ is connected. By Lemma 3.2.11 we can find a sequence $\{\lambda_k\}_{k=1}^{\infty}$ of pairwise distinct complex numbers in Ω and a compact operator $E, \|E\| < \frac{\varepsilon}{16}$ such that $\{\lambda_k\}_{k=1}^{\infty} \subset \sigma_0(A_1)$ and

$$A_1 = B + E \cong \begin{bmatrix} \lambda_1 & & & \star \\ & \lambda_2 & & \\ & & \lambda_3 & \\ 0 & & & \ddots \end{bmatrix} \begin{matrix} \mathbf{C}^n \\ \mathbf{C}^n \\ \mathbf{C}^n \\ \vdots \end{matrix}.$$

Summarizing the arguments above, we can find a compact operator $K_3, \|K_3\| < \frac{\varepsilon}{16}$ such that

$$T + K_1 + K_2 + K_3 \cong \begin{bmatrix} A_1 & \star \\ 0 & \bigoplus\limits_{k=1}^{n} M_-(\Gamma) \end{bmatrix} \begin{matrix} \mathcal{H}_3 \\ \mathcal{H}_4 \end{matrix}.$$

Denote $B_2 = \bigoplus\limits_{k=1}^{n} M_-(\Gamma)$. Note that B_2 is essentially normal and $B_2 \in \mathcal{B}_n(\Phi) \subset \mathcal{B}_n(\Omega)$ [Cowen, M.J. and Douglas, R. (1977)]. By Lemma 3.2.9, we can find a compact operator $F_2, \|F_2\| < \frac{\varepsilon}{32}$ such that

$$A_2 = B_2 + F_2 \in \mathcal{B}_n(\Omega) \cap (SI).$$

Thus there is a compact operator $K_4, \|K_4\| < \frac{\varepsilon}{32}$ such that

$$T + K_1 + K_2 + K_3 + K_4 \cong \begin{bmatrix} A_1 & A_{12} \\ 0 & A_2 \end{bmatrix}.$$

According to Lemma 3.2.7, it is sufficient for us to find a compact operator E_{12} such that $\|E_{12}\| < \frac{\varepsilon}{64}$ and $P_{ker(A_1-\lambda_k)^*}(A_{12}+E_{12})|_{ker(A_1-\lambda_k)}$ is injective. Since

$$\{\lambda_k\}_{k=1}^{\infty} \subset \sigma_p(A_1) \cap \rho_F(A_2)$$

and

$$nul(A_1 - \lambda_k) = n,$$

$$A_2 \cong \begin{bmatrix} \lambda_1 & G_{12} & & \vdots & & \\ & \lambda_2 & & \vdots & \star & \\ & & \ddots & \vdots & & \\ \cdots & \cdots & \cdots & \cdots & \cdots & \cdots \\ & 0 & & \vdots & & A_\infty \end{bmatrix} \begin{matrix} \mathbf{C}^n \\ \mathbf{C}^n \\ \vdots \\ \\ \mathcal{H}_\infty \end{matrix} \cong \begin{bmatrix} G_1 & \star \\ 0 & G_2 \end{bmatrix} \begin{matrix} \oplus \mathbf{C}^n \\ \mathcal{H}_\infty \end{matrix},$$

where

$$G_1 = \begin{bmatrix} \lambda_1 & G_{12} & G_{13} & \cdots \\ & \lambda_2 & G_{23} & \cdots \\ & & \lambda_3 & \ddots \\ 0 & & & \ddots \end{bmatrix} \begin{matrix} \mathbf{C}^n \\ \mathbf{C}^n \\ \mathbf{C}^n \\ \vdots \end{matrix}$$

and $G_2 = A_\infty$. Since $A_2 \in \mathcal{B}_n(\Omega), ker(A_2 - \lambda_k) = ker(G_1 - \lambda_k)$ and

$$G = \begin{bmatrix} A_1 & E_1 & E_2 \\ 0 & G_1 & \star \\ 0 & 0 & G_2 \end{bmatrix} \begin{matrix} \bigoplus_{k=1}^{\infty} \mathbf{C}^n \\ \bigoplus_{k=1}^{\infty} \mathbf{C}^n \\ \mathcal{H}_\infty \end{matrix}.$$

We need only to find a compact operator $F_3, \|F_3\| < \frac{\varepsilon}{64}$, such that

$$P_{ker(A-\lambda_k)^*}(E_1+F_3)|_{ker(G_1-\lambda_k)}$$

is injective. Since $\lambda_k \in \sigma_0(A_1)$ for $k \geq 1$, A_1 has the following expression:

$$
A_1 = \left[
\begin{array}{ccccc|c}
\overline{A}_\infty & \vdots & & \star & & \mathcal{H}_\infty \\
\cdots & \cdots & \cdots & \cdots & \cdots & \cdots \\
& \vdots & \ddots & \vdots & \vdots & \vdots & \vdots \\
& \vdots & & \lambda_3 & C_{23} & C_{13} & \mathbf{C}^n \\
& \vdots & & & \lambda_2 & C_{12} & \mathbf{C}^n \\
& \vdots & & & & \lambda_1 & \mathbf{C}^n
\end{array}
\right].
$$

Thus $\lambda_k \notin \sigma(\overline{A}_\infty)$ for all $k \geq 1$. Under the above decomposition,

$$
E_1 = \left[
\begin{array}{cccccc}
\star & \star & \star & \cdots & \vdots & \star \\
\cdots & \cdots & \cdots & \cdots & \cdots & \cdots \\
\vdots & \vdots & \vdots & \vdots & \vdots & \vdots \\
E_{31} & E_{32} & E_{33} & \cdots & \vdots & \star \\
E_{21} & E_{22} & E_{23} & \cdots & \vdots & \star \\
E_{11} & E_{12} & E_{13} & \cdots & \vdots & \star
\end{array}
\right],
$$

where E_{ij} is an operator from one n-dimensional space to another for $i, j \geq 1$. It is obvious that we can choose F_{11} with $\|F_{11}\| < \frac{\varepsilon}{4}$ such that $E_{11} + F_{11}$ is invertible. Inductively, we can choose F_{jj} with $\|F_{jj}\| < \frac{\varepsilon}{64^j}$ such that $E_{j+1,j+1} + W_j X_j^{-1} V_j + F_{j+1,j+1}$ is invertible, where

$$
W_1 = (G_{j,j+1}, G_{j-1,j+1}, \cdots, G_{1,j+1}, E_{j+1,1}, \cdots, E_{j+1,j}),
$$

$$
X_j = \lambda_j - \left[
\begin{array}{cccccc}
\lambda_j & c_{j,j-1} & \cdots & c_{j,1} & E_{j,1} & \cdots & E_{jj} + F_{jj} \\
& \lambda_{j-1} & \ddots & c_{j-1,1} & E_{j-1,1} & \cdots & E_{j-1,j} \\
& & \ddots & \vdots & \vdots & \vdots & \vdots \\
& & & \lambda_1 & E_{11} + F_{11} & \cdots & E_{1,j} \\
& & & & \lambda_2 & \cdots & G_{1,j} \\
& & & & & \ddots & \\
& & & & & & \lambda_j
\end{array}
\right]
$$

and

$$V_j = \begin{bmatrix} E_{j,j+1} \\ E_{j-1,j+1} \\ \vdots \\ E_{1,j+1} \\ G_{1,j+1} \\ \vdots \\ G_{j,j+1} \end{bmatrix}.$$

Set

$$F_2 = \begin{bmatrix} & & 0 & \vdots & 0 \\ \cdots & \cdots & \cdots & \cdots & \cdots \cdots \\ & & & \cdots & \vdots \\ 0 & & F_{33} & & \vdots \\ & F_{22} & & & \vdots \\ F_{11} & & 0 & & \vdots \end{bmatrix},$$

then F_2 is compact and $\|F_2\| < \frac{\varepsilon}{64}$.

From the construction we can see $P_{ker(A_1-\lambda_k)^*}(E_1 + F_1)|_{ker(A_2-\lambda_k)}$ is injective for all $k \geq 1$. Therefore the proof of the theorem is now complete.

Definition 3.2.12 A sequence $\{P_j : 1 \leq j \leq l\}(1 \leq l \leq \infty)$ of nonzero idempotents in $\mathcal{L}(\mathcal{H})$ is called a spectral family if there exists an invertible operator X such that $\{XP_jX^{-1} : 1 \leq j \leq l\}$ is a sequence of pairwise orthogonal projections and $\sum\limits_{j=1}^{l} P_j = I_{\mathcal{H}}$.

Definition 3.2.13   Let $T \in \mathcal{L}(\mathcal{H})$ and $\mathcal{P} = \{P_j : 1 \leq j \leq l\}(1 \leq l \leq \infty)$ be a spectral family. $\mathcal{P}$ is called a strongly irreducible decomposition of T if the following conditions are satisfied for all j:

(i) $P_jT = TP_j$;
(ii) $T|_{ran P_j} \in (SI)$.

Note, if there exists a strongly irreducible decomposition of T, then we say that T has a strongly irreducible decomposition. In other words, T has a strongly irreducible decomposition means that T is the topological direct sum of at most countably many (SI) operators, or, T is similar to the orthogonal sum of at most countably many (SI) operators.

Definition 3.2.14　Let $\mathcal{P}_1 = \{P_j : 1{\leq}j{\leq}l_1\}$ and $\mathcal{P}_2 = \{Q_j : 1{\leq}j{\leq}l_2\}$ be two strongly irreducible decomposition of T. $\mathcal{P}_1$ and $\mathcal{P}_2$ are said to be similar about T if

　(i)　$l_1 = l_2 = l$;

　(ii)　there exist a permutation π of $\{j : 1{\leq}j{\leq}l\}$ and invertible operator

$$X_i{\in}\mathcal{L}(ranP_j, ranQ_{\pi(j)})$$

such that

$$sup\{\|X_j\|, \|X_j^{-1}\|, 1{\leq}j{\leq}l\} < +\infty$$

and

$$X_jT|_{ranP_j} = T_{ranQ_{\pi(j)}}X_j, 1{\leq}j{\leq}l.$$

In Chapter 4, we will show that $\mathcal{P}_1$ and $\mathcal{P}_2$ are similar about T if and only if there exists an invertible operator $X{\in}\mathcal{A}'(T)$ such that

$$\{XP_jX^{-1} : 1{\leq}j{\leq}l\} = \{Q_j : 1{\leq}j{\leq}l\}.$$

Definition 3.2.15　Suppose that T has strongly irreducible decomposition. T is said to have a *unique strongly irreducible decomposition up to similarity* if any two of the (SI) decomposition of T are similar.

The Jordan canonical theorem in finite dimensional space means essentially that each $n{\times}n$ matrix has a unique (SI) decomposition up to similarity. For operators in $\mathcal{L}(\mathcal{H})$, it is very difficult to obtain a "Jordan canonical theorem". In Chapter 5 of this monograph we will prove the following theorem.

Theorem FJ　*Every Cowen-Douglas operator has a unique (SI) decomposition up to similarity.*

Using Theorem FJ, we obtain the following two approximate Jordan canonical theorems.

Theorem 3.2.16　*Given $T{\in}\mathcal{L}(\mathcal{H})$ and $\varepsilon > 0$, there exists $A{\in}\mathcal{L}(\mathcal{H})$ such that A has a unique (SI) decomposition up to similarity and $\|T - A\| < \varepsilon$.*

Theorem 3.2.17　*Let $T{\in}\mathcal{L}(\mathcal{H})$ be quasitriangular. Assume that $\sigma(T)$ consists of finitely many components and each component intersects $\rho_F(T)$.*

Given $\varepsilon > 0$ there exists a compact operator K with $\|K\| < \varepsilon$ such that $T + K$ has a unique strongly irreducible decomposition up to similarity.

Theorem 3.2.16 can be proved by using of Theorem JW2 in Chapter 1 and Theorem FJ, and Theorem 3.2.17 can be proved by Theorem 3.2.1 and Lemma 3.2.4.

3.3   Open Problems

1. Given $T \in \mathcal{L}(\mathcal{H})$ and $\varepsilon > 0$, does there exist a compact operator K with $\|K\| < \varepsilon$ such that $T + K$ has unique (SI) decomposition up to similarity?
2. Is every operator in $\mathcal{L}(\mathcal{H})$ the sum of two Cowen-Douglas operators of index 1?
3. Given $T \in \mathcal{L}(\mathcal{H})$ and $\varepsilon > 0$, does there exist an integer $p, 1 \leq p < \infty$ and $K \in C^p(\mathcal{H})$ with $\|K\|_p < \varepsilon$ such that $T + K \in (SI)$? where

$$C^p(\mathcal{H}) = \{K \in \mathcal{K}(\mathcal{H}) : \sum_{n=1}^{\infty} \lambda_n^p < \infty, \lambda_n \in \sigma_p(K^*K)^{\frac{1}{2}}\}$$

and

$$\|K\|_p = (\sum_{n=1}^{\infty} \lambda_n^p)^{\frac{1}{p}}.$$

3.4   Remark

The concept of unique strongly irreducible decomposition up to similarity appeared first in [Jiang, C.L. and Wang, Z.Y. (1998)]. Theorem 3.1.1 is given by [Yue, H. (2002)]. Before that, [Jiang, C.L. and Wu, P.Y. (1998)] proved that each operator in $\mathcal{L}(\mathcal{H})$ is the sum of three (SI) operators, and each triangular or compact operator is the sum of two (SI) operators. Theorem 3.1.11 is given by [Jiang, C.L. and Wu, P.Y. (1998)]. Theorem 3.2.1 is due to [Ji, Y.Q. and Jiang, C.L. (2002)]. Theorem 3.2.16 and Theorem 3.2.17 are both proved by [Jiang, C.L.(1)].

Chapter 4

Unitary Invariant and Similarity Invariant of Operators

In this chapter $\mathcal{H}$ always denotes a complex, separable infinite dimensional Hilbert space. One of the basic problems in operator theory is to determine when two operators A and B in $\mathcal{L}(\mathcal{H})$ are unitarily equivalent or similar. In infinite dimensional Hilbert space, this problem has no general solution. What we can do is to find the answer for some special classes of operators. A quantity (quantities) or a property (properties) P is unitary (or similarity) invariant (invariants) if A has P and $A\cong B(A\sim B)$ implies that B has P. For example, reducibility and strong reducibility are unitary invariants while strong reducibility is only the similarity invariant. For a subset R of $\mathcal{L}(\mathcal{H})$, unitary (similarity) invariant (invariants) P is completely unitary (or similarity) invariant (or invariants) if $A\in R$, then $A\cong B$ (or $A\sim B$) if and only if $B\in R$ and A and B have same P. From this point of view, one of the basic problem in operator theory mentioned above is to determine the completely unitary or similarity invariants. We have seen in Chapter 2 that eigenvalues and generalized eigenspaces are completely similarity invariants of $n\times n$ matrices. [Conway, J.B. (1990)] showed that two $*$-cyclic normal (or subnormal) operators A and B in $\mathcal{L}(\mathcal{H})$ are similar if and only if the scalar-valued spectral measures induced by them a equivalent, while they are unitary equivalent if and only if they are similar. Here, an operator A is normal if $A^*A = AA^*$, and A is subnormal if there exist a normal operator N and an invariant subspace $\mathcal{M}$ of N such that $N|_{\mathcal{M}} = A$. For two injective unilateral weighted shifts, the boundedness of the ratios of the products of their weights is the completely similarity invariant [Shields, A.L. (1974)]. There have already been a lot of results on the similarity invariants of operators, especially that of non-adjoint operators, which can be found in, for example, [Herrero, D.A. (1987)], [Herrero, D.A. (1990)], [Conway, J.B. (1990)].

In this chapter, we will discuss further the unitary invariants and similarity invariants of non-self-adjoint operators.

4.1   Unitary Invariants of Operators

We begin with a famous theorem.

Schur theorem   *Each $n \times n$ matrix X is unitary equivalent to the orthogonal direct sum of irreducible matrices.*

For $T \in \mathcal{L}(\mathcal{H})$, let $W^*(T)$ denote the von-Neumann algebra generated by
T. By von-Neumann double commutant theorem we can easily prove the
equivalence of the following conditions.
    (i) T is irreducible;
    (ii) $\mathcal{A}'(W^*(T)) = I$;
    (iii) $W^*(T) = \mathcal{L}(\mathcal{H})$.
The following proposition tells us that the Schur Theorem can not be
generalized to $\mathcal{L}(\mathcal{H})$.

Proposition 4.1.1   *Let $N \in \mathcal{L}(\mathcal{H})$ be a self-adjoint operator with $\sigma_p(N) =
\emptyset$, then N is not orthogonal direct sum of irreducible operators. Further-
more, N is not the topological direct sum of (SI) operators.*

Proof   For the first part of the proposition, if $N = \sum_{i=1}^{l} \oplus N_i, 1 \leq l \leq \infty$,
where $N_i \in (RI)$, then N_i is self-adjoint and the the dimension of the space
$\mathcal{H}_i$, on which N_i acts, is just 1. Thus $\sigma_p(N) \neq \emptyset$. It is a contradiction.

For the second part of the proposition, if $P \in \mathcal{A}'(N)$ is a nontrivial idempotent, then by the spectral theorem of self-adjoint operators, there exists a
orthogonal projection P' such that $ran P = ran P'$ and $P' \in \mathcal{A}'(N)$ [Putnam
I.[1]]. If N is the topological direct sum of sum (SI) operators, then by
the fact stated above, we can find an orthogonal projection $P' \in \mathcal{A}'(N)$ and
$N|_{ran P'}$ is irreducible. By the argument used in the first part we conclude
that $\sigma_p(N) \neq \emptyset$. It is also a contradiction.

Although it is not every operator to be the direct sum of irreducible
operators, we have the following proposition.

Proposition 4.1.2   *Every operator $T \in \mathcal{L}(\mathcal{H})$ is the direct integral of irreducible operators.*

Proof   In fact this is a corollary of Theorem 3.6 of [Azoff, E.A., Fong, C.K.
and Gilfeather., F. (1976)]. The weakly closed algebra $\mathcal{A}(T)$, generated by

T and I, can be expressed as $\int_\Lambda \oplus \mathcal{A}_\lambda d\mu(\lambda)$, where Λ is a separable metric space, μ is a regular σ-finite Borel measure on Λ and $\mathcal{A}_\lambda$ is a weakly closed irreducible operator algebra for almost all $\lambda \in \Lambda$. Recall that an operator algebra is irreducible if it has no nontrivial reducing subspace. Therefore, we have

$$T = \int_\Lambda \oplus T_\lambda d\mu(\lambda), T_\lambda \in \mathcal{A}_\lambda, \lambda \in \Lambda \ \text{a.e.}.$$

Thus $\mathcal{A}(T) \subset \int_\Lambda \oplus \mathcal{A}(T_\lambda) d\mu(\lambda) \subset \int_\Lambda \oplus \mathcal{A}_\lambda d\mu(\lambda) = \mathcal{A}(T)$. This implies that $\mathcal{A}(T_\lambda) = \mathcal{A}_\lambda$ for almost all $\lambda \in \Lambda$. Thus the irreducibility of $\mathcal{A}_\lambda$ implies that $T_\lambda \in (RI)$. Therefore $T = \int_\Lambda \oplus T_\lambda d\mu(\lambda)$ is the asserted irreducible integral decomposition of T.

For C^*-algebra $\mathcal{A}$ and natural number n, let $M_n(\mathcal{A})$ denote the set of all $n \times n$ matrices with entries in $\mathcal{A}$. The following theorem gives the number of reducing subspaces of an operator.

Theorem 4.1.3 *The number of reducing subspaces of any operator $T \in \mathcal{L}(\mathcal{H})$ is either finite or uncountably infinite. The former case occurs if and only if T is the direct sum of finitely many irreducible operators, i.e., $T = T_1 \oplus T_2 \oplus \cdots \oplus T_n$, and T_i, T_j are not unitarily equivalent if $i \neq j$. In this case, the number of reducing subspace is 2^n.*

Theorem 4.1.3 has a similar pattern but in a different context with the following result of [Ong, S.C. (1987)].

Theorem Ong *Given $T \in \mathcal{L}(\mathbf{C}^n)$, the number of invariant subspaces of T is either finite or uncountably infinite. The former case occurs if and only if T has a cyclic vector.*

To prove Theorem 4.1.3, we need four lemmas. The first is a structure theorem for two orthogonal projections, and has been quoted in many literatures before. The reader is referred to [Halmos, P.R. (1968)].

Lemma 4.1.4 *Given two orthogonal projections P and $Q \in \mathcal{L}(\mathcal{H})$, there is a unitary operator U such that*

$$U^* P U = \begin{bmatrix} I_1 & 0 \\ 0 & 0 \end{bmatrix} \oplus I_2 \oplus I_3 \oplus 0 \oplus 0$$

and

$$U^* Q U = \begin{bmatrix} A & B \\ B & I_1 - B \end{bmatrix} \oplus I_2 \oplus 0 \oplus I_4 \oplus 0$$

under the decomposition of the space $\mathcal{H} = \mathcal{H}_1 \oplus \mathcal{H}_2 \oplus \mathcal{H}_3 \oplus \mathcal{H}_4 \oplus \mathcal{H}_5$, *where* $A \in \mathcal{L}(\mathcal{H}_1)$ *is a positive contraction and* $B = [A(I_1 - A)]^{\frac{1}{2}}$. *We may assume that* $0 < A \leq \frac{1}{2}I_1$ *and* A *is unique up to unitary equivalence.*

Using Lemma 4.1.4, we can prove the next lemma.

Lemma 4.1.5 *If* $T \in \mathcal{L}(\mathcal{H})$ *has countably many reducing subspaces, then* $\mathcal{A}'(W^*(T))$ *is abelian.*

Proof Let P, $Q \in \mathcal{A}'(W^*(T))$ be two orthogonal projections represented as in Lemma 4.1.4. Since $PT = TP$ and $QT = TQ$, a simple computation indicates that $T = T_1 \oplus T_2 \oplus \sum_{i=3}^{5} \oplus T_i$ under the decomposition of the space

$$\mathcal{H} = \mathcal{H}_1 \oplus \mathcal{H}_1 \oplus \sum_{i=3}^{5} \oplus \mathcal{H}_i$$

and $T_1 A = A T_1$. For each complex number λ, denote

$$\mathcal{M}_\lambda = \{\lambda BX \oplus X \oplus 0 \oplus 0 \oplus 0 : X \in \mathcal{H}_1\}.$$

It is easily seen that $\mathcal{M}_\lambda$ is a reducing subspace of T, and if $\mathcal{H}_1 \neq \{0\}$, then

$$\mathcal{M}_\lambda \neq \mathcal{M}_{\lambda'} \ (\lambda \neq \lambda').$$

Since T has only countably many reducing subspace, $\mathcal{H}_1 = \{0\}$. Thus

$$P = I_2 \oplus I_3 \oplus 0 \oplus 0$$

and

$$Q = I_2 \oplus 0 \oplus I_4 \oplus 0.$$

This implies that $PQ = QP$. Since von-Neumann algebra $\mathcal{A}'(W^*(T))$ is generated by the projections in it, $\mathcal{A}'(W^*(T))$ is abelian.

Recall that a projection p in a C^*-algebra is minimal if there is no projection q other than 0 and p such that $pq = q$.

Lemma 4.1.6 *Let* $P \in \mathcal{A}'(W^*(T))$ *be a projection, then* P *is minimal if and only if* $T|_{ranP} \in (RI)$.

The proof of this lemma is an easy consequence of the definition of minimal projection.

Lemma 4.1.7 *Let* $A, B \in \mathcal{L}(\mathcal{H}) \cap (\mathcal{RI})$, *then* A *and* B *are unitarily equivalent if and only if there exists a nonzero operator* X *such that* $XA = BX$ *and* $XA^* = B^*X$.

Proof If $XA = BX$ and $XA^* = B^*X$ for some nonzero X, then $kerX$ and $\overline{ranX}$ are reducing subspaces of A and B. If $kerX \neq \{0\}$, by the irreducibility of A $kerX = \mathcal{H}$, i.e., $X = 0$, a contradiction. Thus $kerX = \{0\}$. Similarly, we can conclude that $\overline{ranX} = \mathcal{H}$, i.e., X has a dense range. Let $X = UP$ be the polar decomposition of X, where U is unitary and $P = (X^*X)^{\frac{1}{2}} \geq 0$. Since

$$X^*XA = X^*BX = AX^*X,$$

$$PA = AP.$$

Therefore

$$UAP = UPA = XA = BX = BUP.$$

Since P is also range dense, $UA = BU$. Thus $A \cong B$.

We are now in a position to prove Theorem 4.1.3.

Proof of Theorem 4.1.3 Assume that T has countably infinite many reducing subspaces. By Lemma 4.1.5, $\mathcal{A}'(W^*(T))$ is abelian. Thus $\mathcal{A}'(W^*(T))$ is generated by some Hermitian operator A [Radjaval, H. and Rosenthal, P. (1973)].

Note that $\sigma(A)$ can not be finite. Otherwise,

$$A = \sum_{i=1}^{n} \oplus \lambda_i I_i$$

and

$$W^*(A) = \{\sum_{i=1}^{n} \oplus \alpha_i I_i : \alpha_i \in \mathbf{C}\}.$$

This implies that $W^*(A) = \mathcal{A}'(W^*(T))$ contains only finitely many projections, and contradicts the assumption. Thus $\sigma(A)$ can be decomposed into countably infinitely many pairwise disjoint Borel subsets $\{\sigma_i\}_{i=1}^{\infty}$, each of which has a strictly positive spectral measure. Since $\sigma(A)$ has uncountably many different decompositions, so is the spectral projections of A. Therefore, there are uncountably many orthogonal projections in $W^*(A) = \mathcal{A}'(W^*(T))$. This is also a contradiction. Thus the number of reducing subspaces of T can not be countably infinite.

Now we assume that T has finitely many reducing subspaces. By Lemma 4.1.6 $\mathcal{A}'(W^*(T))$ is abelian. Let $P_1, \cdots, P_n$ be minimal projections in

$\mathcal{A}'(W^*(T))$. Since $P_iP_j = P_jP_i$, it is not difficult to prove that $P_iP_j = 0$ for $i \neq j$, and $\sum\limits_{i=1}^{n} P_i = I$. Let $T = \sum\limits_{i=1}^{n} \oplus T_i$ under the decomposition $\mathcal{H} = \sum\limits_{i=1}^{n} \oplus ranP_i$, where $T_i = T|_{ranP_i}$. By Lemma 4.1.7, $T_i \in (RI)$. We are now to prove that T_i is not unitarily equivalent to T_j for $i \neq j$. Otherwise, if $T_i \cong T_j$, there is a unitary operator U such that $UT_i = T_jU$, where $1 \leq i < j \leq n$. Then for each $\lambda \in \mathbf{C}$, set

$$\mathcal{M}_\lambda = \{0 \oplus \cdots \oplus \underset{i\text{th}}{x} \oplus 0 \oplus \cdots \oplus \underset{j\text{th}}{\lambda Ux} \oplus \cdots \oplus 0 : x \in \mathcal{H}_i\}.$$

It is easy to see that $\mathcal{M}_\lambda \neq \mathcal{M}_{\lambda'}$ if $\lambda \neq \lambda'$ and $\mathcal{M}_\lambda$ is a reducing subspaces of T. Note that there are infinitely many of $\mathcal{M}_\lambda$'s, this contradicts our assumption of T.

Conversely, if $T = \sum\limits_{i=1}^{n} \oplus T_i$ on $\mathcal{H} = \sum\limits_{i=1}^{n} \oplus \mathcal{H}_i$, $T_i \in (RI)$ and T_i is not unitarily equivalent to T_j for $i \neq j$. Let $\mathcal{P} = \{P_{ij}\}_{i,j=1}^{n}$ be orthogonal projections commuting with T. Then $P_{ij}T_j = T_iP_{ij}$ for all $1 \leq i, j \leq n$. Thus

$$P_{ij}T_j^* = P_{ji}^*T_j^* = (T_jP_{ji})^* = (P_{ji}T_i)^* = T_i^*P_{ji}^* = T_i^*P_{ij}.$$

Since T_i and T_j are irreducible and not unitarily equivalent for $i \neq j$, by Lemma 4.1.7, $P_{ij} = 0$ and so $P_{ji} = 0$. Thus P_{ii} is an orthogonal projection commuting with T_i, which implies that $P_{ii} = 0$ or $I_{\mathcal{H}_i}$ $(i = 1, 2, \cdots, n)$. Therefore T has only 2^n reducing subspaces.

In the following, we will characterize the decomposibility of an operator into direct sum of irreducible operators in terms of C^*-algebra language. For $T \in \mathcal{L}(\mathcal{H})$ and an integer $n, 1 \leq n \leq \infty$, let $T^{(n)}$ denote the direct sum of n copies of T.

Theorem 4.1.8 *An operator $T \in \mathcal{L}(\mathcal{H})$ is the direct sum of irreducible operators (i.e., $T = \sum\limits_{i=1}^{n} \oplus T_i^{(n_i)}, 1 \leq n_i \leq \infty, 1 \leq n \leq \infty, T_i \in (RI)$ and are pairwise not unitary equivalent) if and only if $\mathcal{A}'(W^*(T))$ is $*$-isomorphic to $\sum\limits_{i=1}^{n} \oplus M_{n_i}(\mathbf{C})$. Moreover, the (RI)-decomposition of T is unique in the sense of unitary equivalent. Precisely, if $T = \sum\limits_{k=1}^{m} \oplus S_k^{(m_k)}$ is another direct sum of irreducible operators $\{S_k\}_{k=1}^{m}$, which are pairwise not unitary equivalent, then $n = m$ and there exist a permutation π of $\{1, 2, \cdots, n\}$ and a unitary operator U in $\mathcal{A}'(W^*(T))$ such that $n_i = m_{\pi(i)}$ and $UT_i = S_{\pi(i)}U, i =$*

$1, 2, \cdots, n$.

Note that [Takesaki, M. (1979)] showed that every finite dimensional C^*-algebra is *-isomorphic to the direct sum of finitely many full matrix algebras. Thus we have the following corollary.

Corollary 4.1.9 *T is the direct sum of finitely irreducible operators if and only if $dim \mathcal{A}'(W^*(T)) < \infty$.*

To prove Theorem 4.1.8, we need the following lemma.

Lemma 4.1.10 *If T is irreducible on $\mathcal{H}$ and $X \in \mathcal{L}(\mathcal{H})$ satisfying $XT = TX$ and $XT^* = T^*X$, then $X = dI$ for some $d \in \mathbf{C}$.*

Proof Since X^*X commutes with T, T commutes with any spectral projection P of X^*X. Since $T \in (RI), P = 0$ or $P = I$. Thus $\sigma(X^*X)$ is a singleton $\{\alpha\}$ and $X^*X = \alpha I$. On the other hand, it follows from $XT = TX$ and $XT^* = T^*X$ that $ker X$ is a reducing subspace of T. Since $T \in (RI), ker X = \{0\}$ or $\mathcal{H}$. Similarly, $ran \overline{X} = \mathcal{H}$ or $ran X^- = \{0\}$. This implies that $X = 0$ or X is injective with dense range. Thus

$$X = U(X^*X)^{\frac{1}{2}} = \sqrt{\alpha} U,$$

where U is unitary. If $\alpha \neq 0$, then $UT = TU$ and $UT^* = T^*U$. Repeating the arguments above, we have $U = \beta I$ or $X = dI$, where $d = \sqrt{\alpha}\beta$. Thus, the proof is complete.

Proof of Theorem 4.1.8 Let $T = \sum\limits_{i=1}^{n} \oplus T_i^{(n_i)}$ satisfy the conditions of the theorem. For any $X \in \mathcal{A}'(W^*(T))$, by Lemma 4.1.10,

$$X = \sum_{i=1}^{n} \oplus X_i, X_i \in \mathcal{A}'(W^*(T^{(n_i)})).$$

Let

$$X_i = \{Y_{jk}^i\}_{j,k=1}^{n_i},$$

then

$$Y_{jk}^i \in \mathcal{A}'(W^*(T_i)).$$

By Lemma 4.1.10, $Y_{jk}^i = \lambda_{jk}^i I_i$, where I_i is the identity on $\mathcal{H}_i$. Thus

$$X = \sum_{i=1}^{n} \oplus [\lambda_{jk}^i I_i]_{j,k=1}^{n_i}.$$

It is obvious that the mapping $X \mapsto \sum\limits_{i=1}^{n} \oplus [\lambda^i_{jk} I_i]^{n_i}_{j,k=1}$ defines a *-isomorphism from $\mathcal{A}'(W^*(T))$ onto $\sum\limits_{i=1}^{n} \oplus M_{n_i}(\mathbf{C})$.

Conversely, if ϕ is a *-isomorphism from $\mathcal{A}'(W^*(T))$ onto $\mathcal{A} := \sum\limits_{i=1}^{n} \oplus M_{n_i}(\mathbf{C})$ and let E_{ij} denote the element $0 \oplus \cdots \oplus e_{ij} \oplus \cdots \oplus 0$ in $\mathcal{A}$, where e_{ij} is an $n_i \times n_i$ matrix whose (i,j)-entry is 1 and the others are 0. Then $\phi^{-1}(E_{ij}) \in \mathcal{A}'(W^*(T))$ are pairwise orthogonal minimal projections with $\sum\limits_{i,j} \phi^{-1}(E_{ij}) = I_{\mathcal{H}}$. Obviously, by Lemma 4.1.7, $\phi^{-1}(E_{ij})\mathcal{H}$ is one of the reducing subspaces of T with

$$T_{ij} := T|_{\phi^{-1}(E_{ij})\mathcal{H}} \in (RI)$$

and

$$T = \sum_{i,j} \oplus T_{ij}.$$

Since $E_{ij} \cong E_{ik}$ for all (j,k), $T_{ij} \cong T_{ik}$ and $T \cong \sum\limits_{i=1}^{n} \oplus T_i^{(n_i)}$.

In order to prove the uniqueness, let $T = \sum\limits_{k=1}^{m} \oplus S_k^{(m_k)}$ be another expression of T, where $\{S_k\}_{k=1}^{m}$ are pairwise not unitary equivalent and $\mathcal{H} = \sum\limits_{k=1}^{m} \oplus L_k^{(m_k)}$. Let P_{kl} be the orthogonal projection from $\mathcal{H}$ onto the l-th subspace of $L_k^{(m_k)}$, then $F_{kl} = \phi(P_{kl})$ are pairwise orthogonal in $\mathcal{A} := \sum\limits_{i=1}^{n} \oplus M_{n_i}(\mathbf{C})$, and $\sum\limits_{k,l} F_{kl} = I$. Since $T|_{ran P_{kl}} \in (RI)$, P_{kl} is minimal. Thus F_{kl} is minimal in $\mathcal{A}$ and there exists an integer n_i such that $F_{kl} \in M_{n_i}(\mathbf{C})$ with $rank F_{kl} = 1$. Therefore $\sum\limits_{l} F_{kl} = I_{n_i}$ and so $m_k = n_i$. From $\sum\limits_{k,l} F_{kl} = I$, we can conclude that $m = n$. The remainder of the theorem can be proved directly from the *-isomorphism ϕ.

Next we will consider when two operators have isomorphic reducing subspaces lattices. [Conway, J.B. and Gillespie, T.A. (1985)] solved this problem in the case of normal operators. Using their result, we can characterize isomorphism of the reducing subspace lattices of two operators if they can be expressed as direct sums of irreducible operators.

Proposition 4.1.11 *Let*

$$A = \sum_{j=1}^{n} \oplus A_j^{(n_j)}$$

and

$$B = \sum_{k=1}^{m} \oplus B_k^{(m_k)},$$

where $A_j, B_k \in (RI)$, $\{A_j\}$ and $\{B_k\}$ are pairwise not unitarily equivalent, $1 \leq n, m \leq \infty$, $1 \leq n_j, m_k \leq \infty$. Then $RedA$ is isomorphic to $RedB$ if and only if $n = m$ and there exists a permutation π of $\{1, 2, \cdots, n\}$ such that $n_j = m_{\pi(j)}$ for $j = 1, 2, \cdots, n$, where $RedA$ and $RedB$ denote the lattices of reducing subspaces of A and B respectively.

To prove the proposition, we need the following lemma.

Lemma 4.1.12 *If $T \in (RI)$, then $RedT^{(n)}$ is isomorphic to $RedI_n$ for all n, $1 \leq n \leq \infty$, where I_n is the identity on an n-dimensional space.*
Proof If $\mathcal{P} = \{P_{ij}\}_{i,j=1}^{n}$ is a projection in $\mathcal{A}'(T^{(n)})$, by Lemma 4.1.10, $P_{ij} = \lambda_{ij} I$ for some $\lambda_{ij} \in \mathbf{C}, 1 \leq i, j \leq n$. Thus the mapping $P \mapsto \{\lambda_{ij}\}_{i,j=1}^{n}$ induces a lattice isomorphism from $RedT^{(n)}$ onto $RedI_n$.

Proof of Proposition 4.1.11 By Lemma 4.1.7, $RedA$ is isomorphic to $\sum_j \oplus RedA_j^{(n_j)}$, and by Lemma 4.1.12 $\sum_j \oplus RedA_j^{(n_j)}$ is isomorphic to $Red \sum_j \oplus \frac{1}{j} I_{n_j}$. Thus $RedA$ is isomorphic to $Red \sum_j \oplus \frac{1}{j} I_{n_j}$.

Similarly, $RedB$ is isomorphic to $Red \sum_k \oplus \frac{1}{k} I_{n_k}$. Therefore if $RedA$ is isomorphic to $RedB$, then $n = m$ and there is a permutation π of $\{1, 2, \cdots, n\}$ such that $n_j = m_{\pi(j)}$ [Conway, J.B. and Gillespie, T.A. (1985)]. This proves the necessity part. The sufficiency is obvious.

Proposition 4.1.13 *If $T^{(k)}$ is a direct sum of irreducible operators, then so is T.*
Proof Without loss of generality, we assume that

$$T^{(k)} = \sum_{i=1}^{n} \oplus T_i^{(n_i)}, 1 \leq n \leq \infty, 1 \leq n_j \leq \infty$$

$\{T_i\}_{i=1}^{n}$ is a sequence of irreducible operators, which are pairwise unitarily inequivalent. Then there are pairwise orthogonal projections $P_j, j =$

$1, 2, \cdots, n$, each of which commutes with $T^{(k)}$ and $\sum_j P_j = I$, such that $T^{(k)}|_{ranP_j}, j = 1, 2, \cdots, k$ are pairwise unitarily inequivalent. By Lemma 4.1.7,

$$P_j = \sum_i \oplus Q_{ij},$$

where $Q_{ij} \in \mathcal{A}'(T_i^{(n_i)})$, $\{Q_{ij}\}$ are pairwise orthogonal, $\sum_j \oplus Q_{ij} = I_i$ and $T_i^{(n_i)}|_{ranQ_{ij}}, j = 1, 2, \cdots, k$ are pairwise unitarily equivalent. Therefore we need only to prove that if $A^{(k)} \cong B^{(n)}, 1 \leq n \leq \infty, B \in (RI)$, then A is a direct sum of irreducible operators. We may also assume that $n = \infty$. Otherwise $\mathcal{A}'(W^*(A^{(k)})) = M_k(\mathcal{A}'(W^*(A)))$ is finite dimensional by Corollary 4.1.9. This implies that $\mathcal{A}'(W^*(A))$ is also finite dimensional. Therefore A can be expressed as a direct sum of irreducible operators. Since $A^{(k)} \cong B^{(\infty)} = C$, $A^{(k)} \cong C^{(k)}$. It follows that $A \cong C$ by [Kadison, R.V. and Singer, I.M. (1957)] and the proof is complete.

Now we will describe the necessary and sufficient condition for an operator to be sum of irreducible operators in terms of K-theory language. The main result is the following theorem.

Theorem 4.1.14 *Given $T \in \mathcal{L}(\mathcal{H})$. T is the direct sum of irreducible operators if and only if $\bigvee(\mathcal{A}'(W^*(T))) \cong \mathbf{N}_+^{(k_1)} \oplus (\mathbf{N}_+ \cup \{\infty\})^{(k_2)}$ for some $k_1, k_2, 0 \leq k_1, k_2 \leq \infty$, where $\bigvee(\mathcal{A}'(W^*(T)))$ is the semigroup of $\mathcal{A}'(W^*(T))$.*

The proof of Theorem 4.1.14 needs the following lemmas.

Lemma 4.1.15 *Let P and Q be two projections in $\mathcal{A}'(W^*(T))$. If there exists a unitary operator $U \in \mathcal{A}'(W^*(T))$ such that $UP = QU$, then $T|_{ranP} \cong T|_{ranQ}$.*
Proof Set $W = U|_{ranP}$. Then W is a unitary operator from $ranP$ onto $ranQ$ and satisfies $W(T|_{ranP}) = (T|_{ranQ})W$.

Lemma 4.1.16 *Let $T \in \mathcal{L}(\mathcal{H})$ and*

$$\bigvee(\mathcal{A}'(W^*(T))) \cong \mathbf{N}_+^{(k_1)} \oplus (\mathbf{N}_+ \cup \{\infty\})^{(k_2)},$$

$0 \leq k_1, k_2 \leq \infty$ and let $l = k_1 + k_2$, $\{e_i\}_{i=1}^l$ be l free generators in $\bigvee(\mathcal{A}'(W^*(T)))$ and $P \neq 0$ be a projection in $\mathcal{A}'(W^*(T))$. Then $T|_{ranP} \in (RI)$ if and only if $[P] = e_i$ for some i.
Proof Assume that $T|_{ranP}$ is irreducible and $[P] = \sum_{i=1}^l \oplus \alpha_i e_i$, where α_i is

an integer, $0 \leq \alpha_i < \infty$. If there are at least two nonzero α_i's, say $\alpha_1, \alpha_2 \neq 0$. Then $f = \alpha_1 e_1$ and $g = \sum\limits_{i=2}^{l} \oplus \alpha_i e_i$ are nonzero elements in $\bigvee(\mathcal{A}'(W^*(T)))$. Thus there exist a natural number m and mutual orthogonal projections Q and $R \in \mathcal{A}'(W^*(T^{(m)}))$, such that $[Q] = f$ and $[R] = g$. Set $S = Q + R$, then

$$[S] = [Q] + [R] = f + g = \sum_{i=1}^{l} \oplus \alpha_i e_i = [P].$$

Therefore S is unitarily equivalent to $P \oplus 0^{(m-1)}$ in $\mathcal{A}'(W^*(T^{(m)}))$, where 0 denotes the zero operator on $\mathcal{H}$.

By Lemma 4.1.15,

$$T^{(m)}|_{ranS} \cong T^{(m)}|_{ranP \oplus 0^{(m-1)}}$$

$$\cong T^{(m)}|_{ranQ} \oplus T^{(m)}|_{ranR}$$

$$\cong T|_{ranP} \in (RI).$$

This contradicts the fact that $T|_{ranP} \in (SI)$. Thus $[P] = \alpha_i e_i$. Similarly, we can prove $\alpha_i = 1$.

Conversely, assume that $[P] = e_1$ and $T|_{ranP}$ is reducible. Then there are nonzero projections $Q, R \in \mathcal{A}'(W^*(T^{(m)}))$ such that $QR = 0$ and $P = Q + R$. Let

$$[Q] = \sum_{i=1}^{l} \oplus \alpha_i e_i$$

and

$$[R] = \sum_{i=1}^{l} \oplus \beta_i e_i,$$

where $0 \leq \alpha_i, \beta_i < \infty$ for all i. Thus

$$e_1 = [P] = [Q] + [R] = \sum_{i=1}^{l} \oplus (\alpha_i + \beta_i) e_i$$

and

$$\alpha_1 + \beta_1 = 1, \alpha_i + \beta_i = 0$$

for all $i \geq 2$. Therefore $\alpha_1 = 0$ and $\beta_1 = 0$ and $\alpha_i = \beta_i = 0$ for $i \geq 2$. This implies that $[Q] = 0$ or $[R] = 0$. A contradiction. Therefore $T|_{ranP} \in (RI)$.

Lemma 4.1.17 *Let $A \in \mathcal{L}(\mathcal{H})$ be a direct sum of irreducible operators and $B \in \mathcal{L}(\mathcal{K})$ have no reducing subspace on which B is irreducible. If there exists an operator X such that $XA = BX, XA^* = B^*X$, then $X = 0$.*

Proof Without loss of generality, we assume that $A = \sum\limits_{n=1}^{\infty} \oplus A_n$ with respect to $\mathcal{H} = \sum\limits_{n=1}^{\infty} \oplus \mathcal{H}_n$, where $A_n \in (RI)$. Then $X^* = [X_1^*, X_2^*, \cdots]^t$. Now we prove $X_1 = 0$. In fact, from $XA = BX$ and $XA^* = B^*X$, we have $X_1 B = A_1 X_1$ and $X_1 B^* = A_1^* X$. Thus $(X_1 X_1^*)A_1 = A_1(X_1 X_1^*)$ and $(X_1 X_1^*)A_1^* = A_1^*(X_1 X_1^*)$. Since A_1 is irreducible, by Lemma 4.1.10 $X_1 X_1^* = \lambda I_{\mathcal{H}_1}$ for some $\lambda \in \mathbf{C}$. If $\lambda \neq 0$. Let $U = \lambda^{-\frac{1}{2}} X_1$. Then $UU^* = I_{\mathcal{H}_1}$ and $Q := U^*U$ is a projection in $\mathcal{L}(\mathcal{K})$ with $QB = BQ$. Set

$$p = I_{\mathcal{H}_1} \oplus 0, \ q = 0 \oplus Q,$$

then $p, q \in \mathcal{L}(\mathcal{H}_1 \oplus \mathcal{K})$. Set

$$p' = p \oplus 0, q' = q \oplus 0,$$

where

$$p', q' \in \mathcal{L}((\mathcal{H}_1 \oplus \mathcal{K}) \oplus (\mathcal{H}_1 \oplus \mathcal{K})).$$

Set $C = A_1 \oplus B$. We claim that p' and q' are unitarily equivalent in $\mathcal{A}'(W^*(\mathbf{C}^{(2)}))$. To prove this, we define

$$v = \begin{bmatrix} 0 & U \\ 0 & q \end{bmatrix} \in \mathcal{L}(\mathcal{H}_1 + \mathcal{K}).$$

It is easy to see that v is a partial isometry, $vv^* = p$ and $v^*v = q$. Then the assertion follows from the Proposition 5.2.12 of [Wegge-Olsen, N.E. (1993)]. By Lemma 4.1.15, $C^{(2)}|_{ranp'} \cong C^{(2)}|_{ranq'}$. But $C^{(2)}|_{ranp'} = A_1 \in (RI)$. Thus

$$C^{(2)}|_{ranq'} \cong B|_{ranQ} \in (RI),$$

which contradicts our assumption on B. Hence $X_1 = 0$. By the same arguments we can prove that $X_n = 0$ for $n \geq 2$. Thus $X = 0$.

Proof of Theorem 4.1.14 The necessity follows from the analysis above. We only prove the sufficiency. Assume that

$$\bigvee(\mathcal{A}'(W^*(T))) \cong \mathbf{N}_+^{(k_1)} \oplus (\mathbf{N}_+ \cup \{\infty\})^{(k_2)}, \ \ 0 \leq k_1, k_2 \leq \infty.$$

Let P be a projection in $M_k(\mathcal{A}'(W^*(T))) = \mathcal{A}'(W^*(T^{(k)}))$ such that $[P]$ is a free generator of $\bigvee(\mathcal{A}'(W^*(T)))$. By Lemma 4.1.16, $T^{(k)}|_{ranP} \in (RI)$ (here we embed $\mathcal{A}'(W^*(T))$ into $M_k(\mathcal{A}'(W^*(T)))$ with the embedding map $A \mapsto \begin{bmatrix} A & 0 \\ 0 & 0 \end{bmatrix}$). Using Zorn's lemma, we can find a maximal family in $\mathcal{A}'(W^*(T^{(n)}))$ of pairwise orthogonal projections $\{P_j\}_{j=1}^n$, $1 \leq n \leq \infty$ such that $T^{(k)}|_{ranP_j}$ is irreducible, $j = 1, 2, \cdots$. Set $Q = \sum_j P_j$, we will prove that $Q = I^{(k)}$, the identity operator on $\mathcal{H}^{(k)}$. Otherwise, set $T_1 := T^{(k)}|_{ranQ}, T_2 := T^{(k)}|_{ran(I^{(k)}-Q)}$. Since Q is a projection in $\mathcal{A}'(W^*(T^{(k)}))$, T_1 is a direct sum of irreducible operators and T_2 has no reducing subspace $\mathcal{M}$ such that $T_2|_{\mathcal{M}} \in (RI)$. Applying Lemma 4.1.17, we have

$$\mathcal{A}'(W^*(T^{(k)})) = \mathcal{A}'(W^*(T_1)) \oplus \mathcal{A}'(W^*(T_2)).$$

Thus

$$\bigvee(\mathcal{A}'(W^*(T^{(k)}))) \cong \bigvee(\mathcal{A}'(W^*(T_1))) \oplus \bigvee(\mathcal{A}'(W^*(T_2)))$$

(Isomorphism Theorem).

Let R be a projection in $\mathcal{A}'(W^*(T_2^{(m)}))$ for which $[R]$ is a free generator of $\bigvee(\mathcal{A}'(W^*(T_2)))$. By Lemma 4.1.16, $T_2^{(m)}|_{ranR} \in (RI)$. By the similar argument above, we find a nonzero projection $Q_1 \in \mathcal{A}'(W^*(T_2^{(m)}))$ such that $T_3 := T_2^{(m)}|_{ranQ_1}$ is the direct sum of irreducible operators and $T_4 := T_2^{(m)}|_{ran(I-Q_1)}$ has no reducing subspace $\mathcal{M}$ with $T_4|_{\mathcal{M}} \in (RI)$. Using Lemma 4.1.17 again, we have

$$\mathcal{A}'(W^*(T_2^{(m)})) = \mathcal{A}'(W^*(T_3)) \oplus \mathcal{A}'(W^*(T_4)).$$

Thus Q_1 commutes with every operator in $\mathcal{A}'(W^*(T_2^{(m)}))$ and $Q_1 \in \mathcal{A}'(W^*(T_2^{(m)}))$ by von-Neumann double commutant theorem. Therefore $Q = S^{(m)}$, where S is a nonzero projection in $W^*(T_2)$, and

$$T_3 = T_2^{(m)}|_{ranQ_1} = (T_2|_{ranS})^{(m)}.$$

Since T_3 is the direct sum of irreducible operators, so is $T_2|_{ranS}$ by Proposition 4.1.13. This contradicts the assumption on T_2. Thus $Q = I^{(k)}$ and $T^{(k)}$ is a direct sum of irreducible operators. By Proposition 4.1.13, T is also a direct sum of irreducible operators.

By Theorem 4.1.14 and its proof, we have the following theorem.

Theorem 4.1.18 *Let $A, B \in \mathcal{L}(\mathcal{H})$. If A is the direct sum of irreducible operators, i.e., $A = \sum\limits_{i=1}^{l} \oplus A_i^{(n_i)}, 1 \le l \le \infty, 1 \le n_i \le \infty$. Then each two of $\{A_i\}$ are not similar if and only if there exists an isometric isomorphism ϕ such that*

$$\phi(\bigvee(\mathcal{A}'(W^*(A \oplus B)))) = \mathbf{N}_+^{(k_1)} \oplus (\mathbf{N}_+ \cup \{\infty\})^{(k_2)}, \quad 0 \le k_1, k_2 \le \infty$$

and $\phi(I) = 2 \sum\limits_{i=1}^{l} n_i e_i$, where $\{e_i\}$ are the generators of $\bigvee(\mathcal{A}'(W^*(A)))$.

Note that for $A \in \mathcal{L}(\mathcal{H})$, if A is a direct sum of infinitely many irreducible operators, then $K_0(\mathcal{A}'(W^*(A))) = 0$. Thus, in this case, K_0-group can not describe the unitary equivalent relation between operators. But we have the following theorem.

Theorem 4.1.19 *Let $A, B \in \mathcal{L}(\mathcal{H})$. Assume that A is the direct sum of finitely many irreducible operators: $A = \sum\limits_{i=1}^{l} \oplus A_i^{(n_i)}, 1 \le l < \infty, 1 \le n_i < \infty, i = 1, 2, \cdots, l$. Then $A \cong B$ if and only if there exists an isometric isomorphism ϕ such that*

$$\phi(K_0(\mathcal{A}'(W^*(A \oplus B)))) = \mathbf{N}_+^{(l)}$$

and $\phi(I) = 2 \sum\limits_{i=1}^{l} n_i e_i$, where $\{e_i\}$ are the generators of $K_0(\mathcal{A}'(W^*(A)))$.

Note that for $A \in \mathcal{B}_n(\Omega)$. Since A is a direct sum of finitely many irreducible operators, we can get the following theorem by Theorem 4.1.19.

Theorem 4.1.20 *Let $A, B \in \mathcal{B}_n(\Omega)$, then the following statement hold.*

(i) A has a unique (SI) decomposition up to unitary equivalence.

(ii) If $A = \sum\limits_{i=1}^{k} \oplus A_i^{(n_i)}, A_i \in (RI)$ and $\{A_i\}_{i=1}^{k}$ are pairwise unitarily inequivalent. Then $A \cong B$ if and only if there exists an isomorphic map ϕ such that

$$\phi(K_0(\mathcal{A}'(W^*(A \oplus B)))) = \mathbf{Z}^{(k)}$$

and $\phi(I_{\mathcal{A}'(W^*(A \oplus B))}) = 2 \sum n_i e_i$, where $\{e_i\}$ are the generators of $\bigvee(\mathcal{A}'(W^*(A)))$.

4.2 Strongly Irreducible Decomposition of Operators and Similarity Invariant: K_0-Group

It is more convenient in terms of K_0-group to describe the unitary invariant of operators which are direct sum of irreducible operators. The proof of it depends strongly on the tools of C^*-algebra theory. We give, in terms of semigroup theory, a necessary and sufficient condition to that an operator can be expressed as a direct sum of irreducible operators. In this section, we first prove the following theorem.

Theorem 4.2.1 *Given $T \in \mathcal{L}(\mathcal{H})$, the following are equivalent:*

(i) T is similar to $\sum\limits_{i=1}^{k} \oplus A_i^{(n_i)}$ under the decomposition of the space

$\mathcal{H} = \sum\limits_{i=1}^{k} \oplus \mathcal{H}_i^{(n_i)}$, *where $k, n_i < \infty$, $A_i \in (SI)$, $A_i \nsim A_j$ for $i \neq j$ and each $T^{(n)}$ has a unique (SI) decomposition up to similarity;*

(ii) $\bigvee(\mathcal{A}'(T)) \cong \mathbf{N}^{(k)}$, and this isomorphism ϕ sends

$$[I] \longrightarrow n_1 e_1 + n_2 e_2 + \cdots + n_k e_k,$$

where $\{e_i\}_{i=1}^{k}$ are the generators of $\mathbf{N}^{(k)}$ and $\mathbf{N} = \{0, 1, 2, \cdots\}, n_i \neq 0$.

From Theorem 4.2.1, we obtain the following corollary.

Corollary 4.2.2 *Let $T_1, T_2 \in (SI), T = T_1 \oplus T_2$. If $\bigvee(\mathcal{A}'(T)) \cong \mathbf{N}$, then $T_1 \sim T_2$. Moreover, if $T^{(n)}$ has a unique (SI) decomposition for all natural numbers up to similarity, then $T_1 \sim T_2$ if and only if $K_0(\mathcal{A}'(T)) \cong \mathbf{Z}$, where $\mathbf{Z} := \{0, \pm 1, \pm 2, \cdots\}$.*

Before we prove Theorem 4.2.1, the following lemmas are needed.

Lemma 4.2.3 *Given $A, B \in \mathcal{L}(\mathcal{H})$. Assume that φ is an isomorphism from $\mathcal{A}'(A)$ onto $\mathcal{A}'(B)$. $\{P_i\}_{i=1}^{n}$ is an (SI) decomposition of A if and only if $\{\varphi(P_i)\}_{i=1}^{n}$ is an (SI) decomposition of B. In particular, if $A \sim B$, then $\mathcal{A}'(A) \cong \mathcal{A}'(B)$.*

Proof   Since φ is an isomorphism, $0 = \varphi(P_i P_j) = \varphi(P_i)\varphi(P_j)$ for $i \neq j$ and $\sum\limits_{i=1}^{n} \varphi(P_i) = I$. We need only to show that $B|_{\varphi(P_i)\mathcal{H}} \in (SI), i = 1, 2, \cdots, n$. Otherwise, there exist two nonzero idempotents Q_1 and $Q_2 \in \mathcal{A}'(B)$ such that $Q_2 Q_1 = Q_1 Q_2 = 0$ and $Q_1 + Q_2 = \varphi(P_i)$ for some i. Note that $\varphi^{-1}(Q_1), \varphi^{-1}(Q_2)$ are two nonzero idempotents and $P_i = \varphi^{-1}(Q_1) + \varphi^{-1}(Q_2)$. This contradicts $A|_{P_i \mathcal{H}} \in (SI)$, and we proved the first part of the theorem.

If $A \sim B$, there exists an invertible operator X such that $XAX^{-1} = B$. Define $T \mapsto XTX^{-1}$ for all $T \in \mathcal{A}'(A)$. This mapping is an isomorphism from $\mathcal{A}'(A)$ onto $\mathcal{A}'(B)$.

Lemma 4.2.4 *Let $T \in \mathcal{L}(\mathcal{H})$ and $P_1, P_2 \in \mathcal{A}'(T)$ be idempotents. If $P_1 \sim_{\mathcal{A}'(T)} P_2$, then $T|_{P_1\mathcal{H}} \sim T|_{P_2\mathcal{H}}$, where $\sim_{\mathcal{A}'(T)}$ means similarity in $\mathcal{A}'(T)$.*
Proof Since $P_1 \sim_{\mathcal{A}'(T)} P_2$, there is an invertible operator $X \in \mathcal{A}'(T)$ such that $XP_1X^{-1} = P_2$. Thus $X ran P_1 = ran P_2, X ran(I - P_1) = ran(I - P_2)$. Set

$$X_1 = X|_{ran P_1}, \quad X_2 = X|_{ran(I-P_1)}.$$

Then $X = X_1 \dotplus X_2$, the topological sum of X_1 and X_2, and

$$X_1 \in GL(\mathcal{L}(P_1\mathcal{H}, P_2\mathcal{H})), \quad X_2 \in GL(\mathcal{L}((I - P_1)\mathcal{H}, (I - P_2)\mathcal{H})),$$

where $GL(\mathcal{A})$ is the set of invertible elements in the algebra $\mathcal{A}$. Note that

$$T = \begin{bmatrix} T_1 & 0 \\ 0 & T_2 \end{bmatrix} \begin{matrix} P_1\mathcal{H} \\ (I-P_1)\mathcal{H} \end{matrix} = \begin{bmatrix} T'_1 & 0 \\ 0 & T'_2 \end{bmatrix} \begin{matrix} P_2\mathcal{H} \\ (I-P_2)\mathcal{H} \end{matrix},$$

where $T_1 = T|_{P_1\mathcal{H}}, T_2 = T|_{(I-P_1)\mathcal{H}}, T'_1 = T|_{P_2\mathcal{H}}$ and $T'_2 = T|_{(I-P_2)\mathcal{H}}$.
A simple calculation shows that

$$\begin{bmatrix} T'_1 & 0 \\ 0 & T'_2 \end{bmatrix} \begin{bmatrix} X_1 & 0 \\ 0 & X_2 \end{bmatrix} = \begin{bmatrix} X_1 & 0 \\ 0 & X_2 \end{bmatrix} \begin{bmatrix} T_1 & 0 \\ 0 & T_2 \end{bmatrix}.$$

That is $T|_{P_1\mathcal{H}} \sim T|_{P_2\mathcal{H}}$.

Lemma 4.2.5 *Let $T \in \mathcal{L}(\mathcal{H})$ and let $\{P_i\}_{i=1}^n$ and $\{Q_i\}_{i=1}^n$ be two (SI) decompositions of T. If there exist $X_i \in GL(\mathcal{L}(P_i\mathcal{H}, Q_i\mathcal{H}))$ such that*

$$X_i(T|_{P_i\mathcal{H}})X_i^{-1} = T|_{Q_i\mathcal{H}}, i = 1, 2, \cdots, n,$$

then

$$X = X_1 \dotplus X_2 \dotplus \cdots \dotplus X_n \in GL(\mathcal{A}'(T)).$$

Proof Since

$$\mathcal{H} = ran P_1 \dotplus ran P_2 \dotplus \cdots \dotplus ran P_n = ran Q_1 \dotplus ran Q_2 \dotplus \cdots \dotplus ran Q_n,$$

$$T = \begin{bmatrix} T_1 & & 0 \\ & \ddots & \\ 0 & & T_n \end{bmatrix} \begin{matrix} P_1\mathcal{H} \\ \vdots \\ P_1\mathcal{H} \end{matrix} = \begin{bmatrix} T_1' & & 0 \\ & \ddots & \\ 0 & & T_n' \end{bmatrix} \begin{matrix} Q_1\mathcal{H} \\ \vdots \\ Q_1\mathcal{H} \end{matrix} ,$$

where $T_i = T|_{P_i\mathcal{H}}, T_i' = T|_{Q_i\mathcal{H}}, i = 1, 2, \cdots, n$. Clearly, $XT = TX$ and X is invertible.

Lemma 4.2.6 *Let*

$$\{P_1, \cdots, P_m, P_{m+1}, \cdots, P_n\}$$

and

$$\{Q_1, \cdots, Q_m, Q_{m+1}, \cdots, Q_n\}$$

be two spectral families in $\mathcal{A}'(T)$, where $T \in \mathcal{L}(\mathcal{H})$. If there are $X, Y \in \mathcal{A}'(T)$ and a permutation π of $S_n = \{1, 2, \cdots, n\}$ satisfying

(i) $XP_iX^{-1} = Q_i, 1 \le i \le n$;

(ii) $YP_iY^{-1} = Q_{\pi_i}, 1 \le i \le n$.

Then for each $Q_r, m < r \le n$, there exist a $P_{r'}$ with $m < r' \le n$ and Z_r, a product of finitely many Y's and X's, such that $Z_rQ_rZ_r^{-1} = P_{r'}$. Moreover, $\{P_{r'}\}$ is exactly a rearrangement of $\{P_r\}_{r=m+1}^n$.

Proof Given $Q_r, m < r \le n$, it follows from (ii) that exists $P_{j_1}, 1 \le j_1 \le n$, such that $YQ_rY^{-1} = P_{j_1}$. If $m < j_1 \le n$, then we set

$$Z_r = Y$$

and

$$P_{r'} = P_{j_1}.$$

If $1 \le j_1 \le m$, then by (ii) there exists an operator $Q_{j_1}, j_1 \neq r$, such that

$$XYQ_rY^{-1}X^{-1} = Q_{j_1}.$$

By (ii), $YQ_{j_1}Y^{-1} = P_{j_2}$ for some j_2. If $m < j_2 \le n$, then set

$$Z_r = YXY, P_{r'} = P_{j_2}.$$

If $1 \le j_2 \le m$, it is obvious that $j_1 \neq j_2$. Otherwise,

$$Q_{j_1} = Y^{-1}P_{j_2}Y = Y^{-1}P_{j_1}Y = Q_r,$$

which contradicts $m < r \le n$. Using (ii) again, we can find P_{j_3} such that

$$YQ_{j_2}Y^{-1} = P_{j_3}.$$

Similarly, $j_3 \notin \{j_1, j_2\}$. If $m < j_3 \leq n$, then set $Z_r = YXYXY, P_{r'} = P_{j_3}$. Or we can continue the procedure above. Since n is finite, after $s \leq m + 1$ steps we will force $P_s \in \{P_{m+1}, \cdots, P_n\}$. Set $P_r = P_{j_s}, Z_r = YXY \cdots XY$, where X appears S times. Then $Z_r Q_r Z_r^{-1} = P_{j_s}$. We claim that if $r_1 \neq r_2$, where $r_1, r_2 \in \{m + 1, \cdots, n\}$, then $j_{s_1} \neq j_{s_2}$. Otherwise, there exist $Z_{r_1} = YXY \cdots YXY$ (X appears s_1 times) and $Z_{r_2} = YXY \cdots YXY$ (X appears s_2 times) such that

$$Z_{r_1} Q_{r_1} Z_{r_1}^{-1} = Z_{r_2} Q_{r_2} Z_{r_2}^{-1}.$$

Without loss of generality, we may assume that $s_1 \geq s_2$. If $s_1 > s_2$, then

$$Z_{r_2}^{-1} Z_{r_1} Q_{r_1} Z_{r_1}^{-1} Z_{r_2} = Q_{r_2} \in \{Q_{m+1}, \cdots, Q_n\}.$$

Note that $Z_{r_2}^{-1} Z_{r_1} = XY \cdots XY$ (X appears $j_{s_1} - j_{s_2}$ times). Set

$$R = YXY \cdots XY,$$

where X appears $j_{s_1} - j_{s_2} - 1$ times. By the procedure of the choice, we have $RQ_{r_1} R^{-1} \in \{P_1, P_2, \cdots, P_m\}$. Thus

$$XRQ_{r_1} R^{-1} X^{-1} \in \{Q_1, Q_2, \cdots, Q_m\}.$$

But $XRQ_{r_1} R^{-1} X^{-1} = Z_{r_2}^{-1} Z_{r_1} Q_{r_1} Z_{r_1}^{-1} Z_{r_2} = Q_{r_2} \in \{Q_{m+1}, \cdots, Q_n\}$. A contradiction. Thus $s_1 = s_2$. But if $s_1 = s_2$, we can easily prove that $Q_{r_1} = Q_{r_2}$, which is also a contradiction. This completes the proof of our Claim and the lemma.

By the similar argument of Lemma 4.2.6, we can prove the following result.

Lemma 4.2.7 *Let* $T \in \mathcal{L}(\mathcal{H})$ *and let*

$$\{P_1, \cdots, P_{m_1}, \cdots, P_{m_{k-1}-1}, \cdots, P_{m_k}, P_{m_k+1}, \cdots, P_n\}$$

and

$$\{Q_1, \cdots, Q_{m_1}, \cdots, Q_{m_{k-1}-1}, \cdots, Q_{m_k}, Q_{m_k+1}, \cdots, Q_n\}$$

be two spectral families in $\mathcal{A}'(T)$. *If there exist*

$$X_1, X_2, \cdots, X_k, Y \in GL(\mathcal{A}'(T))$$

and a permutation π *of* s_n *such that*

$$X_i P_j X_i^{-1} = Q_j, \ m_i + 1 \leq j \leq m_{i+1}, \ i = 0, 1, \cdots, k - 1, \ m_l = 0$$

and $Y^{-1}P_jY = Q_{\pi(j)}, 1\leq j\leq n$. Then for $\forall r, m_k < r\leq n$, there exists Z_r, a product of finitely many $\{Y, X_1, \cdots, X_k\}$, such that $\{Z_rQ_rZ_r^{-1}\}_{r=m_k+1}^n$ is a rearrangement of $\{P_r\}_{r=m_k+1}^n$.

Lemma 4.2.8 *Let $T\in\mathcal{L}(\mathcal{H})$ and let $\{P_1,\cdots,P_m,P_{m+1},\cdots,P_n\}$ and $\{Q_1,\cdots,Q_m,Q_{m+1},\cdots,Q_n\}$ be two spectral families in $\mathcal{A}'(T)$. If the following conditions are satisfied:*

(i) For each P_i, there is an $X_i\in GL(P_i\mathcal{H}, Q_i\mathcal{H})$ such that

$$X_iT|_{P_i\mathcal{H}}X_i^{-1} = T|_{Q_i\mathcal{H}}, 1\leq i\leq m;$$

(ii) there exist a $Y\in GL(\mathcal{A}'(T))$ and a permutation π of S_n such that

$$Y^{-1}P_iY = Q_{\pi(i)},$$

then given $Q_r, r\in(m+1,\cdots,n)$, we can find $r'\in\{m+1,\cdots,n\}$ and

$$Z_r\in GL(Q_r\mathcal{H}, P_{r'}\mathcal{H})$$

such that

$$Z_r(T|_{Q_r\mathcal{H}})Z_r^{-1} = T|_{P_r\mathcal{H}}.$$

Furthermore, if $r_1 \neq r_2$, then $r_1' \neq r_2'$.

Proof Given $r\in\{m+1,\cdots,n\}$, by (ii) of the lemma there exists an operator $P_{j_1}\in\{P_i\}_{i=1}^n$ such that $YQ_rY^{-1} = P_{j_1}$. If $m < j_1\leq n$, set $Z_r = Y|_{Q_r\mathcal{H}}$. Otherwise it follows from $T|_{(YQ_rY^{-1})\mathcal{H}} = T|_{P_{j_1}\mathcal{H}}$ and (i) that $X_{j_1}T|_{P_{j_1}\mathcal{H}}X_{j_1}^{-1} = T|_{Q_{j_1}\mathcal{H}}$. Using condition (ii) again, we can find $j_2\in\{1,2,\cdots,n\}$ such that $YQ_{j_1}Y^{-1} = Q_{j_2}$. Clearly, $j_1 \neq j_2$. If $j_2\in\{m+1,\cdots,n\}$, set $Z_r = Y|_{Q_j\mathcal{H}}X_{j_1}Y|_{Q_r\mathcal{H}}, P_{r'} = P_{j_2}$. Thus $Z_r(T|_{Q_r\mathcal{H}})Z_r^{-1} = T|_{P_{r'}\mathcal{H}}$. Otherwise, by the similar arguments used in the proof of Lemma 4.2.6, after finite steps, we can find $P_{r'}\in\{P_k\}_{k=m+1}^n$ such that $Z_r(T|_{Q_r\mathcal{H}})Z_r^{-1} = T|_{P_{r'}\mathcal{H}}$. Similarly, we can prove that $r_1' \neq r_2'$ if $r_1 \neq r_2$.

Lemma 4.2.9 *Let $T\in\mathcal{L}(\mathcal{H})$. If T has a unique (SI) decomposition up to similarity, then for each idempotent $P\in\mathcal{A}'(T)$, $T|_{P\mathcal{H}}$ has a unique (SI) decomposition up to similarity.*

Proof Since T has a unique (SI) decomposition up to similarity, each spectral family of $T|_{P\mathcal{H}}$ consists only finite elements, and the (SI) decomposition spectral families consist of same number, say m, of elements.

Let $\{P_i\}_{i=1}^m$ and $\{Q_i\}_{i=1}^m$ be two (SI) decomposition of $T|_{P\mathcal{H}}$, and let $\{P_i\}_{i=m+1}^n$ be an (SI) decomposition of $T|_{(I-P)\mathcal{H}}$. Then

$$\{P_i : 1 \leq i \leq n\}$$

and

$$\{Q_i : 1 \leq i \leq m; \ P_k : m+1 \leq k \leq n\}$$

are two (SI) decomposition of T. By the uniqueness of the decomposition, we can find an operator $Y \in GL(\mathcal{A}'(T))$ such that

$$\{YP_iY^{-1}\} = \{Q_1, \cdots, Q_m, P_{m+1}, \cdots, P_n\}.$$

By Lemma 4.2.6, we can find $Z_i \in GL(\mathcal{L}(Q_i\mathcal{H}, P_i\mathcal{H}))$ and a permutation π of S_n, such that

$$Z_i(T|_{Q_i\mathcal{H}})Z_i^{-1} = T|_{P_{\pi(i)}\mathcal{H}}, \ 1 \leq i \leq m.$$

Set $Z_r = I|_{P_k\mathcal{H}}, k \geq m+1$ and $Z = Z_1 \dotplus \cdots \dotplus Z_n$. By Lemma 4.2.5, $Z \in GL(\mathcal{A}'(T))$ and $Z_{P\mathcal{H}} \in GL(\mathcal{A}'(T)|_{P\mathcal{H}})$. Note that

$$(Z|_{P\mathcal{H}})Q_i(Z|_{P\mathcal{H}})^{-1} = P_{\pi(i)}, 1 \leq i \leq m.$$

The proof of the lemma is complete.

Lemma 4.2.10 *Let $T \in \mathcal{L}(\mathcal{H})$. If T has a unique (SI) decomposition up to similarity. If P and Q are two idempotents in $\mathcal{A}'(T)$, then the following are equivalent:*
   (i) $P \sim_{\mathcal{A}'(T)} Q$;
   (ii) $T|_{P\mathcal{H}} \sim T|_{Q\mathcal{H}}$.
Proof   (i)$\Rightarrow$(ii) is a consequence of Lemma 4.2.4.

(ii)$\Rightarrow$(i).   By Lemma 4.2.9, $T|_{P\mathcal{H}}, T|_{Q\mathcal{H}}, T|_{(I-P)\mathcal{H}}$ and $T|_{(I-Q)\mathcal{H}}$ all have a unique (SI) decomposition up to similarity. Since $T|_{P\mathcal{H}} \sim T|_{Q\mathcal{H}}$, there exists an operator $X \in GL(\mathcal{L}(P\mathcal{H}, Q\mathcal{H}))$ such that

$$X(T|_{P\mathcal{H}})X^{-1} = T|_{Q\mathcal{H}}.$$

Thus if $\{P_1, P_2, \cdots, P_m\}$ is an (SI) decomposition of $T|_{P\mathcal{H}}$, then $(XP_1X^{-1}, XP_2X^{-1}, \cdots, XP_mX^{-1}\}$ is an (SI) decomposition of $T|_{Q\mathcal{H}}$.

Assume that $\{P_{m+1}, \cdots, P_n\}$ and $\{Q_{m+1}, \cdots, Q_n\}$ are (SI) decomposition of $T|_{(I-P)\mathcal{H}}$ and $T|_{(I-Q)\mathcal{H}}$ respectively. Then $\{P_i\}_{i=1}^n$ and

$$\{XP_iX^{-1} : 1 \leq i \leq m; Q_k : m+1 \leq k \leq n\}$$

are two (SI) decompositions of T. By the uniqueness of the decomposition, there exists an operator $Y{\in}GL(\mathcal{A}'(T))$ such that $\{YP_iY^{-1}\}_{i=1}^n$ is a rearrangement of $\{XP_iX^{-1}: 1{\leq}i{\leq}m; Q_k: m+1{\leq}k{\leq}n\}$. By Lemma 4.2.8, for each $r{\in}\{m+1, \cdots, n\}$, we can find $P_{r'}$ with $r'{\in}\{m+1, \cdots, n\}$ and $Z_r{\in}GL(\mathcal{L}(Q_r\mathcal{H}, P_{r'}\mathcal{H}))$ such that

$$Z_r(T|_{Q_r\mathcal{H}})Z_r^{-1} = T|_{P_{r'}\mathcal{H}}$$

and $r'_1 = r'_2$ if and only if $r_1 = r_2$.

Set $Z = Z_1\dot{+}\cdots\dot{+}Z_n{\in}GL(\mathcal{A}'(T))$. Since $ZPZ^{-1} = Q$, by Lemma 4.2.5,

$$P{\sim}_{\mathcal{A}'(T)}Q.$$

Lemma 4.2.11 *Let $T{\in}\mathcal{L}(\mathcal{H})$ and let P and Q be idempotents in $\mathcal{A}'(T)$. If $T|_{P\mathcal{H}}$ is not similar to $T|_{Q\mathcal{H}}$, then $P{\oplus}0_{\mathcal{H}^{(n)}}$ is not similar to $Q{\oplus}0_{\mathcal{H}^{(n)}}$ in $\mathcal{A}'(T^{(n+1)})$ for each natural number n.*

Proof If there is an $X{\in}GL(\mathcal{A}'(T^{(n+1)}))$ satisfying

$$X(P{\oplus}0_{\mathcal{H}^{(n)}})X^{-1} = Q{\oplus}0_{\mathcal{H}^{(n)}}$$

for some $n{\in}\mathbf{N}$, then by Lemma 4.2.4,

$$T^{(n+1)}|_{(P{\oplus}0_{\mathcal{H}^{(n)}})\mathcal{H}^{(n+1)}}{\sim}T^{(n+1)}|_{(Q{\oplus}0_{\mathcal{H}^{(n)}})\mathcal{H}^{(n+1)}}.$$

But we have

$$T^{(n+1)}|_{(P{\oplus}0_{\mathcal{H}^{(n)}})\mathcal{H}^{(n+1)}}{\cong}T|_{P\mathcal{H}}$$

and

$$T^{(n+1)}|_{(Q{\oplus}0_{\mathcal{H}^{(n)}})\mathcal{H}^{(n+1)}}{\cong}T|_{Q\mathcal{H}}.$$

Therefore $T|_{P\mathcal{H}}{\sim}T|_{Q\mathcal{H}}$. A contradiction.

Lemma 4.2.12 *Given $T{\in}\mathcal{L}(\mathcal{H})$. If $T^{(n)}$ has a unique (SI) decomposition up to similarity for each natural number n, then for two idempotents P, Q in $\mathcal{A}'(T)$, $P{\sim}_{\mathcal{A}'(T)}Q$ if and only if $[P] = [Q]$ in $\bigvee(\mathcal{A}'(T))$.*

Proof The "if" part is obvious.

Assume that $[P] = [Q]$, then there is a natural number k such that

$$P{\oplus}0_{\mathcal{H}^{(k)}}{\sim}_{\mathcal{A}'(T^{(k+1)})}Q{\oplus}0_{\mathcal{H}^{(k)}}.$$

By Lemma 4.2.4, $T|_{P\mathcal{H}}{\sim}T|_{Q\mathcal{H}}$. By Lemma 4.2.10 we conclude that $P{\sim}_{\mathcal{A}'(T)}Q$.

Proof of Theorem 4.2.1 (i)$\Rightarrow$(ii).   Let P be the orthogonal projection from $\mathcal{H}$ onto $\mathcal{H}_i$, and let E be an idempotent in $\mathcal{A}'(T^{(n)})$. Since $T^{(n)}$ has a unique (SI) decomposition up to similarity, by Lemma 4.2.9 $T^{(n)}|_{E\mathcal{H}^{(n)}}$ and $T^{(n)}|_{(I-E)\mathcal{H}^{(n)}}$ have unique (SI) decompositions up to similarity. If $\{Q_1, \cdots, Q_a\}$ is an (SI) decomposition of $T^{(n)}|_{E\mathcal{H}^{(n)}}$ and $\{Q_{a+1}, \cdots, Q_b\}$ is an (SI) decomposition of $T^{(n)}|_{(I-E)\mathcal{H}^{(n)}}$, then $\{Q_1, \cdots, Q_b\}$ is an (SI) decomposition of $T^{(n)}$. Since $\{P_1^{(nn_1)}, \cdots, P_k^{(nn_k)}\}$ is also an (SI) decomposition of $T^{(n)}$, by the uniqueness of the decomposition, there is an $X \in GL(\mathcal{A}'(T^{(n)}))$ such that

$$XQ_jX^{-1} = P_i.$$

Since $E = Q_1 + Q_2 + \cdots + Q_a$, $XEX^{-1} = \sum_{i=1}^{k} P_i^{(m_i)}$. Define a mapping

$$\bigvee(\mathcal{A}'(T)) \rightarrow \mathbf{N}^{(k)}$$

by $h([E]) = (m_1, \cdots, m_k)$. Then h is well defined. In fact, if $[E] = [F]$, then by Lemma 4.2.12, $F \sim E \sim \sum_{i=1}^{k} P_i^{(m_i)}$ in $M_\infty(\mathcal{A}'(T))$. If $F \sim \sum_{i=1}^{k} P^{(m_i)}$ in $M_\infty(\mathcal{A}'(T))$, then $h([F]) = h([E])$ and $F \sim E$. Thus h is one to one. For a k-tuple $(m_1, \cdots, m_k)$ of nonnegative integers, we can find a number n such that $m_i \leq nn_i, i = 1, 2, \cdots, k$, which implies that h maps $\sum_{i=1}^{m} P_i^{(m_i)}$ to $(m_1, \cdots, m_k)$ and is onto. Thus

$$\bigvee(\mathcal{A}'(T)) \cong \mathbf{N}^{(k)}.$$

By the construction of h, we know that $h([I]) = (n_1, \cdots, n_k)$.

(ii)$\Rightarrow$(i).   Suppose that $\bigvee(\mathcal{A}'(T)) \cong \mathbf{N}^{(k)}$ and h is the isomorphism. Then there exist a natural number r and k idempotents $Q_1, \cdots, Q_k$ in $\mathcal{A}'(T^{(r)})$ satisfying

$$h([Q_i]) = e_i, 1 \leq i \leq k.$$

Since $\bigvee(\mathcal{A}'(T^{(n)})) \cong \bigvee(\mathcal{A}'(T))$, we need only to prove that T has a unique finite (SI) decomposition up to similarity. First we verify the following assertions:

(a).   For any idempotent P in $\mathcal{A}'(T)$, if $T|_{P\mathcal{H}} \in (SI)$, then there exists a natural number $i, 1 \leq i \leq k$, such that $h([P]) = e_i$.

Assume that $h([P]) = \sum\limits_{i=1}^{k} \lambda_i e_i = \sum \lambda_i h([Q_i]), \lambda_i \in \mathbf{N}$. Set $w = r\sum\limits_{i=1}^{k} \lambda_i$, find a natural number $n > w$ such that

$$P \oplus 0_{\mathcal{H}^{(n-1)}} \sim_{\mathcal{A}'(T^{(n)})} \sum_{i=1}^{k} Q_i^{(\lambda_i)} \oplus 0_{\mathcal{H}^{(n-w)}}.$$

By Lemma 4.2.4,

$$T^{(n)}\big|_{(P \oplus 0_{\mathcal{H}^{(n-1)}})\mathcal{H}^{(n)}} \sim T^{(n)}\big|_{(\sum\limits_{i=1}^{k} Q_i^{(\lambda_i)} \oplus 0_{\mathcal{H}^{(n-w)}})\mathcal{H}^{(n)}}.$$

Thus

$$T|_{P\mathcal{H}} \sim T^{(w)}\big|_{(\sum\limits_{i=1}^{k} Q^{(\lambda_i)})\mathcal{H}^{(w)}}.$$

Note that $T|_{P\mathcal{H}} \in (SI)$, thus only one λ_i equals 1 and the others are zero, i.e.,

$$h([P]) = e_i$$

for some i.

(b). For arbitrary idempotents P and Q in $\mathcal{A}'(T^{(n)})$, if $h([P]) = h([Q])$, then $T|_{P\mathcal{H}} \sim T|_{Q\mathcal{H}}$.

Repeating the arguments in (a), we get (b).

Let $\{P_1, P_2, \cdots, P_m\}$ be a spectral projections of T and assume that

$$h([P_i]) = \sum_{j=1}^{k} \lambda_{ij} e_j, \lambda_{ij} \in \mathbf{N}.$$

Then

$$h(I) = \sum_{i=1}^{m}(\sum P_i) = \sum_{i=1}^{m}\sum_{j=1}^{k} \lambda_{ij} e_j.$$

From $h(I) = \sum\limits_{i=1}^{k} n_i e_i$, we have $\sum\limits_{i=1}^{m}\sum\limits_{j=1}^{k} \lambda_{ij} = \sum\limits_{i=1}^{k} n_i$. Thus $m \le \sum\limits_{i=1}^{k} n_i$. This implies that the number of elements in each set of spectral projections of T is finite and T is the direct sum of only finitely number of (SI) operators.

Furthermore, let $\{P_1, P_2, \cdots, P_l\}$ be an (SI) decomposition of T, then

$$h(\sum_{i=1}^{t}[P_i]) = h([I]) = \sum_{i=1}^{k} n_i e_i.$$

By (a), $t = \sum\limits_{i=1}^{k} n_i$ and for each $i, 1\leq i\leq k$, there exist

$$P_{i_1}, \cdots, P_{i_{n_i}} \in \{P_1, P_2, \cdots, P_t\}$$

such that

$$h([P_{i_1}]) = \cdots = h([P_{i_{n_i}}]) = e_i.$$

By (b), $T|_{P_{i_j}\mathcal{H}} \sim T|_{P_{i_k}\mathcal{H}}, 1\leq j, k\leq n_i$. Denote $A_i = T|_{P_{i_j}\mathcal{H}}, 1\leq j\leq n_i$, we have

$$T \sim \sum_{i=1}^{k} A_i^{(n_i)}.$$

Assume that $\{P_1', P_2', \cdots, P_s'\}$ is another (SI) decomposition of T. Then repeating the above arguments, we have $r = \sum\limits_{i=1}^{k} n_i$ and for each $i, 1\leq i\leq k$, there are n_i idempotents in $\{P_1', P_2', \cdots, P_s'\}$ such that h maps each of them to e_i. By (b), if $h([P_i]) = h([P_j']), 1\leq i, j\leq \sum\limits_{i=1}^{k} n_i$, then $T|_{P_i\mathcal{H}}\sim T|_{P_i'\mathcal{H}}$. By Lemma 4.2.5, T has a unique (SI) decomposition up to similarity. The proof of the theorem now is complete.

Proof of Corollary 4.2.2 Note that if $T^{(n)}$ has a unique (SI) decomposition up to similarity, then by Theorem 4.2.1 $T_1\sim T_2$ if and only if $\bigvee(\mathcal{A}'(T_1\oplus T_2))\cong\mathbf{N}$. Therefore, if $T_1\sim T_2$, then $K_0(\mathcal{A}'(T_1\oplus T_2))\cong\mathbf{Z}$, since $\bigvee(\mathcal{A}'(T_1\oplus T_2))\cong\mathbf{N}$. Conversely, if $K_0(\mathcal{A}'(T_1\oplus T_2))\cong\mathbf{Z}$, by Theorem 3.2.1, $\bigvee(\mathcal{A}'(T_1\oplus T_2))\cong\mathbf{N}^{(k)}, k\leq 2$. Since $K_0(\mathcal{A}'(T_1\oplus T_2))\cong\mathbf{Z}$, $\bigvee(\mathcal{A}'(T_1\oplus T_2))\cong\mathbf{N}$. Thus $T_1\sim T_2$. This completes the proof of the corollary.

The following proposition tells us that besides normal operators, a class of analytic Toeplitz operators whose unitary invariant and similarity invariant are same. In fact, we have stronger result.

Proposition 4.2.13 *Let φ_1 and φ_2 be two univalent analytic functions on the unit disk D. Then the following are equivalent:*
  (i) $T_{\varphi_1}\cong T_{\varphi_2}$;
  (ii) $\ker\tau_{T_{\varphi_1}, T_{\varphi_2}} \neq \{0\}$ and $\ker\tau_{T_{\varphi_2}, T_{\varphi_1}} \neq \{0\}$.
Proof (i)$\Rightarrow$(ii) is obvious.

  (ii)$\Rightarrow$(i). Assume that $\varphi_i(z) = \sum\limits_{j=0}^{\infty} \lambda_j^i z^j, z\in D, i = 1, 2$. Since φ_i is univalent, $\lambda_1^i \neq 0$ for $i = 1, 2$. If there are $X, Y \in \mathcal{L}(H^2)$ such that

$$T_{\varphi_2} Y = Y T_{\varphi_1}, \quad T_{\varphi_1} X = Y T_{\varphi_2}.$$

Denote $\Omega_1 = \varphi_1(D), \Omega_2 = \varphi_2(D)$. Clearly, Ω_1 and Ω_2 are simply connected and $T_{\varphi_1}^* \in \mathcal{B}_1(\Omega_1^*), T_{\varphi_2^*} \in \mathcal{B}_1(\Omega_2^*)$. If $\Omega_1 \neq \Omega_2$, by Lemma 4.2.3

$$ker\tau_{T_{\varphi_1}, T_{\varphi_2}} = ker\tau_{T_{\varphi_2}, T_{\varphi_1}} = \{0\},$$

this contradicts our assumption. Thus we may assume that $\Omega_1 = \Omega_2 = \Omega$ and $\sigma(T_{\varphi_1}) = \sigma(T_{\varphi_2}) = \overline{\Omega}$. Without loss of generality, we assume that $0 \in \Omega$ and $\varphi_1(0) = 0, \varphi_2(z_0) = 0, z_0 \in D$. Then there exists a Möbius transformation

$$\chi : D \to D$$

satisfying $\chi(0) = z_0$. Thus $\varphi_2(\chi(0)) = 0$. This implies that $T_{\varphi_2(z)} \cong T_{\varphi_2(\chi(z))}$. Therefore we may assume that $\varphi_2(0) = 0$. Note that $T_{\varphi_1}^*, T_{\varphi_2}^*$ have the following matrix representations with respect to the ONB $\{1, z, z^2, z^3, \cdots\}$:

$$T_{\varphi_1}^* = \begin{bmatrix} 0 & \lambda_1' & \lambda_2' & \lambda_3' & \cdots \\ & 0 & \lambda_1' & \lambda_2' & \cdot \cdot \\ & & 0 & \lambda_1' & \cdot \cdot \\ 0 & & & & \cdot \cdot \cdot \cdot \end{bmatrix}, \quad T_{\varphi_1}^* = \begin{bmatrix} 0 & \lambda_1^2 & \lambda_2^2 & \lambda_3^2 & \cdots \\ & 0 & \lambda_1^2 & \lambda_2^2 & \cdot \cdot \\ & & 0 & \lambda_1^2 & \cdot \cdot \\ 0 & & & & \cdot \cdot \cdot \cdot \end{bmatrix}.$$

From $T_{\varphi_1}^* Y^* = Y^* T_{\varphi_2}^*$, computation indicates

$$Y^* = \begin{bmatrix} y_{11} & y_{12} & y_{13} & \cdots \\ & y_{22} & y_{23} & \cdots \\ 0 & & & \cdot \cdot \cdot \cdot \end{bmatrix} \quad \text{and} \quad y_{nn} = [\frac{\lambda_1^2}{\lambda_1^1}]^{n-1} y_{11}, \ n = 2, 3, \cdots.$$

We claim that $|\frac{\lambda_1^2}{\lambda_1^1}| \leq 1$. Otherwise, since Y^* is bounded, $y_{11} = 0$ and $y_{nn} = 0$ for $n = 2, 3, \cdots$. Similarly, $y_{ij} = 0$ for $i, j \geq 1$. This contradicts $Y^* \neq 0$. Similarly, $|\frac{\lambda_1^1}{\lambda_1^2}| \leq 1$. Thus $|\lambda_1^1| = |\lambda_1^2| = \lambda$. Denote

$$\theta_j = arg \frac{\lambda_1^j}{\lambda}$$

and

$$U_j = diag(1, e^{i\theta_j}, e^{2i\theta_j}, \cdots), i = 1, 2.$$

Obviously U_j is unitary. Denote

$$R_j = U_j T_{T_j}^* U_j^* = \begin{bmatrix} 0 & \lambda & & & * \\ & 0 & \lambda & & \\ & & 0 & \ddots & \\ 0 & & & \ddots & \end{bmatrix}.$$

Since $U_j T_z^* U_j^* = e^{-i\theta_j} T_z^*$, $R_j \in \mathcal{A}'(T_z^*)$, $j = 1, 2$. Thus, there exists a function $g_j \in H^\infty$ such that $R_j = T_{g_j}^*$. Since $T_{g_j}^* \cong T_{\varphi_j}^*$, $g_j(D) = \varphi_j(D) = \Omega$ and $T_{g_j}^* \in \mathcal{B}_1(\Omega^*)$. Clearly, g_j is a univalent analytic function on D and

$$g_j(0) = 0, g_j'(0) = \lambda > 0, j = 1, 2.$$

It follows from Riemann mapping theorem that $g_1 = g_2$ and therefore $T_{\varphi_1} \cong T_{\varphi_2}$.

In the following we will compute the K_0-groups of some Banach algebras in terms of Theorem 4.2.1.

Theorem 4.2.14 $K_0(H^\infty) \cong \mathbf{Z}$, $\bigvee(H^\infty) \cong \mathbf{N}$.
Proof Consider the analytic Toeplitz operator T_z. It is well known that

$$\mathcal{A}'(T_z) \cong H^\infty.$$

By Theorem 4.2.1, we need only to show that $T = T_z^{(n)}$ has a unique (SI) decomposition up to similarity for each natural number n. Since $T^* \in \mathcal{B}_n(D)$, there are only finitely number of elements in each spectral family of T^*. Thus we need only to prove that if P is a minimal idempotent in $\mathcal{A}'(T)$, then $T|_{P(H^2)^{(n)}} \sim T_z$. Note that T is an isometric isomorphism and $P(H^2)^{(n)}$ is an invariant subspace of T, it follows from the famous von-Neumann-Wold theorem $T|_{P(H^2)^{(n)}} \cong T_z$. This completes the proof of the theorem.

Corollary 4.2.15 $K_0(H^\infty(\Omega)) \cong \mathbf{Z}$ and $\bigvee(H^\infty(\Omega)) \cong \mathbf{N}$, where Ω is a nonempty bounded simply connected domain.
Proof Since Ω is nonempty bounded and simply connected, there exists a univalent analytic function φ such that $\varphi(D) = \Omega$, and $\mathcal{A}'(T_\varphi) \cong H^\infty$. The corollary follows Theorem 4.2.14.

Proposition 4.2.16 Let $\varphi_1, \varphi_2, \cdots, \varphi_n \in H^\infty$ be univalent analytic functions, then there is a natural number $k, k \leq n$, such that

$$\bigvee(\mathcal{A}'(\sum_{i=1}^n \oplus T_{\varphi_i})) \cong \mathbf{N}^{(k)} \quad and \quad K_0(\mathcal{A}'(\sum_{i=1}^n \oplus T_{\varphi_i})) \cong \mathbf{Z}^{(k)}.$$

Furthermore, $T = \sum_{i=1}^{n} \oplus T_{\varphi_i}$ has a unique (SI) decomposition up to similarity.

Proof By Proposition 4.2.13, $\mathcal{A}'(\sum_{i=1}^{n} \oplus T_{\varphi_i}) \cong \sum_{j=1}^{k} \oplus (H^2)^{n_j}$. Thus it follows from the isomorphism of K-groups and Theorem 4.2.14 that

$$\bigvee(\mathcal{A}'(\sum_{i=1}^{n} \oplus T_{\varphi_i})) \cong \sum_{j=1}^{k} \oplus \bigvee((H^2))^{(n_j)} \cong \mathbf{N}^{(k)}$$

and

$$K_0(\mathcal{A}'(\sum_{i=1}^{n} \oplus T_{\varphi_i})) \cong \mathbf{Z}^{(k)}.$$

To summarize the facts discussed above we have the following corollary.

Corollary 4.2.17 *Let $\varphi_1, \varphi_2 \in H^\infty$ be univalent analytic, then the following are equivalent:*

(i) $T_{\varphi_1} \sim T_{\varphi_2}$;

(ii) $T_{\varphi_1} \cong T_{\varphi_2}$;

(iii) $K_0(\mathcal{A}'(T_{\varphi_1} \oplus T_{\varphi_2})) \cong \mathbf{Z}$.

4.3 (*SI*) Decompositions of Some Classes of Operators

Using Lemma 2.2.4 and Theorem 4.2.1, we get the following theorem.

Theorem 4.3.1 *Let $A_1, A_2, \cdots, A_k \in (SI) \cap \mathcal{L}(\mathcal{H})$ satisfying*

$$\mathcal{A}'(A_i)/rad\mathcal{A}'(A_i) \cong \mathbf{C}, \quad i = 1, 2, \cdots, k.$$

Then the following statements hold:

(i) $A_i \sim A_j$ if and only if $K_0(\mathcal{A}'(A_i \oplus A_j)) \cong \mathbf{Z}$;

(ii) Let $T = \sum_{i=1}^{k} \oplus A_i^{(n_i)}$, where $A_i \not\sim A_j, i \neq j$, then

$$\bigvee(\mathcal{A}'(T)) \cong \mathbf{N}^{(k)}, K_0(\mathcal{A}'(T)) \cong \mathbf{Z}^{(k)}.$$

Furthermore, T has a unique (SI) decomposition up to similarity.

A unilateral weighted shift T on Hilbert space $\mathcal{H}$ is an operator that maps each vector in some ONB $\{e_n\}_{n=0}^{\infty}$ into a scalar multiple of the next vector, i.e.,

$$Te_n = \alpha_n e_{n+1}, \alpha_n \in \mathbf{C}$$

for all n. If $\alpha_n \neq 0$ for all n, T is injective.

Lemma 4.3.2 *Let $A \in \mathcal{L}(\mathcal{H})$ be an injective unilateral weighted shift and power bounded, i.e., $\|A^n\| \leq M < \infty$ for a constant $M > 0$ and all $n \geq 1$. Then there is an injective unilateral weighted shift B with $\|B\| \leq 1$ such that $A \sim B$.*

Proof Suppose that

$$A = \begin{bmatrix} 0 & & & 0 \\ \alpha_1 & 0 & & \\ & \alpha_2 & 0 & \\ 0 & & \ddots & \ddots \end{bmatrix} \begin{matrix} e_0 \\ e_1 \\ e_2 \\ \vdots \end{matrix}.$$

It is well known that A is unitarily equivalent to the unilateral weighted shift with weighted sequence $\{|\alpha_n|\}$. Thus we can assume that $\alpha_n \geq 0$ for all n.

If $\alpha_n \leq 1, n = 1, 2, \cdots$, then set $B = A$. Otherwise, denote

$$n_1 = min\{j : \alpha_j > 1\}.$$

If $\prod_{j=n_1}^{k} \alpha_j \geq 1$ for all $k \geq n_1$, then set $m_1 = \infty$. Otherwise, set

$$m_1 = min\{k \geq n_1 : \prod_{j=n_1}^{k} \alpha_j < 1\},$$

then $m_1 < \infty$ and $\alpha_{m_1} < 1$. If $m_1 < \infty$, consider the set $\{k > m_1 : \alpha_k > 1\}$. If it is empty, let $n_2 = \infty$. Otherwise let $n_2 = min\{k > m_1 : a_k > 1\}$. If $\prod_{j=n_2}^{k} a_j \geq 1$ for all $k \geq n_2$, set $m_2 = \infty$.. Otherwise set $m_2 = min\{k \geq n_2 : \sum_{j=n_2}^{k} \alpha_j < 1\}$. Similarly, we define finitely many or countably many n_i's and m_i's inductively satisfying

(1) $m_0 = 0 < n_1 < m_1 < n_2 < m_2 < \cdots < n_k < m_k < n_{k+1} < \cdots$ (the last one is ∞ if the sequence is finite);

(2) $\alpha_j \leq 1$ if $j < n_1$;

(3) $\prod_{i=n_k}^{j} \alpha_i$ if $n_k \leq j \leq m_k, k = 1, 2, \cdots$;

(4) $\alpha_j \leq 1$ if $m_k \leq j \leq n_{k+1}, k = 1, 2, \cdots$.

Define

$$
\chi_j = \begin{cases} 1 & 0 \leq j < n_1 \\[2mm] \displaystyle\prod_{i=n_k}^{j} \alpha_i & n_k \leq j < m_k \\[4mm] 1 & m_k \leq j < n_{k+1}. \end{cases}
$$

Denote $M = \sup\{\|A^k\| : k \geq 1\}$, then $1 \leq \chi_j \leq M, j = 1, 2, \cdots$. Define $X \in \mathcal{L}(\mathcal{H})$ by

$$
X\left(\sum_{j=0}^{\infty} \alpha_j e_j\right) = \sum_{j=0}^{\infty} \chi_j \alpha_j e_j
$$

for all $\sum_{j=0}^{\infty} \alpha_j e_j \in \mathcal{H}$. It is easily seen that X is invertible. Set $B = X^{-1}AX$, then B is still an injective unilateral weighted shift.

If the weighted sequence of B is $\{b_j\}_{j=1}^{\infty}$, then $b_j = \chi_j^{-1} \alpha_j \chi_{j-1}, j \geq 1$. Note that $0 < b_j = \alpha_j \leq 1$ if $j < n_1$; $b_j = \alpha_{n_k}^{-1} \alpha_{n_k} \chi_{n_k-1} = \chi_{n_k-1}$ if $j = n_k, k \geq 1$, since $m_{k-1} \leq n_k - 1 < n_k$; $b_j = (\prod_{i=n_k}^{j} \alpha_i)^{-1} \alpha_j (\prod_{i=n_k}^{j-1} \alpha_i) = 1$ if $n_k < j < m_k, k \geq 1$. If $j = m_k$, $k \geq 1$, $\chi_{j-1} \geq 1$, $\alpha_j < 1$, thus

$$
0 < b_j = 1 \cdot \alpha_j \chi_{j-1}^{-1} \leq 1.
$$

If $m_k < j < n_k, k \geq 1, 0 < b_j = \alpha_j \leq 1$. Therefore $0 < b_j \leq 1$ and $\|B\| \leq 1$.

Lemma 4.3.3 *Let A be an injective unilateral weighted shift with $\|A\| \leq 1$. Then there exists an algebraic homomorphism $\phi : H^\infty \to \mathcal{A}'(A)$ satisfying*
(1) If h is a polynomial, then $\phi(h) = h(A)$;
(2) If $\phi(h) = 0$, then $h = 0$;
(3) Given $h \in H^\infty$, there is a sequence $\{p_n\}_{n=1}^{\infty}$ of polynomials such that $\{p_n(A)\}_{n=1}^{\infty}$ converge in strong operator topology. Denote

$$
\phi(h) = \lim_{\substack{sot \\ n \to \infty}} p_n(A).
$$

(4) For each $h \in H^\infty, \|h(A)\| \leq \|h\|_\infty$, where $\|h\|_\infty$ is the norm of h in H^∞.

Proof Denote $R = \mathcal{H}^{(\infty)}$ and

$$
T = \begin{bmatrix} A & & & 0 \\ (I - A^*A)^{\frac{1}{2}} & 0 & & \\ & I & 0 & \\ 0 & & \ddots & \ddots \end{bmatrix} \begin{matrix} \mathcal{H}_0 \\ \mathcal{H}_1 \\ \mathcal{H}_2 \\ \vdots \end{matrix},
$$

where $\mathcal{H}_i = \mathcal{H}, i = 1, 2, \cdots$. Clearly, $T^*T = I_R$, thus T is an isometry. Since

$$
T^* = \begin{bmatrix} A^* & (I - A^*A)^{\frac{1}{2}} & & 0 \\ & 0 & I & \\ & & 0 & I \\ 0 & & & \ddots \end{bmatrix} \begin{matrix} \mathcal{H}_0 \\ \mathcal{H}_1 \\ \mathcal{H}_2 \\ \vdots \end{matrix}
$$

and since $\{ker(A^*)^k : k \geq 1\} = \mathcal{H}_0, \mathcal{H}_0 \subset \bigvee \{ker(T^*)^k : k \geq 1\}$. If

$$
A = \begin{bmatrix} 0 & & & 0 \\ \alpha_1 & 0 & & \\ & \alpha_2 & 0 & \\ 0 & & \ddots & \ddots \end{bmatrix} \begin{matrix} e_0 \\ e_1 \\ e_2 \\ \vdots \end{matrix}.
$$

For $x_k = 0 \oplus 0 \oplus \cdots \oplus 0 \oplus e_k \oplus 0 \oplus \cdots \in R$, obviously $(T^*)^{k+2} x_k = 0$ for all $k \geq 0$. Hence $\mathcal{H}_1 \subset \bigvee \{ker(T^*)^k : k \geq 1\}$. Similarly, $\mathcal{H}_j \subset \bigvee \{ker(T^*)^k : k \geq 1\}$ for all $j \geq 0$. Therefore T is a pure isometry and for each $h \in H^\infty, h(T)$ is well-defined,

$$
\|h(T)\| = \|h\|_\infty
$$

and there is a sequence $\{p_n\}_{n=1}^\infty$ of polynomials such that

$$
h(T) = SOT - \lim_{n \to \infty} p_n(T).
$$

Let P_0 be the orthogonal projection from R onto $\mathcal{H}_0$. Define

$$
\Phi(h) = P_0 h(T)|_{\mathcal{H}_0}.
$$

Then it is not difficult to see that Φ is a homomorphism from H^∞ to $\mathcal{A}'(A)$ satisfying (1)–(4).

Lemma 4.3.4 *Let A be a power bounded injective unilateral shift with $\sigma_e(A) = \{z : |z| = 1\}$, then given $T \in \mathcal{A}'(A)$, there exists an $h \in H^\infty$ such that $T = h(A)$.*

Proof By Lemma 4.3.2 and Lemma 4.3.3, $h(A)$ is well-defined for each $h \in H^\infty$. Since $0 \notin \sigma_e(A)$ and $ker A = \{0\}$, there exists an injective unilateral weighted shift B such that $A^*B = I$. Since

$$\sigma_e(A) = \{z : |z| = 1\},$$

$$\sigma_e(B) = \{z : |z| = 1\}.$$

Since B has no eigenvalue and is not invertible, $\sigma(B) = \{z : |z| \leq 1\}$. Denote e_0 a unit vector in $ker A^*$. For $|\lambda| < 1$, denote $f_\lambda = \sum_{n=0}^{\infty} \lambda^n B^n e_0$, then $A^* f_\lambda = \lambda f_\lambda$ and $< e_0, f_\lambda > = 1$. It is obvious that f_λ is a vector-valued analytic function for $\lambda \in D$. Given $T \in \mathcal{A}'(A)$, define

$$h(\lambda) = < Te_0, f_{\overline{\lambda}} > = < e_0, T^* f_{\overline{\lambda}} > .$$

Thus h is analytic in D. From $f_{\overline{\lambda}} \neq 0$, $(A^* - \overline{\lambda}) f_{\overline{\lambda}} = 0$ and $dim ker(A^* - \overline{\lambda}) = 1$ for $\lambda \in D$, there exists an $\alpha \in \sigma(T^*)$ such that $T^* f_{\overline{\lambda}} = \alpha f_{\overline{\lambda}}$ because T^* commutes with A^*. Thus

$$h(\lambda) = < Te_0, f_1 > = < e_0, T^* f_{\overline{\lambda}} > = < e_0, \alpha f_{\overline{\lambda}} > = \overline{\alpha}.$$

Since $|h(\lambda)| \leq \|T\|$, $h \in H^\infty$. Note that $h(A)^* f_{\overline{\lambda}} = \overline{h(\lambda)} f_{\overline{\lambda}} = T^* f_\lambda$ for all $\lambda \in D$ and $\bigvee \{ f_{\overline{\lambda}} : |\lambda| < 1 \} = \mathcal{H}$, we have $T = h(A)$.

Using Lemma 4.3.2, Lemma 4.3.3 and Lemma 4.3.4, we have the following proposition and theorem.

Proposition 4.3.5 *If A is a power bounded injective unilateral weighted shift with $\sigma_e(A) = \{z : |z| = 1\}$, then $\mathcal{A}'(A)$ is isomorphic to H^∞.*

Theorem 4.3.6 *If A is a power bounded injective unilateral weighted shift with $\sigma_e(A) = \{z : |z| = 1\}$, then $A^{(n)}$ has a unique (SI) decomposition up to similarity for each natural number n. Furthermore, $\bigvee(\mathcal{A}'(A)) \cong \mathbf{N}$ and $K_0(\mathcal{A}'(A)) \cong \mathbf{Z}$.*

Example 4.3.7 *Let A be an injective unilateral weighted shift with increasing weight sequence $\{w_n\}_{n=1}^{\infty}$ and let $r = \lim_{n \to \infty} |w_n|$. Then A/r is a power bounded injective unilateral weighted shift and $\sigma_e(A/r) = \{z : |z| = 1\}$. By Theorem 4.3.6, $\bigvee(\mathcal{A}'(A)) \cong \mathbf{N}$, $K_0(\mathcal{A}'(A)) \cong \mathbf{Z}$ and $A^{(n)}$ has a unique (SI) decomposition up to similarity for each natural number n.*

Example 4.3.8 *Let T be an injective unilateral weighted shift with decreasing weight sequence $\{w_n\}_{n=1}^{\infty}$ and $\lim_{n \to \infty} w_n = 0$. A simple compu-*

tation shows that $\mathcal{A}'(T)/rad\mathcal{A}'(T)\cong\mathbf{C}$. *By Theorem 4.3.1,* $\bigvee(\mathcal{A}'(T))\cong\mathbf{N}$, $K_0(\mathcal{A}'(T))\cong\mathbf{Z}$ *and* $T^{(n)}$ *has a unique* (SI) *decomposition up to similarity for each natural number* n.

In the following we will discuss the class of bilateral weighted shift. An operator $A\in\mathcal{L}(\mathcal{H})$ is called a bilateral weighted shift if there is an ONB $\{e_n\}_{n=-\infty}^{\infty}$ of the space $\mathcal{H}$ and a sequence $\{\alpha_n\}_{n=-\infty}^{\infty}$ of complex numbers such that $Ae_n = \alpha_n e_{n+1}$. If $\alpha_n \neq 0$ for all n, then A is injective. It is well-known that A is unitarily equivalent to the bilateral weighted shift with weight sequence $\{|\alpha_n|\}_{n=-\infty}^{\infty}$. In what follows we always assume A is an injective bilateral weighted shift with (real) monotone weight sequence and A is not invertible. Denote

$$
A = \begin{bmatrix}
\ddots & & & \vdots & & & \\
0 & \alpha_1 & & \vdots & & & \\
& 0 & \alpha_0 & \vdots & & & \\
& & 0 & \vdots & \alpha_{-1} & & \\
\cdots & \cdots & \cdots & \cdots & \cdots & \cdots & \cdots \\
& & & \vdots & 0 & \alpha_{-2} & \\
& & & \vdots & & 0 & \alpha_{-3} \\
& & & \vdots & & & 0 \\
& & & \vdots & & & \ddots
\end{bmatrix}
\begin{matrix}
\vdots \\ e_2 \\ e_1 \\ e_0 \\ \\ e_{-1} \\ e_{-2} \\ e_{-3} \\ \vdots
\end{matrix}
$$

$$
= \begin{bmatrix} S & R \\ 0 & T \end{bmatrix}
\begin{matrix} \bigvee\{e_j : j\geq 0\} = \mathcal{H}_+ \\ \bigvee\{e_j : j < 0\} = \mathcal{H}_- \end{matrix} .
$$

We assume that $a_j \geq a_{j+1} > 0$ and $\lim\limits_{n\to-\infty} a_n = 1$. Since A is not invertible,

$$
\lim_{n\to+\infty} a_n = 0.
$$

For polynomial $p(z) = \sum\limits_{j=0}^{n} a_j z^j$, denote

$$
p(A) = \begin{bmatrix} p(S) & R_p \\ 0 & p(T) \end{bmatrix} .
$$

Then $\|p(T)\|\leq\|p\|_\infty$ and $\|p(S)\|\leq\|p\|_\infty$. Let Q and P_k be the orthogonal projections from $\mathcal{H}$ onto $\mathcal{H}_-$ and , respectively,

$$\bigvee\{e_k\} := \bigvee\{\lambda e_k : \lambda \in \mathbf{C}\}, k = 0, \pm 1, \pm 2, \cdots .$$

Note that when $k \geq 0$,

$$P_k R_p = \sum_{j=0}^{n} \alpha_j P_k A^j Q = \sum_{k \leq j \leq n} (\alpha_j \prod_{i=k-j}^{k-1} \alpha_i) e_k \otimes e_{k-j}.$$

Thus

$$\|P_k R_p\|_2 \leq (\prod_{i=-1}^{k-1} \alpha_i)(\sum_{j=k+1}^{n} \alpha_j^2)^{\frac{1}{2}} \leq (\prod_{i=-1}^{k-1} \alpha_i)\|p\|_2 \leq (\prod_{i=-1}^{k-1} \alpha_i)\|p\|_\infty,$$

where $\|P_k R_p\|_2$ is the Hilbert-Schmidt norm of $P_k R_p$, $\|p\|_2$ is the norm of p in $H^2(\partial D)$. Therefore,

$$\|R_p\|_2 \leq \sum_{k=0}^{+\infty} \|P_k R_p\|_2 \leq \sum_{k=0}^{+\infty}(\prod_{i=-1}^{k-1} \alpha_i)\|p\|_\infty < +\infty.$$

If $\{p_n\}_{n=1}^\infty$ are uniformly bounded in H^∞ and converges to some $h \in H^\infty$ uniformly in every closed subset of D, then $\{p_n(S)\}_{n=1}^\infty$ and $\{p_n(T)\}_{n=1}^\infty$ converge to $h(S)$ and $h(T)$, respectively, in strong operator topology, and $\{R_{p_n}\}_{n=1}^\infty$ converges to some operator R_h in the Hilbert-Schmidt norm. Thus for each $h \in H^\infty$, we get an operator

$$\begin{bmatrix} h(S) & R_h \\ 0 & h(T) \end{bmatrix},$$

denoted by $h(A)$. Obviously, $\|h(A)\| \geq \|h\|_\infty$ and $h(A) \in \mathcal{A}'(A)$. Furthermore,

$$(h_1 + h_2)(A) = h_1(A) + h_2(A), h_1 h_2(A) = h_1(A) h_2(A)$$

for all $h_1, h_2 \in H^\infty$.

Theorem 4.3.9 $\mathcal{A}'(A)/rad\mathcal{A}'(A) \cong H^\infty$, where A is an injective bilateral weighted shift with decreasing weight sequence $\{\alpha_n\}_{n=-\infty}^{+\infty}$, $\lim\limits_{n \to -\infty} \alpha_n = 1$ and A is not invertible.

Proof By the analysis above, we need only to show that for each $X \in \mathcal{A}'(A)$, there exists an $h \in H^\infty$ such that $X = h(A) + Q$, where

$Q \in rad\mathcal{A}'(A)$. Assume that

$$
X = \begin{bmatrix}
& \vdots & \vdots & \vdots & & & \\
\cdots & X_{11} & X_{10} & \vdots & X_{1\,-1} & X_{1\,-2} & \cdots \\
\cdots & X_{01} & X_{00} & \vdots & X_{0\,-1} & X_{0\,-2} & \cdots \\
\cdots & \cdots & \cdots & \cdots & \cdots & \cdots & \cdots \\
\cdots & X_{-11} & X_{-10} & \vdots & X_{-1\,-1} & X_{-1\,-2} & \cdots \\
\cdots & X_{-21} & X_{-20} & \vdots & X_{-2\,-1} & X_{-2\,-2} & \cdots \\
& \vdots & \vdots & \vdots & & &
\end{bmatrix}
\begin{matrix}
\vdots \\ e_1 \\ e_0 \\ \\ e_{-1} \\ e_{-2} \\ \vdots
\end{matrix}.
$$

For each integer k, denote $\Delta_k(X) = (\varepsilon_{ij}\chi_{ij})_{i,j}$, where $\varepsilon_{ij} = \begin{cases} 1 & i-j = k \\ 0 & i-j \neq k \end{cases}$. That is, $\Delta_k(X)$ is the operator, which reserves the k-th diagonal in the matrix representation of X and let the other entries be zero. It is obvious that X commutes with A if and only if A commutes with every $\Delta_k(X)$. Moreover, computations indicate that $\Delta_0(X) = \chi_{00}I$, $\Delta_k(X)A^{|k|}$ commutes with A and $\Delta_0(\Delta_k(X)A^{|k|}) = \Delta_k(X)A^{|k|}$ for $k < 0$. Thus there is a constant α such that $\Delta_k(X)A^{|k|} = \alpha I$. If $\alpha \neq 0$, we have

$$(\alpha^{-1}\Delta_k(X)A^{|k|-1})A = A(\alpha^{-1}\Delta_k(X)A^{|k|-1}) = I.$$

This contradicts that A is not invertible. Therefore $\alpha = 0$ and $\Delta_k(X) = 0$ for all $k < 0$. Thus

$$
X = \begin{bmatrix} X_1 & X_2 \\ 0 & X_3 \end{bmatrix} \begin{matrix} \mathcal{H}_+ \\ \mathcal{H}_- \end{matrix}
$$

and X commutes with T. A simple computation shows that

$$
\begin{bmatrix} X_1 & X_2 \\ 0 & 0 \end{bmatrix} \in rad\mathcal{A}'(A).
$$

Since $\sigma_e(T) = \{z : |z| = 1\} \cup \{0\}$ and $\|T\| = 1$, we can find $h \in H^\infty$ and $Q \in rad\mathcal{A}'(A)$ such that $X_2 = h(T) + Q$ by Lemma 4.3.4. This completes the proof of the theorem.

Theorem 4.3.10 *Let $A \in \mathcal{L}(\mathcal{H})$ be an injective bilateral weighted shift with monotone weight sequence. If A is not invertible, then*

$$\bigvee(\mathcal{A}'(A)) \cong \mathbf{N}, \quad K_0(\mathcal{A}'(A)) \cong \mathbf{Z}$$

and $A^{(n)}$ has a unique (SI) decomposition up to similarity for each natural number n.

Given two power bounded injective unilateral weighted shifts $A \sim \{\alpha_k\}_{k=1}^{\infty}$ and $B \sim \{\beta_k\}_{k=1}^{\infty}$, by the arguments used in the proofs of Lemma 4.3.3 and Proposition 4.2.13, we have the following proposition.

Proposition 4.3.11 $A \sim B$ *if and only if* $ker\tau_{A,B} \neq \{0\}$ *and* $ker\tau_{B,A} \neq \{0\}$.

Note, if A is not similar to B, then $ker\tau_{A,B} = \{0\}$ or $ker\tau_{B,A} = \{0\}$. Assume that $ker\tau_{A,B} = \{0\}$. Define

$$T = \begin{bmatrix} B & 0 \\ 0 & A \end{bmatrix}.$$

It follows from $ker\tau_{A,B} = \{0\}$ that

$$\mathcal{A}'(T) = \left\{ \begin{bmatrix} T_B & T_{BA} \\ 0 & T_A \end{bmatrix} : T_B \in \mathcal{A}'(B), T_A \in \mathcal{A}'(A) \quad \text{and} \quad T_{BA} \in ker\tau_{B,A} \right\}.$$

By Proposition 4.3.5, $\mathcal{A}'(T)/rad\mathcal{A}'(T) \cong H^\infty \oplus H^\infty$. Thus we have the following proposition.

Proposition 4.3.12 *Let* $A, B \in \mathcal{A}'(T)$ *be two power bounded injective unilateral weighted shifts, then the following are equivalent:*

(i) $A \sim B$ *if and only if* $\mathcal{A}'(A \oplus B)/rad\mathcal{A}'(A \oplus B) \cong M_2(H^\infty)$;

(ii) A *is not similar to* B *if and only if*

$$\mathcal{A}'(A \oplus B)/rad\mathcal{A}'(A \oplus B) \cong H^\infty \oplus H^\infty;$$

(iii) $A \sim B$ *if and only if* $K_0(\mathcal{A}'(A \oplus B)) \cong \mathbf{Z}$;

(iv) A *is not similar to* B *if and only if* $K_0(\mathcal{A}'(A \oplus B)) \cong \mathbf{Z} \oplus \mathbf{Z}$.

Let S be the unilateral shift on H^2 given by $Sf = zf(z), f \in H^2$. Let $\theta \in H^\infty$ be a nonconstant inner function and P_θ denote the projection of H^2 onto $H(\theta) = H^2 \ominus \theta H^2$. The Jordan block $S(\theta)$ is defined by $S(\theta) = P_\theta S|_{H(\theta)}$ [Bercovici, H. (1988)]. In the following, we will prove that for the singular inner function θ with $S(\theta) \in (SI), S(\theta)^{(n)}$ has a unique (SI) decomposition up to similarity. Applying this result we get that $V^{(n)}$ has a unique (SI) decomposition up to similarity.

Theorem 4.3.13 *Let* $\theta \in H^\infty$ *be a singular inner function such that* $S(\theta) \in (SI)$, *then for each* $n \geq 1, S(\theta)^{(n)}$ *has a unique* (SI) *decomposition up to similarity.*

Proof Applying the six-term exact sequence of K-theory [cf. [Taylor, J. (1975)]] and the short exact sequence of Banach algebra

$$0 \longrightarrow \theta H^\infty \overset{i}{\longrightarrow} H^\infty \overset{\pi}{\longrightarrow} H^\infty/\theta H^\infty \longrightarrow 0,$$

we obtain the following exact sequence of groups

$$0 \longrightarrow K_0(H^\infty) \overset{\pi_*}{\longrightarrow} K_0(H^\infty/\theta H^\infty) \overset{\partial}{\longrightarrow} K_1(\theta H^\infty) \overset{i_*}{\longrightarrow} K_1(H^\infty), \quad (4.3.1)$$

where π_* (resp. i_*) is the induced homomorphism of π (resp. i) on $K_0(H^\infty)$ (resp. $K_1(\theta H^\infty)$) and ∂ is the connected homomorphism. It is proved [Tolokommokov, V. (1993)] that $Bar((\theta H^\infty)^+) = 1$, where

$$Bar((\theta H^\infty)^+) = min\{n : (a_1, \cdots, a_{m+1})^T \in Lg_{m+1}((\theta H^\infty)^+)$$

such that

$$(a_1 + b_1, a_{m+1}, \cdots, a_m + b_m a_{m+1})^T \in Lg_m((\theta H^\infty)^+)$$

for some $\{b_i\}_{i=1}^m \subset (\theta H^\infty)^+, \forall\, m \geq n\}$, and

$$Lg_n((\theta H^\infty)^+) = \{(a_1, \cdots, a_n)^T : a_i \in (\theta H^\infty)^+, (i = 1, 2, \cdots, n), \sum_{i=1}^n a_i b_i = 1$$

for some $(b_1, \cdots, b_n)^T, b_i \in (\theta H^\infty)^+ (i = 1, 2, \cdots, n)\}$. From this $i_{(\theta H^\infty)^+}$ is an isomorphism [Wang, Z.Y. and Xue, Y.F. (2000)]. Thus for any $a \in K_1(\theta H^\infty)$ with $i_*(a) = 0$ in $K_1(H^\infty)$, there is an $f = \theta g + 1 \in GL_1((\theta H^\infty)^+)$ for some $g \in H^\infty$ such that $[a] = [f]$ in $K_1(\theta H^\infty)$ and $i_*([f]) = 0$ in $K_1(H^\infty)$. Thus we can find $h \in H^\infty$ such that $f = e^h$ so that $e^{\pi(h)} = \pi(f) = 1$. Noting that $\mathcal{A}'(S(\theta)) \cong H^\infty/\theta H^\infty$ and since $S(\theta) \in (SI)$, we have that $H^\infty/\theta H^\infty$ contains no non-zero idempotent. Therefore $\pi(h) = 2k\pi i$ for some integer k [Taylor, J. (1975)] and hence $h \in (\theta H^\infty)^+$. So $f = e^h \in (\theta H^\infty)^+$. This means that $\partial = 0$ by (4.3.1). Finally, we conclude from (4.3.1) and [Wang, Z.Y. and Xue, Y.F. (2000)] that $K_0(H^\infty/\theta H^\infty) = \{n[1] : n \in \mathbf{Z}\}$.

Let P be a nontrivial idempotent in $\mathcal{A}'(S(\theta)^{(n)}) = M_n(\mathcal{A}'(S(\theta)))$. Since $g \leq r(H^\infty/\theta H^\infty) \leq 1 + Bar(H^\infty/\theta H^\infty) \leq 1 + Bar(H^\infty) = 2$ by [Tolokommokov, V. (1993)] and since

$$\mathcal{A}'(S(\theta)) = \{u(S(\theta)) : u \in H^\infty\} \cong H^\infty/\theta H^\infty,$$

there are $X \in GL_1(\mathcal{A}'(S(\theta)))$ and k with $1 \leq k \leq n - 1$ such that

$$P = X diag(I_k, 0) X^{-1}.$$

Set $T_1 = X|_{diag(I_k,0)H^{(n)}}$ and $T_2 = X^{-1}|_{PH^{(n)}}$. It is easy to verify that

$$T_1 T_2 = I_{PH^{(n)}}$$

and

$$T_2 T_1 = I_{diag(I_k,0)H^{(n)}}$$

and

$$T_2 S(\theta)^{(n)}|_{PH^{(n)}} T_1 = diag(S(\theta)^{(k)}, 0)$$

on $H^{(k)} \oplus 0$ which is similar to $S(\theta)^{(k)}$. Thus $S(\theta)^{(n)}$ has a finite decomposition.

Suppose that $\{P_i\}_{i=1}^m \subset \mathcal{A}'(S(\theta)^{(n)})$ are idempotents,

$$\sum_{i=1}^m P_i = I_{H^{(n)}}, P_i P_j = 0 (i \neq j)$$

and

$$S(\theta)^{(n)}|_{P_i H^{(n)}} \in (SI)(i = 1, 2, \cdots, m).$$

By the above arguments, $S(\theta)^{(n)}|_{P_i H^{(n)}} \in (SI)$ implies that there exists an operator $X_i \in GL_1(\mathcal{A}'(S(\theta)))$ such that

$$P_i = X_i diag(I_{H^2}, 0) X_i^{-1}, (i = 1, 2, \cdots, m).$$

Thus $n[I_{H^2}] = [\sum_{i=1}^m P_i] = \sum_{i=1}^m [P_i] = m[I_{H^2}]$ in $K_0(\mathcal{A}'(S(\theta))) \cong \mathbf{Z}$. Since $[I_{H^2}]$ is the only generator of $K_0(\mathcal{A}'(S(\theta)))$, we get that $n = m$. Therefore we can choose $Y_i \in GL_1(\mathcal{A}'(S(\theta)^{(n)}))$ such that

$$P_i = Y_i diag(0, 0, \cdots, 0, I_{H^2}, 0, \cdots, 0) Y_i^{-1} = Y_i e_i Y_i^{-1}, (i = 1, 2, \cdots, n).$$

Set $Y = \sum_{i=1}^n P_i Y_i$. Then it is easy to check that

$$Y \in GL_1(\mathcal{A}'(S(\theta)^{(n)}))$$

with $Y^{-1} = \sum_{i=1}^n Y_i^{-1} P_i$ and $Y^{-1} P_i Y = e_i$. That is $S(\theta)^{(n)})$ has a unique finite (SI) decomposition up to similarity.

Let V be the Volterra operator on $\mathcal{H} = L^2([0,1])$ defined by

$$Vf(t) = \int_0^t f(s)ds, f \in \mathcal{H}, t \in [0,1].$$

Then $\|V\| = \frac{2}{\pi}, \sigma(V) = \{0\}$ and $\mathcal{A}'(V)$ is the weak closure of the algebra generated by $I_{\mathcal{H}}$ and V. Put $T_V = (I_{\mathcal{H}} - V)(I_{\mathcal{H}} + V)^{-1}$. Then $\mathcal{A}'(T_V) = \mathcal{A}'(T_V)$ and T_V is unitarily equivalent to $S(e)$ by [Bercovici, H. (1987)] and hence $\mathcal{A}'(V) \cong H^\infty / eH^\infty$, where $e(z) = exp(\frac{z+1}{z-1})$.

Corollary 4.3.14 $V \in (SI)$ and $V^{(n)}$ has a unique finite (SI) decomposition up to similarity for each $n \geq 1$.

Proof We have $supp(e) = \{1\}$ and every nontrivial divisor of e has the form $e_t(z) = \lambda exp(t\frac{z+1}{z-1}), 0 < t < 1, |\lambda| = 1, z \in D$. Take

$$z_n = 1 - \frac{1}{n}, n = 1, 2, \cdots .$$

Then $\lim_{n \to \infty} e_t(z_n) = \lim_{n \to \infty} (e/e_t)(z_n) = 0$, i.e., $(e_t, e/e_t)^T \notin Lg_2(H^{(\infty)})$. So $S(e) \in (SI)$, since it is not difficult to see that $S(\theta) \notin (SI)$ if and only if there exists a nonconstant division θ_1 of θ such that $(\theta_1, \theta/\theta_1)^T \in Lg_2(H^{(\infty)})$.

Now suppose that $\{P_i\}_{i=1}^m \subset \mathcal{A}'(V^{(n)})$ is an idempotents, $P_i P_j = 0 (i \neq j)$ and $\sum_{i=1}^m P_i = I$. Let Q be an idempotent in $\mathcal{A}'(T_V^{(n)}|_{P_i \mathcal{H}^{(n)}})$. Then

$$QP_i \in \mathcal{A}'(T_V^{(n)}) = \mathcal{A}'(V^{(n)})$$

is an idempotent and moreover $Q \in \mathcal{A}'(V^{(n)}|_{P_i \mathcal{H}^{(n)}})$. Thus $V^{(n)}|_{P_i \mathcal{H}^{(n)}} \in (SI)$ means that $Q = 0$ or P_i, i.e., $T_V^{(n)}|_{P_i \mathcal{H}^{(n)}} \in (SI)$, so that $\{P_i\}_{i=1}^m$ is an (SI) decomposition spectral family of $T_V^{(n)}$. By Theorem 4.3.13, $n = m$ and there is $Y \in GL_1(\mathcal{A}'(V^{(n)}))$ such that $P_i = Ye_iY^{-1}, (i = 1, 2, \cdots, n)$. The corollary follows.

4.4 The Commutant of Cowen-Douglas Operators

As indicated in Theorem JW2 of Chapter 1, Cowen-Douglas operators have very rich contents, including many of hypernormal operators and subnormal operators. In order to characterize the similarity invariant of Cowen-Douglas operators, we will discuss the commutant of Cowen-Douglas operators in this section, which is the preparation for characterizing the similarity invariant of Cowen-Douglas operators in Chapter 6.

In this section, we always assume that $T \in \mathcal{B}_n(\Omega)$, thus $\bigvee_{z \in \Omega} \ker(T-z) = \mathcal{H}$, where $\mathcal{H}$ is a complex, separable, infinite dimensional Hilbert space. Note that $\mathcal{B}_1(\Omega) \subset (SI)$ and for every $T \in \mathcal{B}_1(\Omega)$ we can easily prove the following result.

Proposition CD [Cowen, M.J. and Douglas, R. (1977)] $\mathcal{A}'(T)$ *is isomorphic to a subalgebra of* $H^\infty(\Omega)$.

Proposition CD indicates that if $T \in \mathcal{B}_1(\Omega)$, then $\mathcal{A}'(T)$ is commutative. In the finite dimensional space $\mathbf{C}^n$, $J_n(\lambda) \in \mathcal{L}(\mathbf{C}^n)$, and $\mathcal{A}'(J_n(\lambda))$ is commutative. For Volterra operator V, $\mathcal{A}'(V)$ is also commutative. In Section 4.5, we will see that for multiplication operator M_f on the Sobolev disk algebra, $M_f^* \in \mathcal{B}_1(D)$ if and only if $\mathcal{A}'(M_f^*)$ is commutative. In Chapter 5, we will show that the set of strongly irreducible operators, whose commutant is commutative, is dense in the set of all Cowen-Douglas operators in the norm topology.

Example 4.4.1 *Let* $\lambda_k = \frac{1}{k}$ *and* $W_k = \lambda_k + J_2 = \begin{bmatrix} \lambda_k & 1 \\ 0 & \lambda_k \end{bmatrix}$. *Define*

$$T = \begin{bmatrix} 0 & W_1 & & 0 \\ & 0 & W_2 & \\ & & \ddots & \ddots \\ 0 & & & \ddots \end{bmatrix} \begin{matrix} \mathbf{C}^2 \\ \mathbf{C}^2 \\ \mathbf{C}^2 \\ \vdots \end{matrix} \in \mathcal{L}(\sum_1^\infty \oplus \mathbf{C}^2).$$

Proposition 4.4.2 *Suppose that* $T \in \mathcal{L}(\mathcal{H})$ *is given in Example 4.4.1, where*
$\mathcal{H} = \oplus \mathbf{C}^2$. *Then*

  (i) $T \in (SI)$;
  (ii) $\mathcal{A}'(T)$ *is not commutative;*
  (iii) $\mathcal{A}'(T)/rad\mathcal{A}'(T)$ *is commutative.*

Proof (i) Suppose that $P \in \mathcal{A}'(T)$, then

$$P = \begin{bmatrix} P_{11} & P_{12} & P_{13} & \cdots \\ & P_{22} & P_{23} & \cdots \\ & & P_{33} & \cdots \\ 0 & & & \ddots \end{bmatrix} \begin{matrix} \mathbf{C}^2 \\ \mathbf{C}^2 \\ \mathbf{C}^2 \\ \vdots \end{matrix} \quad \text{(cf. [Jiang, C.L. and Li, J.X. (2000)])}$$

and $P_{k+1\,k+1} = E_k^{-1} P_{11} E_k$ for all $k > 1$, where $E_k = W_1 W_2 \cdots W_k$. Computation indicates that $E_k = \begin{bmatrix} \frac{1}{k!} & \frac{k+1}{2(k-1)!} \\ 0 & \frac{1}{k!} \end{bmatrix}$. Assume that $P_{11} = \begin{bmatrix} a_{11} & a_{12} \\ a_{21} & a_{22} \end{bmatrix}$.

Then

$$P_{k+1\,k+1} = E_k^{-1} P_{11} E_k$$

$$= \begin{bmatrix} k! & -\frac{k+1}{2(k-1)!}(k!)^2 \\ 0 & k! \end{bmatrix} \begin{bmatrix} a_{11} & a_{12} \\ a_{21} & a_{22} \end{bmatrix} \begin{bmatrix} \frac{1}{k!} & \frac{k+1}{2(k-1)!} \\ 0 & \frac{1}{k!} \end{bmatrix}$$

$$= \begin{bmatrix} a_{11} - a_{21}\frac{k+1}{2(k-1)!}k! & (a_{11} - a_{22})\frac{k+1}{2(k-1)!}k! + a_{12} - a_{21}(\frac{k+1}{2(k-1)!})^2(k!)^2 \\ a_{21} & a_{21}\frac{k+1}{2(k-1)!}k! + a_{22} \end{bmatrix}$$

$$= \begin{bmatrix} a_{11} - a_{21}\frac{k(k+1)}{2} & (a_{11} - a_{22})\frac{k(k+1)}{2} + a_{12} - a_{21}\frac{k^2(k+1)^2}{4} \\ a_{21} & a_{21}\frac{k(k+1)}{2} + a_{22} \end{bmatrix}$$

Since $k(k+1) \longrightarrow \infty$ as $k \longrightarrow \infty$, $a_{21} = 0$ and $a_{11} = a_{22}$.
It is easy to see that if $P^2 = P$, then $P = 0$ or I, i.e., $T \in (SI)$.

(ii)   Denote $A_0 = \begin{bmatrix} 0 & 1 \\ 0 & 0 \end{bmatrix}$. Since $W_k A_0 = A_0 W_k$ for all $k \geq 1$,

$$A := diag(A_0, A_0, \cdots) \in \mathcal{A}'(T).$$

Denote $B_{02} = \begin{bmatrix} 1 & 0 \\ 0 & 0 \end{bmatrix}$. Set $B_{k\,k+2} = E_k^{-1} B_{02} W_3 W_4 \cdots W_{k+2}$ for $k \geq 1$. Then

$$B_{k\,k+2} = \begin{bmatrix} k! & -\frac{k(k+1)}{2}k! \\ 0 & k! \end{bmatrix} \begin{bmatrix} 1 & 0 \\ 0 & 0 \end{bmatrix} \begin{bmatrix} \frac{1}{3\cdot4\cdot5\cdots(k+2)} & \frac{3+4+5+\cdots+(k+2)}{3\cdot4\cdot5\cdots(k+2)} \\ 0 & \frac{1}{3\cdot4\cdot5\cdots(k+2)} \end{bmatrix}$$

$$= \begin{bmatrix} \frac{k!}{3\cdot4\cdot5\cdots(k+2)} & k!\frac{\frac{1}{2}(k+5)\cdot k}{3\cdot4\cdot5\cdots(k+2)} \\ 0 & 0 \end{bmatrix}$$

$$= \begin{bmatrix} \frac{2}{(k+1)(k+2)} & \frac{k(k+5)}{2(k+1)(k+2)} \\ 0 & 0 \end{bmatrix}$$

Thus $\|B_{k\,k+2}\| < 2$ for all $k \geq 1$ and

$$B := \begin{bmatrix} 0 & 0 & B_{02} & 0 & 0 & 0 & 0 & \cdots \\ 0 & 0 & 0 & B_{13} & 0 & 0 & 0 & \cdots \\ 0 & 0 & 0 & 0 & B_{24} & 0 & 0 & \cdots \\ & & & & & \ddots & \end{bmatrix} \in \mathcal{L}(\mathcal{H}).$$

Computation implies $B \in \mathcal{A}'(T)$. But since $A_0 B_{02} \neq B_{02} A_0$, $AB \neq BA$, and therefore $\mathcal{A}'(T)$ is not commutative.

(iii) Suppose that $B \in \mathcal{A}'(A)$, then

$$B = \begin{bmatrix} B_{11} & B_{12} & B_{13} & \cdots \\ & B_{22} & B_{23} & \cdots \\ & & B_{33} & \cdots \\ 0 & & & \ddots \end{bmatrix}.$$

Similar to the proof of (ii), we have

$$B_{ij} = \begin{bmatrix} b_{11}^{ij} & b_{12}^{ij} \\ 0 & b_{22}^{ij} \end{bmatrix}.$$

Let $\{e_k\}_{k=1}^{\infty}$ be an ONB of $H = \oplus \mathbf{C}^{(2)}$ such that

$$A = \begin{bmatrix} 0 & W_1 & 0 & \cdots & \cdots \\ & 0 & 0 & W_2 & 0 & \cdots \\ & & & & & \\ & 0 & 0 & 0 & W_3 & \ddots \\ & \vdots & \vdots & \vdots & \ddots & \ddots \end{bmatrix} \begin{bmatrix} e_1 \\ e_2 \\ e_3 \\ e_4 \\ e_5 \\ e_6 \\ \vdots \end{bmatrix}$$

respect to $\{e_k\}_{k=1}^{\infty}$, where $W_k = \begin{bmatrix} \frac{1}{k} & 1 \\ 0 & \frac{1}{k} \end{bmatrix}$.

We may rearrangement $\{e_k\}_{k=1}^{\infty}$ denoted by $\{f_k\}_{k=1}^{\infty}$ such that

$$A = \begin{bmatrix} A_1 & \overline{S} \\ 0 & \overline{A}_1 \end{bmatrix} \begin{matrix} \mathcal{H}_1 \\ \mathcal{H}_2 \end{matrix},$$

where $\mathcal{H}_1 = \bigvee\limits_{k=1}^{\infty} \{f_{2k}\}, \mathcal{H}_2 = \bigvee\limits_{k=1}^{\infty} \{f_{2k-1}\}$ and

$$
A_1 = \begin{bmatrix} 0 & 1 & 0 & \cdots & \cdots \\ 0 & 0 & \frac{1}{2} & 0 & \cdots \\ 0 & 0 & 0 & \frac{1}{3} & \ddots \\ \vdots & \vdots & \vdots & \vdots & \cdots \end{bmatrix} \begin{matrix} f_2 \\ f_4 \\ f_6 \\ \vdots \end{matrix}, \quad \overline{A}_1 = \begin{bmatrix} 0 & 1 & 0 & \cdots & \cdots \\ 0 & 0 & \frac{1}{2} & 0 & \cdots \\ 0 & 0 & 0 & \frac{1}{3} & \ddots \\ \vdots & \vdots & \vdots & \vdots & \cdots \end{bmatrix} \begin{matrix} f_1 \\ f_3 \\ f_5 \\ \vdots \end{matrix}.
$$

$\overline{S} f_{2k-1} = f_{2k-2}$ and $\overline{S} f_1 = 0$.

Note that

$$
B_{ij} = \begin{bmatrix} b_{11}^{ij} & b_{12}^{ij} \\ 0 & b_{22}^{ij} \end{bmatrix}.
$$

Thus

$$
B = \begin{bmatrix} \overline{B}_1 & \overline{B}_{12} \\ 0 & \overline{B}_2 \end{bmatrix} \begin{matrix} \mathcal{H}_1 \\ \mathcal{H}_2 \end{matrix}.
$$

Without loss of generality, we may assume that $\mathcal{H}_1 = \mathcal{H}_2$. Then

$$
A = \begin{bmatrix} A_1 & S \\ 0 & A_1 \end{bmatrix} \begin{matrix} \mathcal{H} \\ \mathcal{H} \end{matrix},
$$

where S us an injective back shift with power sequence $\{1\}$.

By former argument, we have

$$
B = \begin{bmatrix} B_1 & B_{12} \\ 0 & B_2 \end{bmatrix} \begin{matrix} \mathcal{H} \\ \mathcal{H} \end{matrix}, \quad \forall \, B \in \mathcal{A}'(A),
$$

where $B_i \in \mathcal{A}'(A_1), i = 1, 2$.

Suppose that $B, T \in \mathcal{A}'(A)$. Then

$$
B = \begin{bmatrix} B_1 & B_{12} \\ 0 & B_2 \end{bmatrix} \begin{matrix} \mathcal{H} \\ \mathcal{H} \end{matrix} \quad \text{and} \quad T = \begin{bmatrix} T_1 & T_{12} \\ 0 & T_2 \end{bmatrix} \begin{matrix} \mathcal{H} \\ \mathcal{H} \end{matrix}.
$$

Furthermore,

$$
BT - TB = \begin{bmatrix} 0 & \star \\ 0 & 0 \end{bmatrix} \begin{matrix} \mathcal{H} \\ \mathcal{H} \end{matrix},
$$

it follows that $\mathcal{A}'(A_1)$ is commutative. It shows that $\mathcal{A}'(A)/rad\mathcal{A}'(A)$ is commutative.

In the chapter 6, we will show that the class of operators $\mathcal{L} :=$ $\{T{\in}\mathcal{L}(\mathcal{H}){\cap}(SI) : \mathcal{A}'(T)/rad\mathcal{A}'(T)$ is commutative$\}$ is dense in (SI) operators in norm topology. Therefore we have the following conjecture.

Conjecture   *Given $T{\in}\mathcal{L}(\mathcal{H}){\cap}(SI)$, $\mathcal{A}'(T)/rad\mathcal{A}'(T)$ is commutative.*

If $T{\in}\mathcal{B}_n(\Omega){\cap}(SI)$, we can give a confirmative answer to the conjecture.

Theorem 4.4.3   *Let $A{\in}\mathcal{B}_n(\Omega){\cap}(SI)$, then there exists a natural number $m, 0 < m{\leq}n$ and a bounded connected open set $\Phi{\subset}\sigma(A)$ such that $A{\in}\mathcal{B}_m(\Phi)$ and for each $T{\in}\mathcal{A}'(A), \sigma(T|_{ker(\lambda-A)}) = \{f(\lambda)\}, \lambda{\in}\Phi$, where $f(\lambda)$ is a function analytic in Φ. Furthermore, $\mathcal{A}'(A)/rad\mathcal{A}'(A)$ is commutative.*

Before we prove Theorem 4.4.3, the following definition and proposition are needed.

Definition 4.4.4(Minimal index)   Let

$$T{\in}\mathcal{B}_n(\Omega), A{\in}\mathcal{A}'(T)$$

and

$$A(z) := A|_{ker(T-z)}.$$

If $\sigma(A(z))$ is disconnected at $z_0{\in}\Omega$, then there exists a positive number δ such that $\sigma(A(z))$ is disconnected in each point in $D(z_0,\delta) := \{z : |z-z_0| < \delta\}$. Thus we can find a positive number ε such that $\sigma(A(z)){\cap}\overline{D}(\lambda(z_0),\varepsilon) = \lambda(z_0), z{\in}D(z_0,\delta)$, where $\lambda(z_0)$ is an eigenvalue of $A(z_0)$. Set

$$P(z) = \int_{\partial D(\lambda(z_0),\varepsilon)} (A(z) - \lambda)^{-1}d\lambda.$$

$P(z)$ is said to be a holomorphic idempotent defined on $D(\lambda(z_0), \varepsilon)$ induced by $\mathcal{A}'(T)$. If each holomorphic idempotent $P(z)$ induced by $\mathcal{A}'(T)$ satisfies

$$dimker(T|_{\bigvee_{z{\in}\Phi} ran(P(z)-z_0)}) < n,$$

then n is called the minimal index of T or we say that T has the minimal index n.

Example 4.4.5   *If $T{\in}\mathcal{B}_1(\Omega)$, the minimal index of T is 1.*

Example 4.4.6   *Assume that $f(t) = z(z - \frac{1}{8})$, then $T_f^*{\in}\mathcal{B}_2(\Omega){\cap}(SI), 0{\in}\Omega$. 2 is not the minimal index of T_f^*. But there is a connected open set Ω_1 such that $T_f^*{\in}\mathcal{B}_1(\Omega_1)$.*

Example 4.4.7 *For the analytic Toeplitz operator $T_{z^2}, T_{z^2}^* \in \mathcal{B}_2(D)$. A simple computation shows that $T_{z^2}^*$ has no minimal index.*

By the definition, we have the following proposition.

Proposition 4.4.8 *Given $A \in \mathcal{B}_n(\Omega)$, n is the minimal index of A if and only if there is no $B \in \mathcal{B}_m(\Omega), m < n$, such that $B \in \mathcal{A}'(A)$.*

According to the literature of [Cowen, M.J. and Douglas, R. (1977)] (Chapter 3), we have the following proposition.

Proposition 4.4.9 *$T \in \mathcal{B}_n(\Omega) \cap (SI)$, there is a natural number $m, 0 < m \le n$, and a connected open set Ω_1 such that $T \in \mathcal{B}_m(\Omega)$ and m is the minimal index of T.*

Proposition 4.4.10 *Given $A \in \mathcal{B}_n(\Omega)$ with minimal index n, if $P(z)$ is a holomorphic idempotent defined on a connected open set Φ, induced by $\mathcal{A}'(A)$ and $\dim \operatorname{ran} P(z) = k < n, z \in \Phi$. Denote $\mathcal{H}_1 = \bigvee_{z \in \Phi} \operatorname{ran} P(z)$, then $A|_{\mathcal{H}_1} \in \mathcal{B}_m(\Omega)$ for some $m, k \le m < n$.*

Proof Let P be the orthogonal projection from $\mathcal{H}$ onto $\mathcal{H}_1$ and

$$A_1 = A|_{\mathcal{H}_1}, \quad A_2 = (A^*|_{\mathcal{H}_1^\perp})^*.$$

Then

$$A = \begin{bmatrix} A_1 & A_{12} \\ 0 & A_2 \end{bmatrix} \begin{matrix} \mathcal{H}_1 \\ \mathcal{H}_1^\perp \end{matrix}.$$

By Lemma 1.2 of [Herrero, D.A. (1990)], $\sigma_p(A^*) = \emptyset$. Thus $\dim \mathcal{H}_1^\perp = \infty$. Since $A - z$ is right invertible for $z \in \Omega$, let

$$B = \begin{bmatrix} B_1 & B_{12} \\ 0 & B_2 \end{bmatrix} \begin{matrix} \mathcal{H}_1 \\ \mathcal{H}_1^\perp \end{matrix}$$

be a right inverse of $A - z$, i.e.,

$$(A - z)B = \begin{bmatrix} A_1 - z & A_{12} \\ 0 & A_2 - z \end{bmatrix} \begin{bmatrix} B_1 & B_{12} \\ 0 & B_2 \end{bmatrix}$$

$$= \begin{bmatrix} I_{\mathcal{H}_1} & 0 \\ 0 & I_{\mathcal{H}_1^\perp} \end{bmatrix}.$$

Thus $(A_2 - z)B_2 = I_{\mathcal{H}_1^\perp}$, and $A_2 - z$ is right invertible for $z \in \Omega$.

It is not difficult to prove that $B_2 \in \mathcal{B}_l(\Omega)$ for some $l < n$. Let π be the canonical mapping from $\mathcal{L}(\mathcal{H})$ to $\mathcal{L}(\mathcal{H})/\mathcal{K}(\mathcal{H})$, where $\mathcal{L}(\mathcal{H})/\mathcal{K}(\mathcal{H})$ is the

Calkin algebra. Then we have

$$\pi(B)\pi(A-z) = \begin{bmatrix} \pi(I_{\mathcal{H}_1}) & 0 \\ 0 & \pi(I_{\mathcal{H}_1^{\perp}}) \end{bmatrix}.$$

This implies that $ran(A_1 - z)$ is closed. Denote $(e_1(y), \cdots, e_k(y))$ the holomorphic frame of $ran P(y)$, then for $1 \leq j \leq k$

$$(A_1 - z)e_j(y) = (\lambda(y) - z)e_j(y).$$

Since $\mathcal{H}_1 = \bigvee_{y \in \Phi} ran P(y), ran(A_1 - z) = \mathcal{H}_1$ for each $z \in \Phi$. Thus $A_1 \in \mathcal{B}_{n-l}(\Omega)$. Let $m = n - l$, the proof of the theorem is complete.

Lemma 4.4.11 [[Cowen, M.J. and Douglas, R. (1977)] *Let* $f :$ $\Omega \rightarrow G_r(n, \mathcal{H})$ *be a holomorphic curve, where* Ω *is a connected open set. Let* E_f *be a vector bundle defined on* Ω *and let* $r_1(z), \cdots, r_n(z))$ *be the holomorphic frame of* E_f. *If* $r_1(z_0), \cdots, r_n(z_0)$ *form an ONB of* $f(z_0)$, *then there exists a holomorphic frame* $\overline{r_1}, \cdots, \overline{r_n}$ *of* E_f, *defined on some open set* Δ *containing* z_0, *such that*

$$\overline{r_i}(z_0) = r_i(z_0), i = 1, 2, \cdots, n$$

and

$$< r_i^{(k)}(z_0), r_j(z_0) > = 0, 1 \leq i, j \leq n, k = 1, 2, \cdots.$$

Proof of Theorem 4.4.3 Without loss of generality, we assume that n is the minimal index of A and $D \subset \Omega$. If there is a $B \in \mathcal{A}'(A)$ such that $\sigma(B(0)) = \{\lambda_1, \lambda_2\}, \lambda_1 \neq \lambda_2$, then since $B(z)$ is holomorphic for $z \in \Omega$, there exists an $\varepsilon > 0$ such that

$$\sigma(B(z)) = \{\lambda_1(z), \lambda_2(z)\}, z \in D(0, \varepsilon) = \{z : |z| < \varepsilon\}, \lambda_1(z) \neq \lambda_2(z), \lambda_1(0) = \lambda_1$$

and $\lambda_2(0) = \lambda_2$.

Since $B(z)$ is holomorphic on Ω, we can find an $\varepsilon_1 > 0$ such that

$$D(\lambda_1, \varepsilon_1) \cap \sigma(B(z)) = \overline{D}(\lambda_1, \varepsilon_1) \cap \sigma(B(z)) = \{\lambda_1(z)\}$$

and

$$D(\lambda_2, \varepsilon_1) \cap \sigma(B(z)) = \overline{D}(\lambda_2, \varepsilon_1) \cap \sigma(B(z)) = \{\lambda_2(z)\}.$$

Set

$$P(z) = \int_{\partial D(\lambda_1, \varepsilon_1)} (B(z) - \lambda)^{-1} d\lambda, z \in D(0, \varepsilon_1).$$

Then

$$I - P(z) = \int_{\partial D(\lambda_2, \varepsilon_1)} (B(z) - \lambda)^{-1} d\lambda, z \in D(0, \varepsilon_1).$$

Denote $\mathcal{M} = \bigvee_{z \in D(0, \varepsilon_1)} ran P(z)$. By Proposition 4.4.10, we may assume that

$$A_1 := A|_{\mathcal{M}} \in \mathcal{B}_k(\Omega).$$

Denote $A_2 = (A^*|_{\mathcal{M}^\perp})^*, B_1 = B|_{\mathcal{M}}, B_2 = (B^*|_{\mathcal{M}^\perp})^*$. Note that $\mathcal{M} \in (LatA) \cap (LatB)$ and

$$A = \begin{bmatrix} A_1 & A_{12} \\ 0 & A_2 \end{bmatrix} \begin{matrix} \mathcal{M} \\ \mathcal{M}^\perp \end{matrix}, \quad B = \begin{bmatrix} B_1 & B_{12} \\ 0 & B_2 \end{bmatrix} \begin{matrix} \mathcal{M} \\ \mathcal{M}^\perp \end{matrix},$$

where $LatA$ denotes the lattice of invariant subspaces of A.

Let $\{e_1(z), \cdots, e_k(z)\}$ be a holomorphic frame of E_{A_1} and $\{e_{k+1}(z), \cdots, e_n(z)\}$ be a holomorphic frame of

$$E_2 = \{(x, z) : x \in (I - P(z)) ker(A - z), z \in D(0, \varepsilon)\},$$

then $\{e_1(z), \cdots, e_k(z), e_{k+1}(z), \cdots, e_n(z)\}$ is a holomorphic frame of E_{A_1} on $D(0, \varepsilon)$.

Since $AB = BA, B_1 A_1 = A_1 B_1$. By Proposition 4.4.10 it follows from $A_1 \in \mathcal{B}_k(\Omega)$ that $A_2 \in \mathcal{B}_{n-k}(\Omega)$. Denote

$$\mathcal{H}_1 = kerA, \mathcal{H}_2 = kerA^2 \ominus kerA, \cdots, \mathcal{H}_m = kerA^m \ominus kerA^{m-1}, \cdots.$$

By Theorem CD in Section 1.4, we have

$$\mathcal{H}_1 \oplus \mathcal{H}_2 \oplus \cdots \oplus \mathcal{H}_m = \bigvee \{e_i^{(j)}(0) : 1 \le i \le n, 0 \le j \le m - 1\},$$

and

$$A = \begin{bmatrix} 0 & A_{12} & A_{13} & \cdots \\ & 0 & A_{23} & \cdots \\ & & 0 & \ddots \\ 0 & & & \ddots \end{bmatrix} \begin{matrix} \mathcal{H}_1 \\ \mathcal{H}_2 \\ \mathcal{H}_3 \\ \vdots \end{matrix}, \quad B = \begin{bmatrix} B_{11} & B_{12} & B_{13} & \cdots \\ & B_{22} & B_{23} & \cdots \\ & & B_{33} & \ddots \\ 0 & & & \ddots \end{bmatrix} \begin{matrix} \mathcal{H}_1 \\ \mathcal{H}_2 \\ \mathcal{H}_3 \\ \vdots \end{matrix}.$$

Denote
$\mathcal{L}_1 = kerA_1, \mathcal{L}_2 = kerA_1^2 \ominus kerA_1, \cdots, \mathcal{L}_m = kerA_1^m \ominus kerA_1^{m-1}, \cdots$. By
Theorem CD again, $\mathcal{L}_1 \oplus \cdots \oplus \mathcal{L}_m = \bigvee\{e_i^{(j)}(0) : 1 \le i \le k, 0 \le j \le m-1\}$ and

$$A_1 = \begin{bmatrix} 0 & A'_{12} & A'_{13} & \cdots \\ & 0 & A'_{23} & \cdots \\ & & 0 & \ddots \\ 0 & & & \ddots \end{bmatrix} \begin{matrix} \mathcal{L}_1 \\ \mathcal{L}_2 \\ \mathcal{L}_3 \\ \vdots \end{matrix}, \quad B_1 = \begin{bmatrix} B'_{11} & B'_{12} & B'_{13} & \cdots \\ & B'_{22} & B'_{23} & \cdots \\ & & B'_{33} & \ddots \\ 0 & & & \ddots \end{bmatrix} \begin{matrix} \mathcal{L}_1 \\ \mathcal{L}_2 \\ \mathcal{L}_3 \\ \vdots \end{matrix}.$$

Since $B \in \mathcal{A}'(A), B_1 \in \mathcal{A}'(A_1)$. Therefore

$$B_{k+1,k+1} \sim B_{k,k}$$

and

$$B'_{k+1,k+1} \sim B'_{k,k}, \quad k = 1, 2, \cdots.$$

Since

$$B'_{11} = B|_{\bigvee\{e_1(0),\cdots,e_k(0)\}},$$

$$\sigma(B'_{11}) = \{\lambda_1\} \quad \text{and} \quad \sigma(B'_{kk}) = \{\lambda_1\}, k = 1, 2, \cdots.$$

Set

$$(B_1)_m = \begin{bmatrix} B'_{11} & \cdots & B'_{1m} \\ & \ddots & \vdots \\ 0 & & B'_{mm} \end{bmatrix} \begin{matrix} \mathcal{L}_1 \\ \vdots \\ \mathcal{L}_m \end{matrix},$$

then

$$\sigma((B_1)_m) = \{\lambda_1\}.$$

Since $B_{11} = B|_{kerA}, \sigma(B_{11}) = \{\lambda_1, \lambda_2\}$. Thus

$$\sigma(B_{kk}) = \{\lambda_1, \lambda_2\}, k = 1, 2, \cdots.$$

Set

$$\overline{B}_m = \begin{bmatrix} B_{11} & \cdots & B_{1m} \\ & \ddots & \vdots \\ 0 & & B_{mm} \end{bmatrix} \begin{matrix} \mathcal{H}_1 \\ \vdots \\ \mathcal{H}_m \end{matrix},$$

then

$$\sigma(\overline{B}_m) = \{\lambda_1, \lambda_2\}.$$

Define

$$\overline{P}_m = \int_{\partial D(\lambda_1, \varepsilon_1)} (\overline{B}_m - \lambda)^{-1} d\lambda$$

and

$$P_m = \overline{P}_m \oplus 0 \sum_{k>m} \oplus \mathcal{H}_k.$$

Then P_m is an idempotent and $P_m A_m = A_m P_m$, where

$$A_m = \begin{bmatrix} 0 & A_{12} & \cdots & A_{1m} \\ & 0 & \ddots & \cdots \\ & & \ddots & A_{m-1,m} \\ & & & 0 \end{bmatrix} \begin{matrix} \mathcal{H}_1 \\ \vdots \\ \mathcal{H}_{m-1} \\ \mathcal{H}_m \end{matrix} \oplus 0 \sum_{k>m} \oplus \mathcal{H}_k.$$

Denote $\mathcal{N} = \bigvee\{ran P_m : m = 0, 1, 2, \cdots\}$, then $\mathcal{N} \in (Lat A) \cap (Lat B)$ and $\mathcal{N} \neq \{0\}$.

{Claim 1} $\mathcal{N} = \mathcal{M} = \bigvee\{e_1(z), \cdots, e_n(z) : z \in D(0, \varepsilon_1)\} = \bigvee_{m=1}^{\infty} \mathcal{L}_m.$

Since

$$\bigvee\{e_i^{(j)}(0) : 1 \leq i \leq k, 0 \leq j \leq m-1\} \subset \bigvee\{e_i^{(j)}(0) : 0 \leq i \leq n, 0 \leq j \leq m-1\},$$

$$\mathcal{L}_1 \oplus \cdots \oplus \mathcal{L}_m \subset \mathcal{H}_1 \oplus \cdots \oplus \mathcal{H}_m, \mathcal{L}_1 \oplus \cdots \oplus \mathcal{L}_m \in Lat B$$

and

$$\mathcal{L}_1 \oplus \cdots \oplus \mathcal{L}_m \in Lat B_m.$$

Note that

$$ker(\overline{B}_1 - \lambda)^n = ker(\overline{B}_1 - \lambda_1)^k = ker A_1 = \bigvee\{e_1(0), \cdots, e_k(0)\}.$$

A simple computation indicates that

$$dim ker(\overline{B}_m - \lambda_1)^{mn} = mk.$$

For each $x \in \mathcal{L}_1 \oplus \cdots \oplus \mathcal{L}_m$, we have

$$(\overline{B}_m - \lambda_1)^{mn} x = ((\overline{B_1})_m - \lambda_1)^{mn} x = 0.$$

This implies that

$$\mathcal{L}_1 \oplus \cdots \oplus \mathcal{L}_m \subset ker(\overline{B}_m - \lambda_1)^{mn} = ranP_m.$$

Since $dim(\mathcal{L}_1 \oplus \cdots \oplus \mathcal{L}_m) = mk, \mathcal{L}_1 \oplus \cdots \oplus \mathcal{L}_m = ran\overline{P}_m$ and $\mathcal{N} = \mathcal{M}$.
    Denote

$$\mathcal{K} = \bigvee \{e_{k+1}(z), \cdots, e_n(z) : z \in D(0, \varepsilon_1)\},$$

then

$$\mathcal{K} \in (LatA) \cap (LatB).$$

Define

$$\overline{Q}_m = \int_{\partial D(\lambda_2, \varepsilon_1)} (\overline{B}_m - \lambda)^{-1} d\lambda \text{ and } Q_m = \overline{Q}_m \oplus 0 \underset{k>m}{\sum} \oplus \mathcal{H}_k.$$

Similarly, we can get $\mathcal{K} = \bigvee \{ranQ_m : m = 1, 2, \cdots\}$.
    Since $\{e_1(z), \cdots, e_k(z), e_{k+1}(z), \cdots, e_n(z)\}$ is a holomorphic frame
of E_A, we can obtain an ONB of $kerA$ by Gram-Schmidt orthogonalization. Without loss of generality, we may assume that
$\{e_1(0), \cdots, e_k(0), e_{k+1}(0), \cdots, e_n(0)\}$ is the ONB of $kerA$.
    Since $\lambda_1(z) \neq \lambda_2(z), z \in D(0, \varepsilon_1)$, by Lemma 4.4.11, Theorem JW 1 in
Chapter 1 and the argument of Claim 1, we have the following Claim 2.
    {Claim 2} There are two subspaces E_1^i and E_2^i of $\mathcal{H}_i, i = 1, 2, \cdots$, satis-
fying
    (i) $\mathcal{H}_i = E_1^i + E_2^i, E_1^i \cap E_2^i = \{0\}$;
    (ii) $\sigma(B_{ii}|_{E_1^i}) = \lambda_1(0), \sigma(B_{ii}|_{E_2^i}) = \lambda_2(0)$;
    (iii) $\bigvee\{E_j^i : 1 \leq i < \infty\} \in LatB, j = 1, 2$;
    (iv) $\bigvee\{E_1^i : 1 \leq i < \infty\} = \mathcal{M}$ and $\bigvee\{E_2^i : 1 \leq i < \infty\} = \mathcal{K}$.
    Let $\{g_1^i, \cdots, g_k^i\}$ and $\{{g'}_{k+1}^i, \cdots, {g'}_n^i\}$ be ONB of E_1^i and E_2^i re-
spectively. By Gram-Schmidt orthogonalization, we can get an ONB
$\{g_1^i, \cdots, g_k^i, g_{k+1}^i, \cdots, g_n^i\}$ of $\mathcal{H}_i, i = 1, 2, \cdots$. By (i), (iii) of the claim and
a simple calculation, we have

$$A = \begin{bmatrix} 0 & A_{12} & A_{13} & \cdots \\ & 0 & A_{23} & \cdots \\ & & 0 & \ddots \\ 0 & & & \ddots \end{bmatrix} \begin{matrix} \mathcal{H}_1 \\ \mathcal{H}_2 \\ \mathcal{H}_3 \\ \vdots \end{matrix}, \quad B = \begin{bmatrix} B_{11} & B_{12} & B_{13} & \cdots \\ & B_{22} & B_{23} & \cdots \\ & & B_{33} & \ddots \\ 0 & & & \ddots \end{bmatrix} \begin{matrix} \mathcal{H}_1 \\ \mathcal{H}_2 \\ \mathcal{H}_3 \\ \vdots \end{matrix},$$

where

$$A_{ij} = \begin{bmatrix} a_{11}^{ij} & a_{12}^{ij} \\ 0 & a_{22}^{ij} \end{bmatrix}, \quad B_{ij} = \begin{bmatrix} b_{11}^{ij} & b_{12}^{ij} \\ 0 & b_{22}^{ij} \end{bmatrix}.$$

{Claim 3} $\mathcal{K} + \mathcal{M} = \mathcal{H}$ and $\mathcal{K} \cap \mathcal{M} = \{0\}$.

By Claim 2, we can assume that $\bigvee \{e_j^{(m)}(0) : 1 \leq j \leq n\} = \mathcal{H}_{m+1}$. Set

$$A(d) = \begin{bmatrix} 0 & A_{12} & 0 & \cdots \\ & 0 & A_{23} & \cdots \\ & & 0 & \ddots \\ 0 & & & \ddots \end{bmatrix} \begin{matrix} \mathcal{H}_1 \\ \mathcal{H}_2 \\ \mathcal{H}_3 \\ \vdots \end{matrix}, \quad B(d) = \begin{bmatrix} B_{11} & 0 & 0 & \cdots \\ & B_{22} & 0 & \cdots \\ & & B_{33} & \ddots \\ 0 & & & \ddots \end{bmatrix} \begin{matrix} \mathcal{H}_1 \\ \mathcal{H}_2 \\ \mathcal{H}_3 \\ \vdots \end{matrix},$$

then $A(d) \in \mathcal{B}_n(\Omega_1)$, where $0 \in \Omega_1 \subset \Omega$. Since $AB = BA$,

$$A(d)B(d) = B(d)A(d).$$

Since $\sigma(B_{11}) = \{\lambda_1, \lambda_2\}$ and since $B_{k+1,k+1} \sim B_{k,k}$,

$$\sigma(B(d)) = \{\lambda_1, \lambda_2\}$$

and

$$(B(d) - \lambda_1)^k (B(d) - \lambda_2)^{n-k} = 0.$$

Set $P = \int_{\partial D(\lambda_1, \varepsilon_1)} (B(d) - \lambda)^{-1} d\lambda$. Then P is an idempotent. To complete the proof of Claim 3, we need the following Claim 4.

{Claim 4} $ran P = \mathcal{M}$.

Set

$$\overline{B}_m(d) = \begin{bmatrix} B_{11} & \cdots & 0 \\ & \ddots & \vdots \\ 0 & & B_{mm} \end{bmatrix} \begin{matrix} \mathcal{H}_1 \\ \vdots \\ \mathcal{H}_m \end{matrix}, \quad \overline{P}_m(d) = \int_{\partial D(\lambda_1, \varepsilon_1)} (\overline{B}_m(d) - \lambda)^{-1} d\lambda$$

and $P_m(d) = \overline{P}_m(d) \oplus 0 \sum_{k>m} \oplus \mathcal{H}_k$. Then $\{P_m(d)\}_{m=1}^\infty$ are uniformly bounded. By Banach-Alaoglu Theorem, $P_m(d)$ converge to P in weak operator topology. It is easy to see that

$$ran \overline{P}_1(d) = ran \overline{P}_1 = \{e_1(0), \cdots, e_k(0)\}.$$

Note that $\bigvee\{e_j^{(i)}(0) : 1\leq j\leq k, 0\leq i\leq m\}\in Lat B$, $\bigvee\{e_j^{(m)}(0) : 1\leq j\leq n\} = \mathcal{H}_{m+1}$ and (iv) of Claim 2, we have $ran\overline{P}_2\subset ran\overline{P}_2(d)$. Since

$$dim ran P_2 = 2k = dim ran P_2(d),$$

$$ran\overline{P}_2 = ran P_2(d).$$

Inductively, we can prove that $ran\overline{P}_m(d) = ran P_m$ for all m. Thus $ran P = \mathcal{M}$ and Claim 4 is proved. Similarly, we can prove that $ran(I - P) = \mathcal{K}$. Thus Claim 3 is proved. Since n is the minimal index of A, $\mathcal{M}$ and $\mathcal{K}$ must be nontrivial subspaces of $\mathcal{H}$ and $\mathcal{M} + \mathcal{K} = \mathcal{H}, \mathcal{M}\cap\mathcal{K} = \{0\}$. This contradicts $A\in(SI)$. Thus $\sigma(B(z)) = \{f(z)\}$. Since $B(z)$ is an $\mathcal{L}(\mathbf{C}^m)$-valued holomorphic function, $f(z)$ is an analytic function on Ω. This completes the proof of the first part of Theorem 4.4.3.

For the proof of the second part of Theorem 4.4.3, we assume that $A_1, A_2\in\mathcal{A}'(A)$ and denote $B = A_1 A_2 - A_2 A_1$. Suppose that $\sigma(B(z)) = \{f(z)\}$. Since

$$(B(z) - f(z))ker(A - z)\subset ker(A - z)$$

and

$$dim ker(A - z) = n < +\infty,$$

$$(B(z) - f(z))^n ker(A - z) = 0, z\in\Omega.$$

Since $B(z) = A_1(z)A_2(z) - A_2(z)A_1(z), f(z) = 0$. Thus $B(z)^n ker(A-z) = 0$. It follows from $\bigvee\{ker(A - z) : z\in\Omega\} = \mathcal{H}$ that $B^n = 0$. Therefore $B\in rad\mathcal{A}'(A)$ and the proof of second part of Theorem 4.4.3 is now complete.

By Theorem 4.4.3 and its arguments, we have the following corollary.

Corollary 4.4.12 *Let $A\in\mathcal{B}_n(\Omega)\cap(SI)$, then*

$$rad\mathcal{A}'(A) = \{B\in\mathcal{A}'(A) : B^n = 0\}.$$

Corollary 4.4.13 *Let $A\in\mathcal{B}_n(\Omega)$, then $A\in(SI)$ if and only if $\mathcal{A}'(A)/rad\mathcal{A}'(A)$ is isomorphic to a subalgebra of H^∞.*
Proof Since H^∞ contains no nontrivial idempotent, $\mathcal{A}'(A)/rad\mathcal{A}'(A)$ has no nontrivial idempotent, thus nor does $\mathcal{A}'(A)$. This proves the sufficiency. The necessity follows from Theorem 4.4.3.

By Factorization Theorem $f = \chi Q$ for each function $f \in H^\infty$, where χ is an inner function and Q is an outer function. If χ is a finite Blaschke product, then by the knowledge of Toeplitz operator $T_{\bar{f}}^* \in \mathcal{B}_n(\Omega)$. Thus we have the following corollary.

Corollary 4.4.14 *Let $f \in H^\infty$. If there is a $\lambda_0 \in D$ such that inner part of $f - f(\lambda_0)$ is a finite Blaschke product, then $T_f \in (SI)$ if and only if $\mathcal{A}'(T_f) \cong H^\infty$.*

Proof We first prove that if $T_f \in (SI)$, then $rad\mathcal{A}'(T_f) = \{0\}$. Without loss of generality, we assume that $f = \chi Q$, where χ is a Blaschke product of order n and Q is a outer function. It follows from [Cowen, C.C. (1978)], that $\mathcal{A}'(T_f) \subset \mathcal{A}'(T_\chi)$. But $T_\chi \cong T_z^{(n)}$. Thus $\mathcal{A}'(T_\chi) \cong M_n(\mathcal{A}'(T_z))$. Since

$$radM_n(\mathcal{A}'(T_z)) = \{0\},$$

$$rad\mathcal{A}'(T_f) = \{0\}.$$

By Corollary 4.4.13, $\mathcal{A}'(T_f)$ is commutative. Since $H^\infty \subset \mathcal{A}'(T_f)$, $\mathcal{A}'(T_f) \cong H^\infty$.

4.5   The Sobolev Disk Algebra

Let Ω be an analytic Cauchy domain in the complex plane and let $W^{22}(\Omega)$ denote the Sobolev space $W^{22}(\Omega) = \{f \in L^2(\Omega, dm) :$ the distributional partial derivatives of first and second order of f belong to $L^2(\Omega, dm)\}$, where dm denotes the planar Lebesque measure. For $f, g \in W^{22}(\Omega)$, we define

$$< f, g >:= \sum_{|\alpha| \leq 2} \int D^\alpha f \overline{D^\alpha g} dm,$$

then $W^{22}(\Omega)$ is a Hilbert space and a Banach algebra with identity under an equivalent norm. By Soblov embedding theorem, $f \in W^{22}(\Omega)$ implies that $f \in C(\overline{\Omega})$ and $\|f\|_{C(\overline{\Omega})} \leq M\|f\|_{W^{22}(\Omega)}$ for some M.

For $f \in W^{22}(\Omega)$, the multiplication operator M_f on $W^{22}(\Omega)$ is defined as

$$M_f g = fg, \ g \in W^{22}(\Omega).$$

Let

$$W(\Omega) := \{M_f : f \in W^{22}(\Omega)\},$$

then $W(\Omega)$ is a strictly cyclic operator algebra with strictly cyclic vector $e(s,t) \equiv 1$. It has been proved that $\sigma(M_z) = \sigma_{lre}(M_z) = \overline{\Omega}$ and $\mathcal{A}'(M_z) = W(\Omega)$. Let $\mathcal{A}^a(M_z)$ be the algebra generated by the rational functions of M_z with poles outside $\overline{\Omega}$ and $R(\Omega) := \mathcal{A}^a(M_z)e$. Since $W(\Omega)$ is strictly cyclic, there exist positive numbers N and K such that for each $f \in W^{22}(\Omega)$,

$$N\|f\|_{W^{22}(\Omega)} \leq \|M_f\| \leq K\|f\|_{W^{22}(\Omega)}. \quad \text{[Lambert, A (1971)]}$$

Thus $R(\Omega)$ is the closure in $W^{22}(\Omega)$ of all rational functions with poles outside $\overline{\Omega}$.

Denote

$$M_z(\Omega) := M_z|_{R(\Omega)}.$$

Proposition 4.5.1 *(i)* $\sigma(M_z(\Omega)) = \overline{\Omega}$, $\sigma_e(M_z(\Omega)) = \partial\Omega$, $nul(M_z(\Omega) - z_0) = 0$ *and* $ind(M_z(\Omega) - z_0) = -1$ *for* $z_0 \in \Omega$;

(ii) The maximal ideal space of $\mathcal{A}'(M_z(\Omega))$ *is equivalent to* $\overline{\Omega}$. *The homomorphism* $k_{z_0}^*$ *corresponding to* $z_0 \in \overline{\Omega}$ *is*

$$k_{z_0}^*(M_f(\Omega)) = f(z_0) = <f, k_{z_0}>$$

for some $k_{z_0} \in R(\Omega)$, *and* $k_{z_0} \in ker(M_z(\Omega) - z_0)^*$;

(iii) $\mathcal{A}'(M_z(\Omega)) = \mathcal{A}^a(M_z(\Omega)) = \{M_f(\Omega) : f \in R(\Omega)\}$, *where* $M_f(\Omega) := M_f|_{R(\Omega)}$. *Therefore* $\mathcal{A}'(M_z(\Omega))$ *is strictly cyclic and* $M_z(\Omega)$ *is rational strictly cyclic.*

Proof (i) If $z_0 \notin \overline{\Omega}$, then

$$(z_0 - z)^{-1} \in R(\Omega),$$

$$M_{(z_0-z)^{-1}} = (z_0 - M_z)^{-1}$$

and

$$M_{(z_0-z)^{-1}}(\Omega) = (z_0 - M_z(\Omega))^{-1}.$$

Therefore, $\sigma(M_z(\Omega)) \subset \overline{\Omega}$. Assume that $z_0 \in \overline{\Omega}$ and for all $g \in R(\Omega)$, the function $(z_0 - M_z(\Omega))g = (z_0 - z)g(z)$ vanishes at z_0. Thus $z_0 - M_z(\Omega)$ is not onto and $\sigma(M_z(\Omega)) \subset \overline{\Omega}$.

Let $z_0 \in \Omega$, then for $f \in R(\Omega)$ there exist a number $\delta > 0$ and a function h analytic in Γ such that

$$f(z) - f(z_0) = (z - z_0)h(z), z \in \Gamma,$$

where $\Gamma := \{z \in \mathbf{C} : |z - z_0| < \delta\} \subset \overline{\Gamma} \subset \Omega$. Denote

$$E := \Omega - \frac{1}{3}\overline{\Gamma} = \Omega - \{z \in \mathbf{C} : |z - z_0| \leq \frac{\delta}{3}\}$$

and define

$$g(z) = \begin{cases} h(z) & z \in \Gamma, \\ f(z)(z - z_0)^{-1} & z \in E \end{cases}$$

Then $f(z) - f(z_0) = (z - z_0)g(z)$ for all $z \in \Omega$ and g is analytic in Ω. Note that

$$\int_E |g|^2 dm \leq \frac{9}{\delta^2} \int_E |(z - z_0)g(z)|^2 dm \leq \frac{9}{\delta^2} \|f - f_0\|^2_{W^{22}(\Omega)},$$

i.e.,

$$g \in L^2(E)$$

and

$$\frac{\partial[f(z) - f(z_0)]}{\partial S} = f'(z) = g(z) + (z - z_0)g'(z) \in L^2(E).$$

We have $(z - z_0)g'(z) \in L^2(E)$. Thus

$$\int_E |g'(z)|^2 dm \leq \frac{9}{\delta^2} \int_E |(z - z_0)g'(z)|^2 dm \leq \frac{9}{\delta^2} \|f - f_0\|^2_{W^{22}(\Omega)}.$$

That is $g' \in L^2(E)$. Since

$$\frac{\partial^2[f(z) - f(z_0)]}{\partial s^2} = 2g' + (z - z_0)g'' \in L^2(E),$$

we have $(z - z_0)g'' \in L^2(E)$ and

$$\int_E |g''(z)|^2 dm \leq \frac{9}{\delta^2} \|f - f_0\|^2_{W^{22}(\Omega)}.$$

Therefore, $g \in W^{22}(E)$. For $k = 0, 1, 2$ and $z \in \frac{2}{3}\Gamma = \{z \in \mathbf{C} : |z - z_0| < \frac{2}{3}\delta\}$, by Cauchy formula,

$$|g^{(k)}(z)| = |\frac{1}{2\pi i} \int_{\xi \in \partial\Gamma} \frac{g(\xi)}{(\xi - z)^{k+1}} d\xi| \leq \frac{3^{k+1}}{\delta^k} \max_{\xi \in \partial\Gamma} |g(\xi)|.$$

Therefore, $g, g', g'' \in L^2(\frac{2}{3}\Gamma)$ and $g \in W^{22}(\frac{2}{3}\Gamma)$, Since $g \in W^{22}(E)$ at the same time, we have $g \in W^{22}(\Omega)$. It follows from $f(z) = f(z_0) + (z - z_0)g(z)$ that the codimension of $ran(M_z(\Omega) - z_0)$ is 1. It is obvious that $ker(M_z(\Omega) - z_0) = \{0\}$, thus $z_0 \in \rho_{s-F}(M_z(\Omega)), \rho_{s-F}(M_z(\Omega)) = \Omega$ and $ind(M_z(\Omega) - z_0) = -1$.

(ii) and (iii). Denote $\mathcal{A} := \{M_f(\Omega) : f \in R(\Omega)\}$. It is easy to see that $\mathcal{A}$ is a commutative algebra with identity closed in weak operator topology and $\mathcal{A}e = R(\Omega)$, that is $\mathcal{A}$ is strictly cyclic. By [Lambert, A (1971)], the adjoint space

$$\mathcal{A}^* = \{g^* : \text{ there exists a function } g \in R(\Omega) \text{ such that}$$

$$g^*(M_f(\Omega)) = < M_f(\Omega)e, g >\}.$$

Thus, for homomorphism $k_{z_0}^* : k_{z_0}^*(M_f(\Omega)) = f(z_0), f \in R(\Omega)$, there exists a $k_{z_0} \in R(\Omega)$ such that

$$k_{z_0}^*(M_f(\Omega)) = f(z_0) = < M_f(\Omega)e, k_{z_0} > = < f, k_{z_0} > .$$

On the other hand, if $\rho : \mathcal{A} \to \mathbf{C}$ is a homomorphism, then

$$\rho(M_z(\Omega)) = z_0 \in \sigma(M_f(\Omega)) = \overline{\Omega}.$$

For a rational function $r(z)$ with poles outside $\overline{\Omega}$, $\rho(r(M_z(\Omega))) = r(z_0)$. Since $\{r(M_z(\Omega))\}$ is dense in $\mathcal{A}$, $\rho(M_z(\Omega)) = f(z_0)$ for each $f \in R(\Omega)$.

Let $g \in R(\Omega)$, then $< g, (M_z(\Omega) - z_0)^* k_{z_0} > = < (z - z_0)g, k_{z_0} > = 0$. Therefore, $k_{z_0} \in ker(M_z(\Omega) - z_0)^*$.

Assume that $T \in \mathcal{A}'(M_z(\Omega))$, then for $z_0 \in \Omega$,

$$M_z(\Omega)^* T^* k_{z_0} = T^* M_z(\Omega)^* k_{z_0} = \overline{z_0} T^* k_{z_0}.$$

Since $nul(M_z(\Omega) - z_0)^* = 1$, there is a complex number $t(z_0)$ such that

$$T^* k_{z_0} = \overline{t(z_0)} k_{z_0}.$$

Thus

$$(Tf)(z) = < Tf, k_z > = < f, T^* k_z > = t(z)f(z), \; f \in R(\Omega), \; z \in \Omega.$$

Because $t(z) = (Te)(z) \in R(\Omega), T = M_t(\Omega)$. This proved that

$$\mathcal{A}'(M_z(\Omega)) = \mathcal{A} = \mathcal{A}^a(M_z(\Omega)).$$

Proposition 4.5.2 *Let Ω be a bounded simply connected Cauchy domain, then the set of polynomials is dense in $R(\Omega)$.*

Proof Given a positive number ε and an $f \in R(\Omega)$, let $r(z)$ be a rational function with poles outside $\overline{\Omega}$ such that $\|f - r\|_{W^{22}(\Omega)} < \varepsilon$. For this r, there is a bounded Cauchy domain Ω_1 such that $\overline{\Omega}_1 \supset \Omega_1 \supset \overline{\Omega} \supset \Omega$ and the poles of $r(z)$ are outside $\overline{\Omega}_1$. By Mergelyan theorem, there exist polynomials p_n such that p_n converge to p uniformly on $\overline{\Omega}_1$. If follows from Cauchy formula that $p'_n \to r'$ and $p''_n \to r''$ uniformly on Ω. In conclusion, $p_n \to r$ in $W^{22}(\Omega)$ and there is an integer N such that $\|p_N - r\|_{W^{22}(\Omega)} < \varepsilon$. Therefore, $\|p_N - f\|_{W^{22}(\Omega)} < 2\varepsilon$.

Because of the special definition of the inner product in Sobolov space for general Ω, it is very difficult to discuss the properties of the space $R(\Omega)$ and the multiplication operators further. But if we choose a better Ω, many results can be obtained.

When $\Omega = D$, the unit disc, we call $R(D)$ Sobolev disk algebra. For simplicity of symbols, in what follows we will denote $M_f(D)$, the multiplication operator with symbol $f \in R(D)$, by M_f.

Proposition 4.5.3 *(i) Hilbert space $R(D)$ has an ONB $\{e_n\}_{n=0}^{\infty}$, where $e_n = \beta_n z^n, \beta_n = [\frac{n+1}{(3n^4 - n^2 + 2n + 1)\pi}]^{\frac{1}{2}}, n = 0, 1, 2, \cdots$;*
(ii) $R(D)$ is a functional Hilbert space, the reproducing kernel of which is given by $k(u,v) = \sum\limits_{n=0}^{\infty} \beta_n^2 u^n \overline{v}^n$. For $z_0 \in \overline{\Omega}$,

$$k_{z_0} = \sum_{n=0}^{\infty} (\beta_n^2 \overline{z}_0^n) z^n;$$

(iii) If $f(z) = \sum\limits_{n=0}^{\infty} f_n z^n$ is analytic in D, then $f \in R(D)$ if and only if

$$\sum_{n=0}^{\infty} \frac{|f_n|^2}{\beta_n^2} < +\infty.$$

If $f \in R(D)$, then $S_n = \sum\limits_{k=0}^{n} f_k z^k$ converge to f uniformly in $\overline{D}$;
(iv) M_z is an essentially normal unilateral weighted shift, $M_z e_n = \alpha_n e_{n+1}, \alpha_n = \frac{\beta_n}{\beta_{n+1}}, n = 0, 1, 2, \cdots$, and $\|M_z\| = \sqrt{\frac{98}{15}}$;
(v) Assume that $f(z) = \sum f_n z^n \in R(D)$, then

$$M_f = \begin{bmatrix} f_0 & & & & 0 \\ f_1 \frac{\beta_0}{\beta_1} & f_0 & & & \\ f_2 \frac{\beta_0}{\beta_2} & f_1 \frac{\beta_1}{\beta_2} & f_0 & & \\ f_3 \frac{\beta_0}{\beta_3} & f_2 \frac{\beta_1}{\beta_3} & f_1 \frac{\beta_2}{\beta_3} & f_0 & \\ \vdots & \vdots & \vdots & \vdots & \ddots \end{bmatrix} \begin{matrix} e_0 \\ e_1 \\ e_2 \\ e_3 \\ \vdots \end{matrix}.$$

Proof (i) Computation shows that $z^n = (s+it)^n, n = 0, 1, 2, \cdots$, is an orthogonal system and $\|z^n\|^2 = \frac{(3n^4 - n^2 + 2n + 1)\pi}{n+1}$. By Proposition 4.5.2, $\{e_n\}_{n=0}^{\infty}$ forms an ONB of $R(D)$.

(ii) For $z_0 \in \overline{D}, |f(z_0)| \leq \|f\|_{C\overline{D}} \leq M\|f\|_{W^{22}(D)}$ for each $f \in R(D)$. Then $R(D)$ is a functional Hilbert space with reproducing kernel

$$k(u, v) = \sum_{n=0}^{\infty} e_n(u)\overline{e_n(v)} = \sum_{n=0}^{\infty} \beta_n^2 u^n \overline{v}^n.$$

Since $< f, \sum \beta_n^2 \overline{z_0}^n z^n > = f(z_0)$ for $f \in R(D)$, $k_{z_0} = \sum(\beta_n^2 \overline{z_0}^n)z^n$.

(iii) Assume that $f(z)$ is a function analytic in D with the Taylor expansion $f = \sum f_n z^n$, then $f = \sum \frac{f_n}{\beta_n} e_n \in R(D)$ if and only if $\sum |\frac{f_n}{\beta_n}|^2 < +\infty$. If $f \in R(D)$, then $s_n(z) = \sum_{k=0}^{n} f_k z^k = \sum_{k=0}^{n} \frac{f_k}{\beta_k} e_k$ converge to f in $W^{22}(D)$. By Sobolev imbedding theorem, $s_n(z)$ converges to $f(z)$ uniformly on $\overline{D}$.

(iv) Since

$$M_z e_n = z\beta_n z^n = \frac{\beta_n}{\beta_{n+1}} e_{n+1} = \alpha_n e_{n+1}, n = 0, 1, 2, \cdots,$$

$$M_z^* M_z - M_z M_z^* = diag(\alpha_0^2, \alpha_1^2 - \alpha_0^2, \alpha_2^2 - \alpha_1^2, \cdots).$$

Since $\alpha_{n+1}^2 - \alpha_n^2 \longrightarrow 0$, $M_z \in C^1$.

(v) For $f = \sum f_n z^n = \sum \frac{f_n}{\beta_n} e_n \in R(D)$,

$$< M_f e_n, e_m > = < \sum_k f_k \frac{\beta_n}{\beta_{n+k}} e_{n+k}, e_m > = \begin{cases} f_{m-n} \frac{\beta_n}{\beta_m} & m \geq n \\ 0 & m < n. \end{cases}$$

Thus M_f can be represented as a matrix in (v).

Proposition 4.5.4 *Let f be analytic in D, then $f \in R(D)$ if and only if $f' \in H^2, f'' \in L^2(D)$.*

Proof If $f \in R(D)$ and $f(z) = \sum f_n z^n$. By Proposition 4.5.3, $\sum \frac{|f_n|^2}{\beta_n^2} < +\infty$. Thus $\sum n^2 |f_n|^2 < +\infty$ and $f' \in H^2$. From the definition of $R(D), f'' \in L^2(D)$.

Conversely, if $f' \in H^2, f'' \in L^2(D)$. A theorem in [Duren, P.L. (1970)](Theorem 3.11) asserts that function f is analytic in D, continuous to $C = \partial D$ and absolutely continuous on C if and only if $f' \in H^1$. Thus $f \in C(\overline{D})$ and $f \in R(D)$.

For $f \in R(D)$, denote $f_r := f(rz), z \in \overline{D}, 0 < r \leq 1$.

Proposition 4.5.5 *(i) f_r converge to f in $R(D)$ as $r \to 1$;*
(ii) $f_r(M_z) \longrightarrow M_f$ as $r \to 1$.

Proof (i) Since $f(z)$ is uniformly continuous in $\overline{D}$, $f(rz) \longrightarrow f(z)$ uniformly in $\overline{D}$. Thus

$$\int_D |f(rz) - f(z)|^2 dm \longrightarrow 0 \quad (r \longrightarrow 1).$$

Given a positive ε, since f' and $f'' \in L^2(D)$, we can choose r_0 such that $0 < r_0 < 1$ and $\int_{D \backslash (2r_0 - 1)D} |f'(z)|^2 dm < \varepsilon$ and $\int_{D \backslash (2r_0 - 1)D} |f''(z)|^2 dm < \varepsilon$.

Thus

$$\int_D |rf'(rz) - f'(z)|^2 dm$$

$$= \int_{D \backslash r_0 D} |rf'(rz) - f'(z)|^2 dm + \int_{r_0 D} |rf'(rz) - f'(z)|^2 dm$$

$$\leq 2[\int_{D \backslash r_0 D} |rf'(rz)|^2 dm + \int_{D \backslash r_0 D} |f'(z)|^2 dm] + \int_{r_0 D} |rf'(rz) - f'(z)|^2 dm.$$

Similarly,

$$\int_D |r^2 f''(rz) - f''(z)|^2 dm$$

$$\leq 2[\int_{D \backslash r_0 D} |r^2 f''(rz)|^2 dm + \int_{D \backslash r_0 D} |f''(z)|^2 dm] + \int_{r_0 D} |r^2 f''(rz) - f''(z)|^2 dm.$$

Since $rf'(rz)$ and $r^2 f''(rz)$ uniformly converge to $f'(z)$ and $f''(z)$ respectively in any closed subset of D, we can find a r_1 such that when $r > r_1$,

1. If $|z| \leq r_0$, then $|rf'(rz) - f'(z)| < \varepsilon$ and $|r^2 f''(rz) - f''(z)| < \varepsilon$;
2. If $z \in D \backslash r_0 D$, then $rz > 2r_0 - 1$.

Thus, if $r > r_1$, we have

$$\int_D |rf'(rz) - f'(z)|^2 dm < 4\varepsilon + \pi \varepsilon^2$$

and

$$\int_D |r^2 f''(rz) - f''(z)|^2 dm < 4\varepsilon + \pi \varepsilon^2.$$

Therefore, $f_r \longrightarrow f$ $(r \longrightarrow 1)$ in $R(D)$.

(ii) Given $g \in R(D)$,

$$f_r(M_z)g = \frac{1}{2\pi i} \int\limits_{|\xi|=\frac{1}{r}} f(r\xi)(\xi - M_z)^{-1}g(z)d\xi$$

$$= \frac{1}{2\pi i} \int\limits_{|\xi|=\frac{1}{r}} f(r\xi)(\xi - z)^{-1}g(z)d\xi$$

$$= f(rz)g(z), \quad z \in \overline{D}.$$

Assume that $f(z) = \sum f_n z^n$, then $f_r(z) = \sum f_n r^n z^n$. Since

$$\sum |\frac{f_n r^n}{\beta_n}|^2 \leq \sum |\frac{f_n}{\beta_n}|^2 < +\infty.$$

By Proposition 4.5.3, $f_r \in R(D)$. Thus $f_r(M_z) = M_{f_r}$. Since $\mathcal{A}'(M_z)$ is strictly cyclic, $f_r \longrightarrow f$ in $R(D)$ is equivalent to $M_{f_r} = f_r(M_z) \longrightarrow M_f$.

Theorem 4.5.6 *Assume that $f \in R(D)$, then*
(i) $\sigma(M_f) = f(\overline{D})$;
(ii) $\sigma_e(M_f) = \sigma_{lre}(M_f) = f(C)$. If $z_0 \in D$ and $f(z_0) \notin f(C)$, then

$$ind(M_f - f(z_0)) = -nul(M_f - f(z_0))^* = -n,$$

where n is the number of zeros of $f(z) - f(z_0)$ in D, including multiplicity.
Proof (i) Let $z_0 \notin \overline{D}$, then the value of the functions in the range of $M_f - f(z_0)$ is zero at z_0 and $M_f - f(z_0)$ is not onto. This implies that $\sigma(M_f) \supset f(\overline{D})$. On the other hand, if $w_0 \notin f(\overline{D})$, it is easy to see that

$$[f(z) - w_0]^{-1} \in R(D)$$

and

$$[M_f - w_0]M_{(f-w_0)^{-1}} = M_{(f-w_0)^{-1}}[M_f - w_0] = I.$$

Thus $\sigma(M_f) = f(\overline{D})$.
(ii) By Proposition 4.5.1, $\sigma_e(M_z) = \sigma_{lre}(M_z) = C$. By [Conway, J.B., Herrero, D.A. and Morrel, B.B. (1989)],

$$\sigma_e(M_{f_r}) = \sigma_{lre}(M_{f_r}) = \sigma_e(f_r(M_z)) = f_r(C) = \{f(z) : |z| = r\}.$$

If $z_0 \in C$, $f(rz_0) \in \sigma_{lre}(M_{f_r})$. Since

$$M_{f_r} - f(rz_0) \longrightarrow M_f - f(z_0) \quad \text{as} \quad r \longrightarrow 1,$$

$$f(z_0) \in \sigma_{lre}(M_f) \quad \text{and} \quad f(C) \subset \sigma_{lre}(M_f).$$

Conversely, if there is a $z_0 \in D$ such that $f(z_0) \notin f(C)$, then $f(z) - f(z_0)$ has only finitely many zeros in $\overline{D}$, denoted by $\{z_0, z_1, \cdots, z_n\} \subset \overline{D}$. From the proof of Proposition 4.5.1(i), we know that

$$f(z) - f(z_0) = (z - z_0)^{k_0}(z - z_1)^{k_1} \cdots (z - z_n)^{k_n} g(z), \quad g \in R(D)$$

and $g(z) \neq 0$ for $z \in \overline{D}$. By Proposition 4.5.1,

$$M_f - f(z_0) = (M_{z-z_0})^{k_0} \cdots (M_{z-z_n})^{k_n} M_g$$

is a Fredholm operator and $ind(M_f - f(z_0)) = -\sum_{i=0}^{n} k_i = -n$. It is easily seen that $nul(M_f - f(z_0)) = 0$ and therefore $nul(M_f - f(z_0))^* = n$.

Proposition 4.5.7 *Let $f \in R(D)$, then*
(i) If $z_0 \in D$ and $f(z_0) \notin f(C)$, then $M_f^* \in \mathcal{B}_n(\Omega)$, where Ω is a component of $\rho_{s-F}(M_f)$ containing $f(z_0)$ and n is the zeros of $f(z) - f(z_0)$ in D;
(ii) M_f is an essentially normal operator.

Proof (i) It needs only to prove

$$\bigvee \{ker[(M_f - f(z_0))^*]^k : k \geq 1\} = R(D).$$

In fact, if

$$y \in R(D) \ominus \bigvee \{ker[(M_f - f(z_0))^*]^k : k \geq 1\},$$

then for each k,

$$y \in [ker[(M_f - f(z_0))^*]^k]^\perp = ran(M_f - f(z_0))^k.$$

Thus we can find a function $h \in R(D)$ such that

$$y = [M_f - f(z_0)]^k h = (z - z_0)^{k_0 k} \cdots (z - z_n)^{k_n k} g^k h.$$

Note that z_0 is the zero of y of order kk_0. Since $k \geq 1$ can be any natural number, z_0 is an essentially singular point of y. It is a contradiction.

(ii) By Proposition 4.5.3, there is a sequence of polynomials p_n converging to f in $R(D)$. Since $\mathcal{A}'(M_z) = \{M_g : g \in R(D)\}$ is a strictly cyclic algebra,

$$M_{p_n} \longrightarrow M_f \ (n \longrightarrow \infty).$$

By Proposition 4.5.3, M_z is essentially normal, and $\pi(M_z)$ is normal in the Calkin algebra $\mathcal{L}(R(D))/\mathcal{K}(R(D))$, where π is the canonical mapping from

$\mathcal{L}(R(D))$ to $\mathcal{L}(R(D))/\mathcal{K}(R(D))$. Therefore $\pi(p_n)$ are normal elements and so is

$$\pi(M_f) = \lim_{n\to\infty} \pi(M_{p_n}),$$

i.e., M_f is an essentially normal operator in $\mathcal{L}(R(D))$.

Example 4.5.8 $f(z) = z^3 + z^2$, $g(z) = \frac{\alpha-z}{1-\bar{\lambda}z} \cdot \frac{\beta-z}{1-\bar{\beta}z}$, $|\alpha| < 1$, $|\beta| < 1$. $f(C)$ *separates* $f(\overline{D})$ *into three components:* Ω_1, Ω_2 *and* $\Omega_3, 1\in\Omega_1, -1\in\Omega_2$ *and* $\frac{1}{4}\in\Omega_3$. $M_f^*\in\mathcal{B}_1(\Omega_1)\cap\mathcal{B}_2(\Omega_2)\cap\mathcal{B}_3(\Omega_3)$. $M_g^*\in\mathcal{B}_2(D)$.

In the following, we will discuss the commutant of the multiplication operators.

Lemma 4.5.9 *Let* $f\in R(D)$, $M_f^*\in\mathcal{B}_n(\Omega)$, $z_0\in D_1 := f^{-1}(\Omega)$ *and*

$$f(z) - f(z_0) = (z - z_0)^{h_1}(z - z_1)^{h_2}\cdots(z - z_l)^{h_{l+1}}g_{z_0}(z),$$

where $\{z_i\}_{i=1}^l\subset D_1$ *are pairwise distinct,* $\sum\limits_{i=1}^{l+1} h_i = n, g_{z_0}(z) \neq 0, z\in\overline{D}$. *Then there exist* n *linearly independent vectors*

$$K_{z_0} := \{k_{z_0}, k_{z_0}^1, \cdots, k_{z_0}^{h_1-1}, k_{z_1}, \cdots, k_{z_1}^{h_2-1}, \cdots, k_{z_l}^{h_{l+1}-1}\}$$

such that

$$\ker M_{f-f(z_0)}^* = \bigvee K_{z_0}.$$

Proof Choose $k_{z_0}^1, k_{z_0}^2, \cdots, k_{z_0}^{h_1-1}\in R(D)$ such that

$$M_{z-z_0}^* k_{z_0}^1 = k_{z_0}, M_{z-z_0}^* k_{z_0}^2 = k_{z_0}^1, \cdots, M_{z-z_0}^* k_{z_0}^{h_1-1} = k_{z_0}^{h_1-2}.$$

Then for every $u\in R(D)$ and $0\leq j\leq h_1 - 1$, here let $k_{z_0}^0 = k_{z_0}$, we have

$$< u, M_{f-f(z_0)}^* k_{z_0}^j > = < (f - f(z_0))u, k_{z_0}^j >$$

$$= < (z - z_0)^{h_1-j}(z - z_1)^{h_2}\cdots(z - z_l)^{h_{l+1}}g_{z_0}u, k_{z_0} >$$

$$= 0.$$

Thus $k_{z_0}^j\in\ker M_{f-f(z_0)}^*$. Similarly, choose $k_{z_i}^1, \cdots, k_{z_i}^{h_{i+1}-2}\in R(D)$ such that

$$M_{z-z_i}^* k_{z_i}^j = k_{z_i}^{j-1}, 0 < i\leq l, 1\leq j\leq h_{i+1}^{k_{i+1}-2}.$$

Then $k_{z_i}^j\in\ker M_{f-f(z_0)}^*$.

If there is a sequence of complexes $\{c_i^j : 0 \leq i \leq l, 0 \leq j \leq h_i - 1\}$ satisfying that $\sum_{i,j} c_i^j k_{z_i}^j = 0$, apply on both sides by $M^*_{(z-z_0)^{h_1-1}\cdots(z-z_l)^{h_l+1}}$, we get $c_0^{h_1-1} = 0$. Similarly, $c_i^j = 0, (0 \leq i \leq l, 0 \leq j \leq h_i - 1)$. Thus k_{z_0} is linearly independent and

$$\ker M^*_{f-f(z_0)} = \bigvee K_{z_0}.$$

Given $f \in R(D), M_f^* \in \mathcal{B}_n(\Omega), D_1 = f^{-1}(\Omega)$. For $z_0 \in D_1$,

$$f(z) - f(z_0) = (z - z_0)(z - z_1)\cdots(z - z_{n-1})g_{z_0}(z), \quad g_{z_0}(z) \neq 0 \text{ for } z \in \overline{D}. \tag{4.5.1}$$

Let $N_{z_0} = \{z_i\}_{i=0}^{n-1}$, the numbers in N_{z_0} can repeat. Denote

$$\Gamma = \cup\{N_{z_0} : z_0 \in D_1, \quad \text{there is at least one } z_i (0 \leq i \leq n-1) \quad \text{such that}$$

$$f'(z_i) = 0\}.$$

Lemma 4.5.10 Γ *is an at most countable subset of* D_1 *and* $z_0 \in D_1 \backslash \Gamma$ *if and only if the numbers in* D_{z_0} *are pairwise distinct.*
Proof If $z_k \in N_{z_0}$ with $f'(z_k) = 0$ for some $k, 0 \leq k \leq n-1$. Then

$$f'(z) = (z - z_1)\cdots(z - z_{n-1})g_{z_0}(z) + \cdots$$

$$+ (z - z_0)\cdots(z - z_{k-1})(z - z_{k+1})\cdots(z - z_{n-1})g_{z_0}(z) + \cdots$$

$$+ (z - z_0)\cdots(z - z_{n-1})g'_{z_0}(z).$$

Let $z = z_k$, we have

$$(z_k - z_0)\cdots(z_k - z_{k-1})(z_k - z_{k+1})\cdots(z_k - z_{n-1})g_{z_0}(z_k) = 0.$$

Thus there exists at least an $i \neq k$ with $z_i = z_k$. The converse is obvious.

Lemma 4.5.11 *Let* $f \in R(D), M_f^* \in \mathcal{B}_n(\Omega)$, *then there exist an open subset* $\Lambda \subset f^{-1}(\Omega)$ *and analytic functions*

$$\alpha_1(z), \alpha_2(z), \cdots, \alpha_n(z)$$

and

$$Z_1(z), Z_2(z), \cdots, Z_{n-1}(z)$$

on Λ such that for each $A \in \mathcal{A}'(M_f)$,

$$(Ag)(z) = \alpha_1(z)g(z) + \alpha_2(z)g(Z_1(z)) + \cdots + \alpha_n(z)g(Z_{n-1}(z)), z \in \Lambda, g \in R(D).$$

Proof Denote $D_1 = f^{-1}(\Omega)$. If the set $\{z \in D_1 : f'(z) = 0\}$ is finite, then Γ is finite. Let $\Lambda = D_1 \backslash \Gamma$. If the set $\{z \in D_1 : f'(z) = 0\}$ is a countably many set. Choose an open subset $D_2 \subset D_1$, then $D_2 \cap \Gamma$ is finite. Let $\Lambda = D_2 \backslash \Gamma$.

For $z \in \Lambda$, by Lemma 4.5.9 and Lemma 4.5.10,

$$ker M^*_{f-f(z)} = \bigvee \{k_z, k_{z_1}, \cdots, k_{z_{n-1}}\}$$

and the numbers in N_z are pairwise distinct. Since $A \in \mathcal{A}'(M_f)$,

$$A^* k_z \in ker M^*_{f-f(z)}.$$

Thus there exist n complex numbers $\overline{\alpha_1(z)}, \overline{\alpha_2(z)}, \cdots, \overline{\alpha_n(z)}$ such that

$$A^* k_z = \overline{\alpha_1(z)} k_z + \overline{\alpha_2(z)} k_{z_1} + \cdots + \overline{\alpha_n(z)} k_{z_{n-1}}.$$

Therefore,

$$\begin{aligned}(Ag)(z) &= <Ag, k_z> \\ &= <g, A^* k_z> \\ &= \alpha_1(z)g(z) + \alpha_2(z)g(z_1) + \cdots + \alpha_n(z)g(z_{n-1}). \quad (4.5.2)\end{aligned}$$

For $N_z = \{z, z_1, \cdots, z_{n-1}\}$, we choose $u_1, u_2, \cdots, u_{n-1} \in D_1$ such that

(i) the n open balls $B(z, \varepsilon), \cdots, B(z_k, \varepsilon)(1 \le k \le n-1)$ are pairwise disjoint in D_1 for some $\varepsilon > 0$;

(ii) $\{u_k\}_{k=1}^{n-1}$ are pairwise distinct and

$$\{u_k\}_{k=1}^{n-1} \cap [\overline{B}(z, \varepsilon) \cup (\bigcup_{k=1}^{n-1} \overline{B}(z_k, \varepsilon))] = \emptyset.$$

Set

$$f_k(Z, Z_1, \cdots Z_{n-1}) = (u_k - Z)(u_k - Z_1) \ldots (u_k - Z_{n-1}) g_Z(u_k) - f(u_k) + f(Z),$$

where $(k = 1, 2, \cdots, n-1)$ and $Z \in \Lambda$. Computations show that

$$|\Delta| = |det[\frac{\partial f_k}{\partial Z_i}|_{(Z, Z_1, \cdots, Z_{n-1}) = (z, z_1, \cdots, z_{n-1})}]|$$

$$= |[\prod_{i \ne j}(u_i - u_j)][\prod_{i \ne j}(z_i - z_j)][\prod_{i=1}^{n-1}(u_i - z)][\prod_{i=1}^{n-1} g_Z(u_i)]| \ne 0$$

By the implicit holomorphic function theorem [Griffiths, P. (1985)] (Theorem 9.6), there exists a $\delta > 0$ and analytic functions $Z_k = Z_k(v)$, $(k = 1, 2, \cdots, n-1)$ in $B(z, \delta)$ such that for $v \in B(z, \delta)$,

$$f(u_k) - f(v) = (u_k - v)(u_k - Z_1(v)) \cdots (u_k - Z_{n-1}(v)) g_v(u_k), \ \ k = 1, 2, \cdots, n-1.$$

Thus for $v \in B(z, \delta)$, $v, Z_1(v), \cdots, Z_{n-1}(v)$ satisfying (4.5.1) and the analytic functions $Z_1(v), \cdots, Z_{n-1}(v)$ can be extended analytically to Λ satisfying

$$f(u) - f(z) = (u - z)(u - Z_1(z)) \cdots (u - Z_{n-1}(z)) g_z(u).$$

From (4.5.2), we get

$$(Ag)(z) = \alpha_1(z)g(z) + \alpha_2(z)g(Z_1(z)) + \cdots + \alpha_n(z)g(Z_{n-1}(z)), \ \forall \ g \in R(D), z \in \Lambda. \tag{4.5.3}$$

Set $g = e, z, z^2, \cdots, z^{n-1}$ respectively and denote

$$h_1 = Ae, h_2 = Az, \cdots, h_n = Az^{n-1},$$

then

$$\begin{cases} \alpha_1(z) + \qquad \alpha_2(z) + \cdots + \qquad \alpha_n(z) = h_1(z) \\[2mm] z\alpha_1(z) + \quad Z_1(z)\alpha_2(z) + \cdots + Z_{n-1}(z)\alpha_n(z) = h_2(z) \\[2mm] \qquad\qquad\qquad\qquad\vdots \\[2mm] z^{n-1}\alpha_1(z) + Z_1^{n-1}(z)\alpha_2(z) + \cdots + Z_{n-1}^{n-1}(z)\alpha_n(z) = h_n(z) \end{cases} \quad z \in \Lambda$$

Since the coefficient determinant $J(z)$ is a Vandermonde determinant,

$$J(z) = \begin{vmatrix} 1 & 1 & \cdots & 1 \\ z & Z_1 & \cdots & Z_{n-1} \\ \vdots & \vdots & \vdots & \vdots \\ z^{n-1} & Z_1^{n-1} & \cdots & Z_{n-1}^{n-1} \end{vmatrix} \neq 0.$$

By Cramer's rule, $\alpha_1(z), \cdots, \alpha_n(z)$ are analytic in Λ. Since Ag is analytic, (4.5.3) determines the vector Ag and describes $A \in \mathcal{A}'(M_f)$.

Corollary 4.5.12 *If $M_f^* \in \mathcal{B}_1(\Omega)$, then $\mathcal{A}'(M_f) = \{M_g : g \in R(D)\}$.*

Let $n \geq 2$ and w be the n-th root of 1, i.e., $w \in C, w^n = 1$. Let Δ_n denote the Vandermonde determinant of order n:

$$\Delta_n = \begin{vmatrix} 1 & 1 & \cdots & 1 \\ 1 & w & \cdots & w^{n-1} \\ \vdots & \vdots & \vdots & \vdots \\ 1 & w^{n-1} & \cdots & w^{(n-1)^2} \end{vmatrix}.$$

For $1 \leq i, j \leq n$, the (i,j)-cofactor will be denoted by Δ_{ij}.

Proposition 4.5.13 *An operator $A \in \mathcal{A}'(M_{z^n})$ if and only if for each $g \in R(D)$,*

$$(Ag)(z) = \sum_{i=1}^{n} \alpha_i(z) g(w^{i-1} z), z \neq 0,$$

where $\alpha_i(z) = \frac{1}{\Delta_n}(\sum_{j=1}^{n} \Delta_{ij} \frac{h_j}{z^{j-1}})$ and $\{h_j\}_{j=1}^{n}$ are n functions in $R(D)$.

Proof For $f(z) = z^n, M_{z^n}^* \in \mathcal{B}_n(D), \Lambda = \{z \in D : z \neq 0\}$. If

$$z, z_0 \in \Lambda, z^n - z_0^n = (z - z_0)(z - wz_0) \cdots (z - w^{n-1} z_0),$$

then

$$Z_i(z_0) = w^{i-1} z_0.$$

Let

$$A \in \mathcal{A}'(M_{z^n}),$$

denote

$$h_k = Az^{k-1} \ (k = 1, 2, \cdots, n).$$

Using Theorem 4.5.11, computations indicate that

$$\alpha_i(z) = \frac{1}{\Delta_n z^{\frac{n(n+1)}{2}}} [\sum_{j=1}^{n} \Delta_{ij} z^{\frac{n(n+1)}{2} - j + 1} h_j(z)] = \frac{1}{\Delta_n} [\sum_{j=1}^{n} \Delta_{ij} \frac{h_j(z)}{z^{j-1}}].$$

On the other hand, for arbitrary $h_1, h_2, \cdots, h_n \in R(D)$, set

$$\alpha_i(z) = \frac{1}{\Delta_n} \sum_{j=1}^{n} \Delta_{ij} \frac{h_j(z)}{z^{j-1}}, z \neq 0, 1 \leq i \leq n.$$

For $g \in R(D)$, formally, set

$$(Ag)(z) = \sum_{i=1}^{n} \alpha_i(z)g(w^{i-1}z) \quad (z \neq 0).$$

{Claim} A is bounded.

For each integer m, let $\overline{m} \equiv m \pmod{n}$ and $0 \leq \overline{m} \leq n-1$. Assume that

$$g(z) = \sum_{m=0}^{\infty} g_m z^m,$$

then

$$(Ag)(z) = \sum_{i=1}^{n} \alpha_i(z)g(w^{i-1}z)$$

$$= \frac{1}{\Delta_n} \sum_{i=1}^{n} [(\sum_{j=1}^{n} \Delta_{ij}\frac{h_j(z)}{z^{j-1}})(\sum_{m=0}^{\infty} g_m w^{m(i-1)} z^m)]$$

$$= \frac{1}{\Delta_n} \sum_{i=1}^{n} [\sum_{m=0}^{\infty} g_n(\sum_{j=1}^{n} \Delta_{ij} w^{\overline{m}(i-1)} h_j z^{m-j+1})]$$

$$= \frac{1}{\Delta_n} \sum_{m=0}^{\infty} g_m [\sum_{i=1}^{n} (\sum_{j=1}^{n} \Delta_{ij} w^{\overline{m}(i-1)} h_j z^{m-j+1})]$$

$$= \frac{1}{\Delta_n} \sum_{m=0}^{\infty} g_m [\sum_{j=1}^{n} (\sum_{i=1}^{n} \Delta_{ij} w^{\overline{m}(i-1)} h_j z^{m-j+1})].$$

It is easy to see that

$$\sum_{i=1}^{n} \Delta_{ij} w^{\overline{m}(i-1)} = \begin{cases} \Delta_n & \overline{m} = j-1 \\ 0 & \overline{m} \neq j-1. \end{cases}$$

Therefore,

$$(Ag)(z) = \sum_{k=0}^{\infty} (g_{kn}z^{kn}h_1(z) + g_{kn+1}z^{kn}h_2(z) + \cdots + g_{kn+n-1}z^{kn}h_n)$$

$$= \sum_{i=0}^{n} (\sum_{k=0}^{\infty} g_{kn+i-1}z^{kn}h_i).$$

For each $i, 1 \leq i \leq n$,

$$\| \sum_{k=0}^{\infty} g_{kn+i-1} z^{kn} h_i \|^2$$

$$\leq M \|h_i\|^2 \| \sum_{k=0}^{\infty} g_{kn+i-1} z^{kn} \|^2$$

$$= M \|h_i\|^2 \sum_{k=0}^{\infty} | \frac{g_{kn+i-1}}{\beta_{kn+i-1}} \frac{\beta_{kn+i-1}}{\beta_{kn}} e_{kn} |^2$$

$$= M \|h_i\|^2 \sum_{k=0}^{\infty} | \frac{g_{kn+i-1}}{\beta_{kn+i-1}} |^2 | \frac{\beta_{kn+i-1}}{\beta_{kn}} |^2$$

$$\leq N \|g\|^2.$$

Thus $\|Ag\|^2 \leq nN\|g\|^2$ and A is a bounded linear operator. It is easy to show that $A \in \mathcal{A}'(M_{z^n})$.

For an operator $T \in \mathcal{L}(\mathcal{H})$, there exists a Banach reducing decomposition to each idempotent P in $\mathcal{A}'(T)$, i.e., $\mathcal{H} = \mathcal{H}_1 \dotplus \mathcal{H}_2$, where

$$\mathcal{H}_i \in LatT, \quad \mathcal{H}_1 = ranP$$

and

$$\mathcal{H}_2 = ran(I - P).$$

The following proposition characterizes all of the Banach reducing decompositions of M_{z^2}.

Proposition 4.5.14 *If $P \in \mathcal{A}'(M_{z^2})$, the following statements are equivalent:*

(i) $P^2 = P$;

(ii) $(Pg)(z) = \alpha_1(z)g(z) + \alpha_2(z)g(-z)$ $(z \neq 0)$ for $g \in R(D)$ and one of the following two must be true: (1) $\alpha_2(z) \equiv 0$ and $P = I$ or 0;

(2) $\alpha_1(z) = \frac{a_{-1}}{z} + \frac{1}{2} + \sum_{k=0}^{\infty} a_{2k+1} z^{2k+1}$ and $\alpha_1(z)\alpha_1(-z) = \alpha_2(z)\alpha_2(-z)$;

(iii) If $h_1, h_2 \in R(D), h_1(z) = \sum_{n=0}^{\infty} b_n z^n$, $h_2(z) = \sum_{n=0}^{\infty} c_n z^n$, then one of the following two must be true: (1) $zh_1(z) = h_2(z)$ and $P = I$ or 0;

(2) $b_0 + c_1 = \frac{1}{2}, b_{2k} = -c_{2k+1}$ $(k \geq 1)$ and $h_1(z)h_2(-z) = h_1(-z)h_2(z)$.

Proof   (i)$\Rightarrow$(ii).   If $\alpha_1(z)$ and $\alpha_2(z)$ correspond to P, then $Z_1(z) = -z$ $(z \neq 0)$.

Since $P^2 = P$,

$$\alpha_1^2(z)g(z) + \alpha_1(z)\alpha_2(z)g(-z) + \alpha_2(z)\alpha_1(-z)g(-z) + \alpha_2(z)\alpha_2(-z)g(z)$$

$$= \alpha_1(z)g(z) + \alpha_2(z)g(-z) \ (z \neq 0)$$

for each $g \in R(D)$.

When $g = e$,

$$\alpha_1^2(z) + \alpha_1(z)\alpha_2(z) + \alpha_2(z)\alpha_1(-z) + \alpha_2(z)\alpha_2(-z) = \alpha_1(z) + \alpha_2(z).$$

When $g = z$,

$$\alpha_1^2(z) - \alpha_1(z)\alpha_2(z) - \alpha_2(z)\alpha_1(-z) + \alpha_2(z)\alpha_2(-z) = \alpha_1(z) - \alpha_2(z).$$

Simple computations show that

$$\alpha_1^2(z) + \alpha_2(z)\alpha_2(-z) = \alpha_1(z) \tag{4.5.4}$$

$$[\alpha_1(z) + \alpha_1(-z)]\alpha_2(z) = \alpha_2(z) \tag{4.5.5}$$

Since α_1 and α_2 are analytic functions, $\alpha_2(z) \equiv 0$ or $\alpha_1(z) + \alpha_1(-z) \equiv 1$. If $\alpha_2(z) \equiv 0$, then $\alpha_1(z) \equiv 1$ or 0 and $P = I$ or $P = 0$. If $\alpha_1(z) + \alpha_1(-z) \equiv 1$, if follows from Proposition 4.5.13 that $\alpha_1(z)$ is analytic in $D \backslash \{0\}$ with a pole of order 1 at $z = 0$. Therefore, $\alpha_1(z)$ can be expressed as $\alpha_1(z) = \frac{a_{-1}}{z} + \frac{1}{2} + \sum_{k=0}^{\infty} a_{2k+1} z^{2k+1}$. By (4.5.4),

$$\alpha_2(z)\alpha_2(-z) = \alpha_1(z)[1 - \alpha_1(z)]$$

$$= \alpha_1(z)[-\frac{a_{-1}}{z} + \frac{1}{2} - \sum_{k=0}^{\infty} a_{2k+1} z^{2k+1}]$$

$$= \alpha_1(z)\alpha_1(-z).$$

(ii)$\Rightarrow$(i).

$$(P^2 g)(z) = \alpha_1^2(z)g(z) + \alpha_2(z)\alpha_2(-z)g(z) + \alpha_2(z)g(-z)[\alpha_1(z) + \alpha_1(-z)]$$

$$= \alpha_1^2(z)g(z) + \alpha_1(z)\alpha_1(-z)g(z) + \alpha_2(z)g(-z)$$

$$= \alpha_1(z)g(z)[\alpha_1(z) + \alpha_1(-z)] + \alpha_2(z)g(-z)$$

$$= \alpha_1(z)g(z) + \alpha_2(z)g(-z) = (Pg)(z), \qquad g \in R(D).$$

Similar computation shows that (ii)$\Leftrightarrow$(iii).

Example 4.5.15 *Operator P is defined by*

$$(Pg)(z) = (\frac{1}{2} + sinz)g(z) - (\frac{1}{2} + sinz)g(-z), \ z \neq 0, g \in R(D).$$

Then $P \in \mathcal{A}'(M_{z^2})$ and $P^2 = P$.

In the following we will discuss the commutants of the multiplication operators on $R(D)$ with the symbol of the form $f(z) = z^n h(z)$.

Proposition 4.5.16 *Let $f \in R(D), f'(0) \neq 0, T \in \mathcal{A}'(M_f)$, then $T = M_\varphi$ for some $\varphi \in R(D)$ if and only if T admits a lower triangular matrix representation with respect to the ONB $\{e_n\}_{n=0}^{\infty}$.*

Proof If $T \in \mathcal{A}'(M_f)$ admits a lower triangular matrix representation

$$T = (t_{ij})_{ij}, t_{ij} = 0 \, (j > i).$$

Let $f(z) = \sum_{n=0}^{\infty} f_n z^n$ with $f_1 \neq 0$. Compare the (2.1) entries of $TM_f = M_f T$,

$$f_1 t_{22} \frac{\beta_0}{\beta_1} = f_1 t_{11} \frac{\beta_0}{\beta_1}.$$

Since $f_1 \neq 0, t_{11} = t_{22}$. Compare the (3.2) entries, we get $t_{22} = t_3$. In general, $t_{11} = t_{22} = \cdots = c_0$. Compare the (3.1) entries, we get

$$f_1 t_{32} \frac{\beta_0}{\beta_1} + f_2 t_{33} \frac{\beta_0}{\beta_2} = f_2 t_{11} \frac{\beta_0}{\beta_2} + f_1 t_{21} \frac{\beta_1}{\beta_2}.$$

Since $t_{33} = t_{11}, t_{32} \frac{\beta_0}{\beta_1} = t_{21} \frac{\beta_1}{\beta_2}$. Thus $t_{32} = c_1 \frac{\beta_1}{\beta_2}$, where $c_1 = \frac{\beta_0}{\beta_1} t_{21}$. Suppose that $t_{l+1,l} = c_1 \frac{\beta_{l-1}}{\beta_l}$ $(l \leq k-1)$. Compare the $(k+1, k-1)$ entries, we get

$$f_1 t_{k+1,k} \frac{\beta_{k-2}}{\beta_{k-1}} + f_2 t_{k+1,k+1} \frac{\beta_{k-2}}{\beta_k} = f_2 t_{k-1,k-1} \frac{\beta_{k-2}}{\beta_k} + f_1 t_{k,k-1} \frac{\beta_{k-1}}{\beta_k}.$$

It follows from $t_{k+1,k+1} = t_{k-1,k-1}$ that

$$t_{k+1,k}\frac{\beta_{k-2}}{\beta_{k-1}} = t_{k,k-1}\frac{\beta_{k-1}}{\beta_k} = c_1\frac{\beta_{k-2}}{\beta_{k-1}} \cdot \frac{\beta_{k-1}}{\beta_k}$$

and $t_{k+1,k} = c_1\frac{\beta_{k-1}}{\beta_k}$. By the mathematical deduction, $t_{k+1,k} = c_1\frac{\beta_{k-1}}{\beta_k}$ for all $k \geq 1$.

Similarly, by the same arguments we can prove that if $t_{k,1} = c_{k-1}\frac{\beta_0}{\beta_{k-1}}$, then $t_{k+i,1+i} = c_{k-1}\frac{\beta_i}{\beta_{k-1+i}}$, $i = 0,1,2,\cdots$. Therefore,

$$T = \begin{bmatrix} c_0 & & & 0 \\ c_1\frac{\beta_0}{\beta_1} & c_0 & & \\ c_2\frac{\beta_0}{\beta_2} & c_1\frac{\beta_1}{\beta_2} & c_0 & \\ \vdots & \vdots & \vdots & \ddots \end{bmatrix}.$$

Since $\varphi = Te_0 \in R(D)$, by Proposition 4.5.3, $\varphi = \sum_{n=0}^{\infty} c_n z^n$ and $T = M_\varphi$.

Conversely, if $T = M_\varphi$, by Proposition 4.5.3, T admits a lower triangular matrix representation with respect to the ONB $\{e_n\}_{n=0}^{\infty}$.

Lemma 4.5.17[Deddens, J.A. and Wong, T.K. (1973), Lemma 2] *Let* N *be a nilpotent on* $\mathcal{H}$, $X_0 = \lambda + N, 0 \neq \lambda \in \mathbf{C}$. *If* $B, A_0, A_1, \cdots \in \mathcal{L}(\mathcal{H})$ *satisfying* $\|A_k\| \leq M$ *and* $A_k X_0 = X_0 A_{k-1} + B$ $(k = 1,2,\cdots)$, *then* $A_0 = A_1 = A_2 = \cdots$.

Lemma 4.5.18 *Let* $T \in \mathcal{A}'(M_{z^n}) \cap \mathcal{A}'(M_f), f \in R(D), f = z^r g, 1 \leq r < n$ *and* $g(0) \neq 0$. *Then* $T \in \mathcal{A}'(M_{z^s})$, *where* $s = (n,r)$ *denotes the maximal common divisor of* n *and* r.

Proof By Proposition 4.5.3,

$$M_g = \begin{bmatrix} G_{11} & & 0 \\ G_{21} & G_{22} & \\ \vdots & \vdots & \ddots \end{bmatrix} \quad \text{and} \quad M_{z^r} = \begin{bmatrix} 0 & & & 0 \\ W_1 & 0 & & \\ 0 & W_2 & 0 & \\ \vdots & \vdots & \vdots & \ddots \end{bmatrix}$$

with respect to the ONB $\{e_k\}_{k=0}^{\infty}$, where G_{ij}, W_k are $r \times r$ matrices. W_k is invertible and since $g(0) \neq 0, G_{ii}$ is also invertible $(i,j,k = 1,2,\cdots)$. Since

$$TM_{z^n} = M_{z^n}T,$$

$$T = \begin{bmatrix} T_{11} & 0 \\ T_{21} & T_{22} \\ \vdots & \vdots & \ddots \end{bmatrix}$$

with respect to the ONB $\{e_k\}_{k=0}^{\infty}$, where T_{ij} is $n \times n$ matrix $(i, j = 1, 2, \cdots)$.

{Case 1} If r is a divisor of n, $n = pr$, $r > 1$ and $s = r$. Suppose that

$$T_{kk} = \begin{bmatrix} V_{11}^k & V_{12}^k & \cdots & V_{1p}^k \\ V_{21}^k & V_{22}^k & \cdots & V_{2p}^k \\ \cdots & \cdots & \cdots & \cdots \\ V_{p1}^k & V_{p2}^k & \cdots & V_{pp}^k \end{bmatrix},$$

where V_{ij}^k is a $r \times r$ matrix $(k, i, j = 1, 2, \cdots)$. Denote $m = (k-1)p, k = 1, 2, \cdots$. It follows from $TM_{z^r g} = M_{z^r g}T$ that $T_{kk}F_{kk} = F_{kk}T_{kk}$ and F_{kk} equals to

$$\begin{bmatrix} 0 & & & & 0 \\ W_{m+1}G_{m+1,m+1} & 0 & & & \\ W_{m+2}G_{m+2,m+1} & W_{m+2}G_{m+2,m+2} & 0 & & \\ \vdots & \vdots & \vdots & \ddots & \\ W_{m+p-1}G_{m+p-1,m+1} & W_{m+p-1}G_{m+p-1,m+2} & \cdots & W_{m+p-1}G_{m+p-1,m+p-1} & 0 \end{bmatrix}.$$

Compare the $(1, p-1)$ entries of $T_{kk}F_{kk} = F_{kk}T_{kk}$, we get

$$V_{1p}^k W_{m+r-1}G_{m+p-1,m+p-1} = 0.$$

Since W_{m+r-1} and $G_{m+p-1,m+p-1}$ are invertible, $V_{1p}^k = 0$. Similarly, $V_{ij}^k = 0$ if
$j > i, k = 1, 2, \cdots$. Thus, T_{kk} admits a block lower triangular matrix representation and

$$T = \begin{bmatrix} V_{11} & 0 \\ & V_{22} \\ \star & & \ddots \end{bmatrix},$$

where V_{kk} is a $r \times r$ matrix.

By the arguments used in the proof of Proposition 4.5.18, $T \in \mathcal{A}'(M_{z^r})$.

{Case 2} If r is not a divisor of n. Find positive integers p and q such that $qr - pn = s$. Since $TM_{z^{qr}g^q} = M_{z^{qr}g^q}T, TM_{z^{s+np}g^q} = M_{z^{s+np}g^q}T$. Because

$$TM_{z^{pn}} = M_{z^{pn}}T,$$

$$TM_{z^s g^q} = M_{z^s g^q} T.$$

Note that s is a divisor of n and $g^q(0) \neq 0$. Repeating the proof of Case 1, we get $T \in \mathcal{A}'(M_{z^s})$ and complete the proof.

Using Lemma 4.5.17 and Lemma 4.5.18, we get the following proposition.

Proposition 4.5.19 Let $f(z) = z^n h(z) \in R(D), h(z) \neq 0$ for $z \in \overline{D}$ and $n \geq 1$. Then $\mathcal{A}'(M_f) = \mathcal{A}'(M_{z^n}) \cap \mathcal{A}'(M_h) = \mathcal{A}'(M_{z^s})$, where

$$s = (n, n_1, n_2, \cdots), h(z) = a_0 + a_1 z^{n_1} + a_2 z^{n_2} + \cdots.$$

Proof Assume that $f(z) = h_n z^n + h_{n+1} z^{n+1} + \cdots$. Set

$$B_k = \begin{bmatrix} \beta_{nk-n} & & & 0 \\ & \beta_{nk-n+1} & & \\ & & \ddots & \\ 0 & & & \beta_{nk-1} \end{bmatrix}$$

and

$$H_k = \begin{bmatrix} h_{nk} & h_{nk-1} & \cdots & h_{nk-(n-1)} \\ h_{nk+1} & h_{nk} & \cdots & h_{nk-(n-2)} \\ \cdots & \cdots & \cdots & \cdots \\ h_{nk+n-1} & h_{nk+n-2} & \cdots & h_{nk} \end{bmatrix},$$

where $h_i = 0$ if $i < n$. Then

$$M_f = \begin{bmatrix} 0 & & & 0 \\ F_{21} & 0 & & \\ F_{31} & F_{32} & 0 & \\ \vdots & \vdots & \vdots & \ddots \end{bmatrix},$$

where $F_{k+i,k} = B_{k+i}^{-1} H_i B_k$.
Since $h(z) \neq 0$ for $z \in \overline{D}$,

$$ker(M_f^*)^k = \bigvee_{i=1}^{nk-1} \{e_i\}.$$

Suppose that $T \in \mathcal{A}'(M_f)$. It follows from $T^* M_f^* = M_f^* T^*$ that $T^*(\bigvee_{i=0}^{nk-1} \{e_i\}) \subset ker(M_f^*)^k = \bigvee_{i=0}^{nk-1} \{e_i\}$ and hence T^* admits a block upper triangular matrix representation with respect to the ONB $\{e_i\}_{i=0}^{\infty}$, i.e.,

$$T = \begin{bmatrix} T_{11} & 0 \\ T_{21} & T_{22} \\ \vdots & \vdots & \ddots \end{bmatrix}.$$

Compare the $(2, 1)$ entries of $TM_f = M_f T$, we get $F_{21}T_{11} = T_{22}F_{21}$. It follows from $F_{21} = B_2^{-1}H_1B_1$ and $B_2^{-1}H_1B_1T_1 = T_{22}B_2^{-1}H_1B_1$ that

$$H_1B_1T_{11}B_1^{-1} = B_2T_{22}B_2^{-1}H_1.$$

Compare the $(i, i-1)$ entries, we get

$$T_{ii}F_{i,i-1} = F_{i,i-1}T_{i-1,i-1},$$

i.e.,

$$T_{ii}B_i^{-1}H_1B_{i-1} = B_i^{-1}H_1B_{i-1}T_{i-1,i-1}.$$

Thus

$$H_1B_{i-1}T_{i-1,i-1}B_{i-1}^{-1} = B_iT_{ii}B_i^{-1}H_1.$$

By Lemma 4.5.17,

$$B_{i-1}T_{i-1,i-1}B_{i-1}^{-1} = B_iT_{ii}B_i^{-1},$$

i.e.,

$$B_i^{-1}B_{i-1}T_{i-1,i-1} = T_{ii}B_i^{-1}B_{i-1}.$$

Thus

$$T_{i+1,i+1}W_i = W_iT_{ii},$$

where $W_i = B_{i+1}^{-1}B_i$. Similarly, by mathematical deduction we can prove that

$$B_{k+i-1}T_{k+1+i,i+1}B_{i+1}^{-1} = B_{k+i}T_{k+i,i}B_i^{-1},$$

i.e.,

$$T_{k+1+i,i+1}W_i = W_{k+i}T_{k+i,i}, k = 0, 1, 2, \cdots, i = 1, 2, 3, \cdots.$$

This implies that

$$TM_{z^n} = M_{z^n}T.$$

Since

$$TM_{z^n}M_h = M_{z^n}M_hT,$$

$$M_{z^n}TM_h = M_{z^n}M_hT.$$

This means that

$$z^n(TM_hg)(z) = z^n(M_hTg)(z)$$

for $g \in R(D)$ and $z \in D$. Thus

$$TM_h = M_hT$$

and

$$T \in \mathcal{A}'(M_{z^n}) \cap \mathcal{A}'(M_h).$$

Conversely, $\mathcal{A}'(M_{z^n}) \cap \mathcal{A}'(M_h) \subset \mathcal{A}'(M_f)$ is obvious. Thus,

$$\mathcal{A}'(M_f) = \mathcal{A}'(M_{z^n}) \cap \mathcal{A}'(M_h).$$

If $T \in \mathcal{A}'(M_f) = \mathcal{A}'(M_{z^n}) \cap \mathcal{A}'(M_h)$. Suppose that

$$h(z) = a_0 + a_1 z^{n_1} + a_2 z^{n_2} + \cdots,$$

where $a_i \neq 0$ $(i = 0, 1, 2, \cdots)$. Then $h - h(0) = z^{nk_1}h_1$, where $n_1 = nk_1 + r_1$ for $k_1, r_1 \in \mathbf{N}$ and $r_1 < n$ and $h_1 = z^{r_1}g_1$. Since $T \in \mathcal{A}'(M_h)$,

$$TM_{z^{nk_1}}M_{h_1} = M_{z^{nk_1}}M_{h_1}T.$$

Since $T \in \mathcal{A}'(M_{z^n})$,

$$M_{z^{nk_1}}TM_{h_1} = M_{z^{nk_1}}M_{h_1}T,$$

i.e.,

$$z^{nk_1}(TM_{h_1})g = z^{nk_1}(M_{h_1}T)g$$

for $g \in R(D)$. Thus

$$TM_{h_1} = M_{h_1}T.$$

By Lemma 4.5.18, $T \in \mathcal{A}'(M_{z^{s_1}})$, where $s_1 = (n, r_1) = (n, n_1)$.
Similarly, if $n_2 = nk_2 + r_2$ for $k_2, r_2 \in \mathbf{N}$ and $r_2 < n$, then

$$a_2 z^{n_2} + a_3^{n_3} + \cdots = z^{nk_2}h_2,$$

where $h_2 = z^{r_2}g_2$. By the same argument $T \in \mathcal{A}'(M_{z^{s_2}})$ for $s_2 = (n, n_1, n_2)$. In general, $T \in \mathcal{A}'(M_{z^s})$, where $s = (n, n_1, n_2, \cdots)$.

Conversely, if $T \in \mathcal{A}'(M_{z^s})$, then it is obvious that $T \in \mathcal{A}'(M_f)$. Thus

$$\mathcal{A}'(M_f) = \mathcal{A}'(M_{z^n}) \cap \mathcal{A}'(M_h) = \mathcal{A}'(M_{z^s}), s = (n, n_1, n_2, \cdots).$$

Corollary 4.5.20 *Suppose that* $f \in R(D), f(z) = z^n h(z), h(z) \neq 0$ *for* $z \in \overline{D}$ *and for each* $k > 0$, $h(z)$ *is not a function of* z^k, *then*

$$\mathcal{A}'(M_f) = \mathcal{A}'(M_z) = \{M_g : g \in R(D)\}.$$

In the following, we will discuss the strong irreducibility of multiplication operator M_f on $R(D)$.

Theorem 4.5.21 *Given* $f \in R(D)$, *the following are equivalent.*
 (i) $M_f^* \in \mathcal{B}_1(\Omega)$;
 (ii) $\mathcal{A}'(M_f) = \{M_g : g \in R(D)\}$;
 (iii) $\mathcal{A}'(M_f)$ *is commutative;*
 (iv) *If* $M_f^* \in \mathcal{B}_n(\Omega_1)$, *then for each* $A \in \mathcal{A}'(M_f)$,

$$(Az)(z) = zh_1(z), (Az^2)(z) = z^2 h_1(z), \cdots, (Az^{n-1})(z) = z^{n-1}h_1(z),$$

where $h_1 = Ae_1$;
 (v) $M_f \in (SI)$.
Proof (i)$\Rightarrow$(ii) Corollary 4.5.12.
 (ii)$\Rightarrow$(iii). Obvious.
 (iii)$\Rightarrow$(iv). If there an operator $A \in \mathcal{A}'(M_f)$ such that $(Az^k)(z_0) \neq z_0^k h_1(z_0)$ for some k, $1 < k \leq n-1$ and $z_0 \in D_1 = f^{-1}(\Omega), z_0 \neq 0$. Then

$$(M_{z^k}Ae)(z_0) = z_0^k h_1(z_0) \neq (Az^k)(z_0) = (AM_{z^k}e)(z_0).$$

Thus $M_{z^k}A \neq AM_{z^k}$. But A and M_{z^k} are in $\mathcal{A}'(M_f)$. A contradiction.
 (iv)$\Rightarrow$(v). For $z \in \Lambda$, since $\alpha_1(z) + \alpha_2(z) + \cdots + \alpha_n(z) = h_1(z)$ and

$h_{k+1} = z^k h_1$ for $1 \le k \le n-1$, by Theorem 4.5.11 and (iv)

$$
\left\{
\begin{array}{l}
(Z_1(z) - z)\alpha_2(z) + (Z_2(z) - z)\alpha_3(z) + \cdots + (Z_{n-1}(z) - z)\alpha_n(z) = 0 \\[2mm]
(Z_1^2(z) - z^2)\alpha_2(z) + (Z_2^2(z) - z^2)\alpha_3(z) + \cdots + (Z_{n-1}^2(z) - z^2)\alpha_n(z) = 0 \\[2mm]
\qquad\qquad\qquad\qquad\vdots \\[2mm]
(Z_1^{n-1}(z) - z^{n-1})\alpha_2(z) + (Z_2^{n-1}(z) - z^{n-1})\alpha_3(z) \\[2mm]
\qquad + \cdots + (Z_{n-1}^{n-1}(z) - z^{n-1})\alpha_n(z) = 0
\end{array}
\right.
$$

Computations indicate that the coefficient determinant is still a Vandermonde determinant:

$$
V =
\begin{vmatrix}
Z_1 - z & Z_2 - z & \cdots & Z_{n-1} - z \\
Z_1^2 - z^2 & Z_2^2 - z^2 & \cdots & Z_{n-1}^2 - z^2 \\
\vdots & \vdots & \vdots & \vdots \\
Z_1^{n-1} - z^{n-1} & Z_2^{n-1} - z^{n-1} & \cdots & Z_{n-1}^{n-1} - z^{n-1}
\end{vmatrix}
$$

$$
= (-1)^{n-1}
\begin{vmatrix}
1 & 1 & \cdots & 1 & 1 \\
Z_1 & Z_2 & \cdots & Z_{n-1} & z \\
Z_1^2 & Z_2^2 & \cdots & Z_{n-1}^2 & z^2 \\
\vdots & \vdots & \vdots & \vdots & \vdots \\
Z_1^{n-1} & Z_2^{n-1} & \cdots & Z_{n-1}^{n-1} & z^{n-1}
\end{vmatrix}
\neq 0.
$$

Therefore $\alpha_2(z) = \alpha_3(z) = \cdots = \alpha_n(z) = 0$ and $\mathcal{A}'(M_f) = \{M_g : g \in R(D)\}$.

Since $\mathcal{A}'(M_f)$ does not contain nontrivial idempotent, $M_f \in (SI)$.

(v)$\Rightarrow$(i). Suppose that the minimal index of M_f^* is n and $n \ge 2$. By Theorem 4.5.11, $Ag = \alpha_1(z)g(z) + \cdots + \alpha_n(z)g(Z_{n-1}(z))$ for $A \in \mathcal{A}'(M_f)$, $g \in R(D)$ and $z \in \Lambda$. Since $M_z^* \in \mathcal{A}'(M_f^*)$, $M_z^* k_u = \bar{u} k_u$ and $M_z^* k_{Z_1(u)} = \overline{Z_1(u)} k_{Z_1(u)}$ for $u \in \Lambda$. By Theorem 4.4.3, the spectrum of $M_z^*|_{ker(M_f^* - f(u))}$ is connected. Thus $u = Z_1(u)$. This contradicts $z \neq Z_1(z)$ when $z \in \Lambda$. Therefore $n = 1$ and $M_f^* \in \mathcal{B}_1(\Omega)$.

Proposition 4.5.22 $M_{z^n} \in (SI)$ *if and only if* $n = 1$.

Proof If $n \ge 2$, define an operator P by

$$
Pf = \frac{1}{n}[f(z) + f(\omega z) + \cdots + f(\omega^{n-1}z)] \quad f \in R(D).
$$

By Proposition 4.5.13, $P \in \mathcal{A}'(M_{z^n})$. For each i, $1 \le i \le n-1$,

$$Pf(w^i z) = \tfrac{1}{n}[f(\omega^i z) + f(\omega^{i+1} z) + \cdots + f(\omega^{n-1+i} z)]$$

$$= \tfrac{1}{n}[f(z) + f(\omega z) + \cdots + f(\omega^{n-1} z)].$$

Thus

$$P^2 f = \tfrac{1}{n^2}[n(f(z) + f(\omega z) + \cdots + f(\omega^{n-1} z))]$$

$$= \tfrac{1}{n}[f(z) + f(\omega z) + \cdots + f(\omega^{n-1} z)] = Pf.$$

It is obvious that P is nontrivial and $M_{z^n} \notin (SI)$.

Proposition 4.5.23 *Let $f \in R(D)$ with $f(z) = z^n h(z^m)$, $h(z) \ne 0$, $z \in \overline{D}$ and $n \ge 1$. If for each $k > m$, $h(z^m)$ is not a function of z^k, then $M_f \in (SI)$ if and only if $(n, m) = 1$.*
Proof By proposition 4.5.19 $\mathcal{A}'(M_f) = \mathcal{A}'(M_{z^s})$, where $s = (m, n)$. By Proposition 4.5.22, $\mathcal{A}'(M_{z^s})$ contains no nontrivial idempotent if and only if $s = 1$. Thus $M_f \in (SI)$ if and only if $s = 1$.

Proposition 4.5.23 requires $n \ge 1$. One may asks the question: Is the conclusion of Proposition 4.5.23 true when $n = 0$? In fact the answer in general is negative unless f is an integral function.

Example 4.5.24 *Let $f(z) = 2z - \sum\limits_{n=1}^{\infty} \frac{3}{2^n} z^n$, then $M_f \notin (SI)$.*
Proof It is easy to see that $f(z)$ is analytic in $2D = \{z \in \mathbf{C} : |z| < 2\}$. Thus $f \in R(D)$. Define an operator P as follows:

$$(Pg)(z) = \frac{1}{2} g(z) + \frac{1}{2} g(2 + \frac{3}{z-2}), \quad g \in R(D).$$

Then

$$(Pg)(2 + \frac{3}{z-3}) = \frac{1}{2} g(z) + \frac{1}{2} g(2 + \frac{3}{z-2}).$$

Simple computations show that $P^2 = P$ and $f(z) = f(2 + \frac{3}{z-2})$. Thus

$$PM_f g(z) = \tfrac{1}{2} f(z) g(z) + \tfrac{1}{2} f(2 + \tfrac{3}{z-2}) g(2 + \tfrac{3}{z-2})$$

$$= \tfrac{1}{2} f(z) g(z) + \tfrac{1}{2} f(z) g(2 + \tfrac{3}{z-2})$$

$$= M_f Pg(z), \quad \text{for all} \quad g \in R(D).$$

Set $g(z) = 10 + f(z)$, then g satisfies the requirement of Proposition 4.5.23 with $n = 0$. But $M_g \notin (SI)$.

Proposition 4.5.25 $\mathcal{A}'(M_{z^2})/rad\mathcal{A}'(M_{z^2})$ *is noncommutative.*

Proof By Theorem 4.5.21 $\mathcal{A}'(M_{z^2})$ is not commutative. Assume that R is a left ideal of $\mathcal{A}'(M_{z^2})$. Set $R_1 = \{Ae : A \in R\}, R_2 = \{Az : A \in R\}$.
{**Claim(i)**} R_1, R_2 are ideals of $R(D)$, and if $h_1 \in R_1, h_2 \in R_2$, then

$$h_1(-z) \in R_1, \frac{h_1(z) - h_1(-z)}{z} \in R_1, h_2(-z) \in R_2$$

and

$$\frac{h_2(z) - h_2(-z)}{z} \in R_2.$$

Let $A \in R$ and $h_1 = Ae$. Denote $h_2 = Az \in R_2$. For arbitrary $f_1, f_2 \in R(D)$, denote $B \in \mathcal{A}'(M_{z^2})$ determined by f_1, f_2. For $g \in R(D)$,

$$(BA)g(z)$$

$$= [\frac{f_1(z)\frac{h_1(z)+h_1(-z)}{2} + f_2(z)\frac{h_1(z)-h_1(-z)}{2z}}{2} + \frac{f_1(z)\frac{h_2(z)+h_2(-z)}{2} + f_2(z)\frac{h_2(z)-h_2(-z)}{2z}}{2z}]g(z)$$

$$+ [\frac{f_1(z)\frac{h_1(z)+h_1(-z)}{2} + f_2(z)\frac{h_1(z)-h_1(-z)}{2z}}{2} - \frac{f_1(z)\frac{h_2(z)+h_2(-z)}{2} + f_2(z)\frac{h_2(z)-h_2(-z)}{2z}}{2z}]g(-z).$$

Choose $f_2 = zf_1$, i.e., $B + M_{f_1}$, then

$$(BA)g(z) = [\frac{f_1(z)h_1(z)}{2} + \frac{f_1(z)h_2(z)}{2z}]g(z) + [\frac{f_1(z)h_1(z)}{2} - \frac{f_1(z)h_2(z)}{2z}]g(-z),$$

which means $f_1h_1 \in R_1$ and R_1 is an ideal of $R(D)$.
Set $f_1 = e, f_2 = 0$, i.e., $(Bg)(z) = \frac{1}{2}[g(z) + g(-z)]$, then

$$(BA)g(z) = [\frac{h_1(z)+h_1(-z)}{4} + \frac{h_2(z)+h_2(-z)}{4z}]g(z)$$

$$+ [\frac{h_1(z)+h_1(-z)}{4} - \frac{h_2(z)+h_2(-z)}{4z}]g(-z),$$

which means $h_1(z) + h_1(-z) \in R_1$ and $h_1(-z) \in R_1$.
Set $f_1 = 0, f_2 = e$, i.e., $(Bg)(z) = \frac{1}{2z}[g(z) - g(-z)]$, then

$$(BA)g(z) = [\frac{\frac{h_1(z)-h_1(-z)}{2z}}{2} + \frac{\frac{h_2(z)-h_2(-z)}{2z}}{2z}]g(z)$$

$$+ [\frac{\frac{h_1(z)-h_1(-z)}{2z}}{2} - \frac{\frac{h_2(z)-h_2(-z)}{2z}}{2z}]g(-z),$$

which means $\frac{h_1(z)-h_1(-z)}{z}\in R_1$. Similarly, we can prove that R_2 has the same properties.

{**Claim(ii)**} For arbitrary $z_1, z_2 \in D, z_1 \neq 0, z_2 \neq 0$, Set

$$R_1(z_1) = (z^2 - z_1^2)R(D), R_2(z_2) = (z^2 - z_2^2)R(D),$$

then $R_1(z_1)$ is a "maximal" ideal which has the properties: if $h_1 \in R_1(z_1)$ then $h_1(-z) \in R_1(z_1)$ and $\frac{h_1(z)-h_1(-z)}{z}\in R_1(z_1)$. Similar properties hold for $R_2(z_2)$. For each $h_1 \in R_1(z_1)$ and $h_2 \in R_2(z_2)$, find $A \in \mathcal{A}'(M_{z^2})$ with

$$(Ag)(z) = (\frac{h_1(z)}{2} + \frac{h_2(z)}{2z})g(z) + (\frac{h_1(z)}{2} - \frac{h_2(z)}{2z})g(-z), z \neq 0.$$

Let $R(z_1, z_2)$ be the set of all such operator A. Simple computation shows that $R(z_1, z_2)$ is an ideal of $\mathcal{A}'(M_{z^2})$. If there is an ideal $\mathcal{S}$ of $\mathcal{A}'(M_{z^2})$ such that $R(z_1, z_2) \subset \mathcal{S}$, then $\mathcal{S}_1 := \{Ae : A \in \mathcal{S}\} \supset R_1(z_1)$ is an ideal and by (i) and the "maximality" of $R_1(z_1)$, $\mathcal{S}_1 = R_1(z_1)$. Similarly, $\mathcal{S}_2 := \{Az : A \in \mathcal{S}\} = R_2(z_2)$. Hence $\mathcal{S} = R(z_1, z_2)$, i.e., $R(z_1, z_2)$ is a maximal ideal of $\mathcal{A}'(M_{z^2})$.

{**Claim(iii)**} $rad\mathcal{A}'(M_{z^2}) = \{0\}$.

Let

$$R = \cap\{R(z_1, z_2) : z_1, z_2 \in D, z_1 \neq 0, z_2 \neq 0\},$$

then

$$R_1 = \{Ae : A \in R\} = \cap\{R_1(z_1) : z_1 \in D, z_1 \neq 0\}$$

$$= \{h \in R(D) : h(z) = h(-z) = 0, z \in D\} = \{0\},$$

$$R_2 = \{Az : A \in R\} = \cap\{R_2(z_2) : z_2 \in D, z_2 \neq 0\}$$

$$= \{h \in R(D) : h(z) = h(-z) = 0, z \in D\} = \{0\}.$$

Thus $R = \{0\}$ and $rad\mathcal{A}'(M_{z^2}) = \{0\}$. Since $\mathcal{A}'(M_{z^2}) = \{0\}$ is noncommutative, $\mathcal{A}'(M_{z^2})/rad\mathcal{A}'(M_{z^2})$ is noncommutative.

In the last part of this section we will consider in invariant subspace of M_z on $R(D)$.

Proposition 4.5.26 *Let $f \in R(D)$.*
(i) If $B_a(z) = \frac{z-a}{1-\bar{a}z}, a \in D$, then $M_{f \circ B_a} \sim M_f$;
(ii) $M_f \sim M_z$ if and only if $f(z) = \lambda\frac{z-a}{1-\bar{a}z}, |\lambda| = 1$ and $a \in D$.

Proof (i) It is obvious that $f \circ B_a \in R(D)$. Define an operator S_a as follows: $S_a f = f \circ B_a$, $f \in R(D)$. Computation indicates that there are positive numbers M_1, M_2, M_3 and M_4 such that

$$\int_D |f \circ B_a|^2 dm \leq M_1 \int_D |f|^2 dm, \quad \int_D |(f \circ B_a)'|^2 dm \leq M_2 \int_D |f'|^2 dm$$

and

$$\int_D |(f \circ B_a)''|^2 dm \leq M_3 \int_D |f'|^2 dm + M_4 \int_D |f''|^2 dm.$$

Thus $\|f \circ B_a\| \leq M \|f\|$ for some number M and all $f \in R(D)$, i.e., S_a is bounded. Note that for each $g \in R(D)$,

$$S_a M_f g = S_a(fg) = f(B_a) g(B_a)$$

and

$$M_{f \circ B_a} S_a g = M_{f \circ B_a} g(B_a) = f(B_a) g(B_a).$$

Thus $S_a M_f = M_{f \circ B_a} S_a$. Since B_a is an invertible analytic function, S_a is one to one and onto. Thus S_a is invertible and $M_{f \circ B_a} = S_a M_f S_a^{-1}$.

(ii) If $f(z) = \lambda \frac{z - a}{1 - \bar{a} z}$, then $M_f \sim M_z$ by (i).

On the other hand, if $M_f \sim M_z$, by Theorem 3.5.6,

$$f(\overline{D}) = \sigma(M_f) = \sigma(M_z) = \overline{D}.$$

Thus $f(D) \subset \overline{D}$. Since f is continuous on $\overline{D}$, by maximal module theorem $f(D) \subset D$. For arbitrary $\lambda \in D$, since $\lambda \in \sigma(M_z) \backslash \sigma_e(M_z) = f(\overline{D}) \backslash f(C)$, $\lambda \in f(D)$ and $D \subset f(D)$. Therefore f maps D onto D. Note that for each $\lambda \in D$,

$$nul(\lambda - M_f^*) = nul(\lambda - M_z^*) = 1,$$

this implies that $f(z) - \lambda$ has only one zero. Thus f maps D one to one onto D. It must be a Möbius transformation up to a coefficient of module one.

Lemma 4.5.27 *Assume that $f_1, f_2, \cdots, f_n$ are n functions in $R(D)$ without common zeroes in $\overline{D}$. Then there are functions $g_1, g_2, \cdots, g_n$ in $R(D)$ such that $\sum\limits_{k=1}^{n} f_k g_k = e$.*

Proof Denote $\mathcal{J} = \{g_1 f_1 + g_2 f_2 + \cdots + g_n f_n : g_1, g_2, \cdots, g_n \in R(D)\}$.

It is obvious that $\mathcal{J}$ is an ideal of $R(D)$. If $\mathcal{J}$ is a nontrivial ideal, it must be contained in a maximal ideal. But this is impossible, because the maximal ideal space of $R(D)$ is $\overline{D}$ and $f_1, f_2, \cdots, f_n$ have no common zeroes in $\overline{D}$. Thus $e \in \mathcal{J}$ and $\sum_{k=1}^{n} f_k g_k = e$ for some $g_1, g_2, \cdots, g_n \in R(D)$.

Theorem 4.5.28 *(i) Subspace M with finite common zeroes $\{z_1, z_2, \cdots, z_n\}$ in D, multiplicity included, is an invariant subspace of M_z if and only if*

$$M = (z - z_1) \cdots (z - z_n) R(D);$$

(ii) If M is an invariant subspace in (i), then the projection onto $\mathcal{M}$ is $P_M = M_\chi (M_\chi^* M_\chi)^{-1} M_\chi^*$, where $\chi = \prod_{i=1}^{n} \frac{z - z_i}{1 - \bar{z}_i z}$.

Proof (i) If $M = (z - z_1) \cdots (z - z_n) R(D)$, it is obvious M is an invariant subspace of M_z with common zeroes $\{x_i\}_{i=1}^{n}$ in D.

Conversely, if $M \in Lat M_z$ with common zeroes $\{z_i\}_{i=1}^{n}$. Denote

$$N := \{g \in R(D) : (z - z_1) \cdots (z - z_n) g \in M\}.$$

Then it is not difficult to see that N is an invariant subspace of M_z. For each $w \in \overline{D}$, since $\{z_i\}_{i=1}^{n} \subset D$, there is a function $f_w \in N$ such that $f_w(w) \neq 0$. Since f_w is continuous in $\overline{D}$, choose a neighborhood $U(w, \varepsilon)$ such that f_w does not equal zero in $U(w, \varepsilon)$. Thus we can find a finite open cover $U(w_1, \varepsilon_1), \cdots, U(w_k, \varepsilon_k)$ of $\overline{D}$ and functions $f_{w_1}, f_{w_2}, \cdots, f_{w_k}$ in N such that f_{w_i} has no zero in $U(w_i, \varepsilon_i)$ $(1 \leq i \leq k)$. Thus $\{f_{w_i}\}_{i=1}^{k}$ have no common zero in $\overline{D}$.

By Lemma 4.5.27, there are $g_1, g_2, \cdots, g_k \in R(D)$ such that

$$\sum_{i=1}^{k} f_{w_i} g_i = e.$$

Note that since polynomials are dense in $R(D)$, each invariant subspace of M_z in fact is an ideal. Thus $e \in N$ and $N = R(D)$. Thus

$$(z - z_1) \cdots (z - z_n) R(D) \subset M.$$

Since $M \subset (z - z_1) \cdots (z - z_n) R(D)$ is obvious, $M = (z - z_1) \cdots (z - z_n) R(D)$.

(ii) Since M_χ^* is a Fredholm and $ker M_\chi = \{0\}$, $M_\chi^* M_\chi$ is invertible. It is obvious that $M_\chi (M_\chi^* M_\chi)^{-1} M_\chi^*$ is a self-adjoint idempotent. Thus it is an orthogonal projection. By Proposition 4.5.7, M_χ^* is a Cowen-Douglas operator of index n, and $ran M_\chi^* = R(D)$. Thus

$$ran[M_\chi (M_\chi^* M_\chi)^{-1} M_\chi^*] = ran M_\chi = \chi R(D) = M,$$

i.e., P_M is the projection onto M.

It is well-known that in Hardy space H^2 M_z similar to the restriction of it on any nontrivial invariant subspace M. The following result indicates that this statement is valid in Sobolev disk algebra if and only if M has only finitely many common zeroes in D.

Proposition 4.5.29   *Let* $M \in LatM_z$, *then* $M_z \sim M_z|_M$ *if and only if*

$$M = (z - z_1) \cdots (z - z_n) R(D),$$

where $\{z_i\}_{i=1}^n \subset D$.

Proof   Assume that $M = (z - z_1) \cdots (z - z_n) R(D)$. Denote

$$p = (z - z_1) \cdots (z - z_n).$$

Define $T_p : R(D) \longrightarrow M$ by $T_p f = pf, f \in R(D)$. It is easy to see that T_p maps $R(D)$ one to one onto M. Since

$$T_p^{-1} M_z|_M T_p f = zf = M_z f, f \in R(D),$$

$$M_z = T_p^{-1} M_z|_M T_p.$$

On the other hand, if there is an $W : R(D) \longrightarrow M$ such that

$$M_z = W^{-1} M_z|_M W,$$

then

$$W M_z = M_z|_M W.$$

Denote $h = We$. Computations indicate that $Wz^n = z^n h$. Thus $Wp = ph = M_h p$ for each polynomial p. By Proposition 4.5.2, $Wf = M_h f$ for all $f \in R(D)$. This implies that $W = M_h$ and M_h has a closed range. By Theorem 4.5.6, $0 \notin h(C)$. Since M_h is essentially normal, h has finitely many zeroes in D. By Theorem 4.5.28, $M = (z - z_1) \cdots (z - z_n) R(D)$ for $\{z_i\}_{i=1}^n \subset D$.

Theorem 4.5.28 and Proposition 4.5.29 describe the structure of invariant subspaces of M_z with finitely many common zeroes in D. The following example is an invariant subspace of infinitely many common zeroes.

Example 4.5.30   *Let* $f(z) = (\prod_{n=1}^{\infty} \frac{a_n - z}{1 - a_n z})(z - 1)^5$, *where* $a_n = 1 - \frac{1}{n^2}$. *Then* $f \in R(D)$.

Proof By a result in [Gardner, B.J. (1989)], $f(z)$ is continuous in $\overline{D}$.

Let $B(z) = \prod_{k=1}^{\infty} b_{a_k}$, where $b_{a_k} = \frac{a_k - z}{1 - a_k z}$.

{Claim} $B'(z) = \sum_{n=1}^{\infty} (\frac{a_n^2 - 1}{(1 - a_n z)^2} \prod_{k \neq n} b_{a_k}(z))$.

Let $G = \{z \in \mathbf{C} : |z| < r < 1\}$, choose r so that $z \in G$. Assume that $a_n > r$ when $n > N$. $B(z) = (\prod_{k=1}^{N} b_{a_k})(\prod_{k=N+1}^{\infty} b_{a_k})$. Thus

$$B'(z) = (\prod_{k=1}^{N} b_{a_k})'(\prod_{k=N+1}^{\infty} b_{a_k}) + (\prod_{k=1}^{N} b_{a_k})(\prod_{k=N+1}^{\infty} b_{a_k})'$$

$$= (\prod_{k=1}^{N} b_{a_k})' B_N(z) + (\prod_{k=1}^{N} b_{a_k}) B_N'(z), \quad \text{where} \quad B_N(z) = \prod_{k=N+1}^{\infty} b_{a_k}.$$

Note that $B_N(z) \neq 0$, $z \in G$. Thus

$$ln(B_N(z)) = \sum_{k=N+1}^{\infty} ln(\frac{a_k - z}{1 - a_k z}).$$

But

$$\frac{B_N'(z)}{B_N(z)} = \sum_{k=N+1}^{\infty} \frac{a_k^2 - 1}{(a_k - z)(1 - a_k z)}$$

converges uniformly in G. Thus

$$B_N'(z) = \sum_{k=N+1}^{\infty} [\frac{a_k^2 - 1}{(1 - a_k z)^2} \prod_{n \neq k, n \geq N+1} b_{a_n}(z)].$$

Note that

$$|B'(z)(z - 1)^5| = |\sum_{k=1}^{\infty} \frac{(a_k^2 - 1)(z - 1)^5}{(1 - a_k z)^2} \prod_{n \neq k} b_{a_n}(z)|$$

$$\leq 16 \sum_{k=1}^{\infty} (1 - a_k)|\frac{1-z}{1-a_k z}|^2$$

$$\leq 64 \sum_{k=1}^{\infty} (1 - a_k) < \infty,$$

thus $B'(z)(z - 1)^5$ is bounded on $\overline{D}$. It is easy to see that $B(z)(z - 1)^4$ is bounded on $\overline{D}$. Therefore $(B(z)(z - 1)^5)'$ is bounded.

Similarly, $(B(z)(z-1)^5)''$ is bounded on $\overline{D}$, these imply $f(z) \in R(D)$.

Denote $M\{a_n\} := \{g \in R(D) : g(a_n) = 0, n = 1, 2, \cdots\}$. It is obvious that $M\{a_n\}$ is an invariant subspace of M_z with infinitely many common zeroes and $f \in M\{a_n\}$.

For a given set M, $[M]$ denotes the minimal invariant subspace of M_z containing M. If $M \neq \{0\}$ is an invariant subspace of M_z on Hardy space H^2, Beurlig Theorem asserts that $dim(M \ominus zM) = 1$ and $[M \ominus zM] = M$. Comparing to this, the invariant subspaces of M_z on Bergman space L_a^2 are more complicated. [Apostal, C., Bercobici, H., Foias, C. and Pearcy, C. (1985)] proved that if n is any positive integer or $+\infty$, there is an invariant subspace M in L_a^2 such that

$$dim(M \ominus zM) = n.$$

For Sobolev disk algebra $R(D)$, we have the following result.

Proposition 4.5.31 *Let $M \neq \{0\}$ is an invariant subspace of M_z in $R(D)$, then $dim(M \ominus zM) = 1$.*

Proof [Richter, S. (1987)] studied a Banach space $\mathcal{B}$ of analytic functions, which satisfies the following conditions:

(i)   For each $\lambda \in D$, the point evaluation functional is continuous;

(ii) If $f \in \mathcal{B}$, then $zf \in \mathcal{B}$;

(iii) If $f \in \mathcal{B}$ and $f(\lambda) = 0$, then $f = (z - \lambda)g$ for some $g \in \mathcal{B}$.

S. Richter also proved that if the above Banach space $\mathcal{B}$ is an algebra and M is its closed ideal, then $dim(M \ominus zM) = 1$.

It is obvious that Sobolev disk algebra and each nonzero invariant subspace M of M_z satisfy these conditions.

Thus $dim(M \ominus zM) = 1$.

4.6   The Operator Weighted Shift

Let $\{W_k\}_{k=1}^\infty$ be a sequence of uniformly bounded operator on $\mathbf{C}^n$. An operator S in $\mathcal{L}(\bigoplus_{k=0}^\infty \mathbf{C}^n)$ is called a unilateral operator weighted shift with weighted sequence $\{W_k\}_{k=1}^\infty$, denoted by $S \sim \{W_k\}_{k=1}^\infty$, if

$$S(x_0, x_1, \cdots, x_k, \cdots) = (0, W_1 x_0, \cdots, W_{k+1} x_k, \cdots)$$

for all $(x_k) \in \mathcal{H}_+ = \overset{+\infty}{\underset{k=0}{\bigoplus}} \mathbf{C}^n$. Similarly, an operator S on $\mathcal{H} = \overset{+\infty}{\underset{k=-\infty}{\bigoplus}} \mathbf{C}^n$ is called a bilateral operator weighted shift with weighted sequence $\{W_k\}_{k=-\infty}^{+\infty}$, denoted by $S \sim \{W_k\}_{k=-\infty}^{+\infty}$, if

$$S(\cdots, x_{-1}, \widehat{x_0}, x_1, \cdots) = (\cdots, W_{-1}x_{-2}, \widehat{W_0 x_{-1}}, W_1 x_0, \cdots)$$

for all $(x_k) \in \mathcal{H}$ In general, unilateral and bilateral operator weighted shifts are both called operator weighted shift, denoted by $S \sim \{W_k\}$, and n is called the multiplicity of S.

When $n = 1$, S is the scalar-valued weighted shift, from which the operator weighted shift is naturally generalized. We must point out here that it is not just a formal generalization. For example, using operator weighted shift, [Pearcy, C. and Petrovic, S. (1994)] proved that an n-normal operator is power bounded if and only if it is similar to a contraction operator.

In this section, we will discuss injective operator weighted shift, that is, each W_k is invertible. Given an operator weighted shift $S \sim \{W_k\}$, since S and $e^{i\theta}S$ are unitarily equivalent for each $\theta \in [0, 2\pi]$ [Lambert, A (1971)], $\sigma(S), \sigma_e(S)$ and $\sigma_r(S)$ have circular symmetry. It is easily seen that

$$r(S) = \lim_{k \to \infty} (\sup_i \|W_{i+k} \cdots W_{i+1}\|)^{\frac{1}{k}},$$

$$r_1(S) = \lim_{k \to \infty} (\inf_i \|W_{i+k} \cdots W_{i+1}\|)^{\frac{1}{k}},$$

where $r(S)$ is the spectral radius of S and

$$r_1(S) = \lim_{k \to \infty} (m(S^k))^{\frac{1}{k}},$$

$$m(S) := \inf\{\|Sx\| : \|x\| = 1\}$$

is the lower bound of S.

Since $dim ker(S - \lambda) \le n < \infty$ and $dim ker(S - \lambda)^* \le n < \infty$ for each $\lambda \in \mathbf{C}$, $\rho_{s-F}(S) = \rho_F(S)$. For unilateral operator weighted shift $S \sim \{W_k\}_{k=1}^{\infty}$, we have

$$\sigma_l(S) = \sigma_\pi(S) = \sigma_e(S) \subset \{\lambda : r_1(S) \le |\lambda| \le r(S)\}.$$

Clearly, $\sigma_p(S) = \emptyset$, thus $\sigma_r(S) = \sigma(S) = \{\lambda : |\lambda| \le r(S)\}$ [Lambert, A (1971)], where $\sigma_\pi(S)$ is the approximate point spectrum of S.

An operator weighted shift $S \sim \{W_k\}$ is said to be upper triangular, if there is an ONB $\{e_i\}_{i=1}^{n}$ of $\mathbf{C}^n$ for each k such that W_k admits an upper triangular matrix representation with respect to this ONB. By the basic matrix theory, we have the following proposition.

Proposition 4.6.1 *Every operator weighted shift* $S \sim \{W_k\}$ *is unitarily equivalent to an upper triangular operator weighted shift* $\widehat{S} \sim \{\widehat{W_k}\}$.

Theorem 4.6.2 *Let* $S \sim \{W_k\}_{k=1}^{\infty}$ *be a unilateral operator weighted shift, if* $\sigma_e(S)$ *is disconnected then* $S \notin (SI)$.

To prove this theorem we need the following lemmas.

Lemma 4.6.3[Herrero, D.A. (1990)] *Let* $A, B \in \mathcal{L}(\mathcal{H})$, *then*

$$\sigma_\pi(AB) = \sigma_l(\tau_{AB}) = \sigma_l(A) - \sigma_r(B).$$

Lemma 4.6.4 *Let* $A \in \mathcal{L}(\mathcal{H}_1)$, $B \in \mathcal{L}(\mathcal{H}_2)$ *and*

$$dim \mathcal{H}_1 = dim \mathcal{H}_2.$$

If $\sigma_l(A) \cap \sigma_r(B) = \emptyset$, *then* $G := \{Y \in ran\tau_{AB} : \|Y\| \leq 1\}$ *is closed in weak operator topology (WOT)*.

Proof Without loss of generality, we assume that $\mathcal{H} = \mathcal{H}_1 = \mathcal{H}_2$. Let $\{Y_\alpha\}_{\alpha \in \Lambda}$ be a net in G and $Y = (WOT) - \lim_\alpha Y_\alpha$. Then there exists $X_\alpha \in \mathcal{L}(\mathcal{H})$ such that $Y_\alpha = \tau_{AB}(X_\alpha)$. By Lemma 4.6.3, $0 \notin \sigma_l(\tau_{AB})$. Thus there exists $T \in \mathcal{L}(\mathcal{L}(\mathcal{H}))$, such that $T\tau_{AB} = I$. Therefore,

$$\|X_\alpha\| = \|T(Y_\alpha)\| \leq \|T\| \|Y_\alpha\| \leq \|T\|$$

for each α. Since each bounded closed set in $\mathcal{L}(\mathcal{H})$ is compact in weak operator topology [Conway, J.B. (1990)], we can find an operator $X \in \mathcal{L}(\mathcal{H})$ such that $X = WOT - \lim_\alpha X_\alpha$. Therefore, for arbitrary $x, y \in \mathcal{H}$,

$$\lim < (\tau_{AB}X_\alpha)x, y > = \lim < AX_\alpha x, y > - \lim < X_\alpha Bx, y >$$

$$= < AXx, y > - < XBx, y > = < (\tau_{AB}X)x, y > .$$

Thus $Y = WOT - \lim \tau_{AB}X_\alpha = \tau_{AB}X$. Clearly $\|Y\| \leq 1$, therefore $Y \in G$.

Lemma 4.6.5 *Let* $A \sim \{A_k\}_{k=1}^{\infty}$ *and* $B \sim \{B_k\}_{k=1}^{\infty}$ *be unilateral operator weighted shifts with multiplicities* n *and* m *respectively. If* $\sigma_l(A) \cap \sigma_r(B) = \emptyset$

and if C is an operator of the form

$$
C = \begin{bmatrix} 0 & & & 0 \\ c_1 & 0 & & \\ & c_2 & 0 & \\ 0 & & c_3 & 0 \\ & & & & \ddots \end{bmatrix} \begin{matrix} \mathbf{C}^m \\ \mathbf{C}^m \\ \mathbf{C}^m \\ \mathbf{C}^m \\ \vdots \end{matrix} ,
$$

$$\mathbf{C}^n \ \ \mathbf{C}^n \ \ \mathbf{C}^n \ \cdots$$

then $C \in ran\tau_{A,B}$.

Proof   Let $M = \sup\limits_{k}\{\|c_k\|\}$,

$$
F_k = \begin{bmatrix} 0 & & & & & 0 \\ c_1 & 0 & & & \\ & c_2 & 0 & & \\ & & \ddots & \ddots & \\ & & & c_k & 0 \\ & & & & 0 & 0 \\ & & & & & \ddots & \ddots \end{bmatrix} \quad \text{and} \quad E_k = \begin{bmatrix} 0 & & & & & 0 \\ 0 & 0 & & & \\ & 0 & 0 & & \\ & & \ddots & \ddots & \\ & & & c_k & 0 \\ & & & & 0 & 0 \\ & & & & & \ddots & \ddots \end{bmatrix}.
$$

Then $\|F_k\| \leq M$, $F_k = \sum\limits_{j=1}^{k} E_j$ and $C = WOT - \lim F_k$. By Lemma 4.6.4 we need only to prove that each $E_k \in ran\tau_{A,B}$. Set

$$
\begin{aligned}
X_{k-1} &= A_k^{-1} c_k, \\
X_{k-2} &= (A_k A_{k-1})^{-1} c_k B_{k-1}, \\
&\vdots \\
X_1 &= (A_k \cdots A_2)^{-1} c_k B_{k-1} \cdots B_2, \\
X_0 &= (A_k \cdots A_1)^{-1} c_k B_{k-1} \cdots B_1.
\end{aligned}
$$

Define

$$
X = \begin{bmatrix} X_0 & & & & 0 \\ & \ddots & & & \\ & & X_{k-1} & & \\ 0 & & & 0 & \\ & & & & \ddots \end{bmatrix} \begin{matrix} \mathbf{C}^m \\ \mathbf{C}^m \\ \mathbf{C}^m \\ \mathbf{C}^m \\ \vdots \end{matrix} .
$$

$$\mathbf{C}^n \ \ \mathbf{C}^n \ \ \mathbf{C}^n \ \ \mathbf{C}^n \ \cdots$$

A straightforward computation shows that $AX - XB = E_k$.

Let $\mathcal{H}_k$ be the k-th subspace in the orthogonal direct sum $\mathcal{H}_+ = \bigoplus_{k=0}^{\infty} \mathbf{C}^n$. Then for each unilateral operator weighted shift $S \sim \{W_k\}_{k=1}^{\infty}$, W_k can be regarded as an invertible operator from $\mathcal{H}_{k-1}$ to $\mathcal{H}_k$. In what follows we will always identify $x \in \mathcal{H}_k$ with $(0, \cdots, 0, x, 0, \cdots) \in \mathcal{H}_+$, the k-th of which is x.

Lemma 4.6.6 *There exists an ONB $\{e_i^{(k)}\}_{i=1}^n$ of $\mathcal{H}_k$ ($k = 0, 1, 2, \cdots$), such that the weighted W_k of S is of the form*

$$W_k = \begin{bmatrix} W_1^{(k)} & & \star \\ & \ddots & \\ & & W_n^{(k)} \\ e_1^{(k)} & \cdots & e_n^{(k)} \end{bmatrix} \begin{matrix} e_1^{(k-1)} \\ \vdots \\ e_n^{(k-1)} \end{matrix} \tag{4.6.1}$$

and $d_i \geq d_{i+1}$ ($i = 1, 2, \cdots, n-1$), where

$$d_i = \overline{\lim_{k \to \infty}} \left(\prod_{j=1}^k |W_i^{(j)}| \right)^{\frac{1}{k}}. \tag{4.6.2}$$

Proof For convenience denote $M_k = W_k W_{k-1} \cdots W_1$ and $M_0 = I$. Choose an ONB $\{e_1^{(0)}, e_2^{(0)}, \cdots, e_n^{(0)}\}$ of $\mathcal{H}_0$. Set

$$d_1 = \max\{\overline{\lim_{k \to \infty}} \|M_k e_i^{(0)}\|^{\frac{1}{k}} : 1 \leq i \leq n\}.$$

Without loss of generality, we may assume that $d_1 = \overline{\lim_{k \to \infty}} \|M_k e_1^{(0)}\|^{\frac{1}{k}}$. Set

$$e_1^{(k)} = \frac{M_k e_1^{(0)}}{\|M_k e_1^{(0)}\|} \in \mathcal{H}_k.$$

Then

$$W_k e_1^{(k-1)} = \frac{\|M_k e_1^{(0)}\|}{\|M_{k-1} e_1^{(0)}\|} e_1^{(k)} := \omega^{(k)} e_1^{(k)} \tag{4.6.3}$$

and $B_1 := \{e_1^{(k)}\}_{k=0}^{\infty}$ is ONB of $\mathcal{H}_+$. Let $P_1^{(k)}$ be the projection from $\mathcal{H}_k$ onto the subspace $[\bigvee\{e_1^{(k)}\}]^{\perp}$ and $P_1^{(0)} = I$. By (4.6.3),

$$P_1^{(k)} W_k P_1^{(k-1)} = P_1^{(k)} W_k \tag{4.6.4}$$

Suppose that we have found an orthogonal set B_j of $\mathcal{H}_+$,

$$B_j = \{e_j^{(k)} : 1 \leq j \leq i, k \geq 0\}$$

and $d_j = \varliminf\limits_{k \to \infty} \|P_{j-1}^{(k)} M_k e_j^{(0)}\|^{\frac{1}{k}}$, where $P_j^{(k)}$ $(1 \leq j \leq i, k \geq 0)$ is the projection from $\mathcal{H}_+$ onto $(\bigvee\{e_1^{(k)}, \cdots, e_j^{(k)}\})^{\perp}$ and $P_j^{(0)} = I$. They satisfy the following properties:

(i) $d_1 \geq d_2 \geq \cdots \geq d_i$;

(ii) $P_{j-1}^{(k)} W_k e_j^{(k-1)} = W_j^{(k)} e_j^{(k)}$, $(j = 1, 2, \cdots, i)$. $\quad$ (4.6.5)

(iii) $P_j^{(k)} W_k P_j^{(k-1)} = P_j^{(k)} W_k$, $(j = 1, 2, \cdots, i)$. $\quad$ (4.6.6)

Without loss of generality, we may assume that

$$d_{i+1} = \varliminf\limits_{k \to \infty} \|P_i^{(k)} M_k e_{i+1}^{(0)}\|^{\frac{1}{k}}.$$

Thus $d_i \geq d_{i+1}$. Set

$$e_{i+1}^{(k)} = \frac{P_i^{(k)} M_k e_{i+1}^{(0)}}{\|P_i^{(k)} M_k e_{i+1}^{(0)}\|} \in \mathcal{H}_k.$$

By (4.6.6),

$$P_i^{(k)} M_k e_{i+1}^{(k-1)} = \frac{P_i^{(k)} W_k P_i^{(k-1)} M_{k-1} e_{i+1}^{(0)}}{\|P_i^{(k-1)} M_{k-1} e_{i+1}^{(0)}\|}$$

$$= \frac{\|P_i^{(k)} M_k e_{i+1}^{(0)}\|}{\|P_i^{(i-1)} M_{k-1} e_{i+1}^{(0)}\|} e_{i+1}^{(k)} \qquad (4.6.7)$$

$$= \omega_{k+1}^{(k)} e_{i+1}^{(k)}.$$

Besides, $B_{i+1} = \{e_j^{(k)} : 1 \leq j \leq i+1, k \geq 0\}$ is still an orthogonal set of $\mathcal{H}_+$. Let $P_{i+1}^{(k)}$ be the projection from $\mathcal{H}_k$ onto the subspace $(\bigvee\{e_1^{(k)}, \cdots, e_{i+1}^{(k)}\})^{\perp}$ and $P_{i+1}^{(0)} = I$. It follows from (4.6.6) and (4.6.7) that $P_{i+1}^{(k)} W_k P_{i+1}^{(k-1)} = P_{i+1}^{(k)} W_k$.

Repeating the procedure above, we can find $\{d_i\}_{i=1}^n$ and ONB $\{e_i^{(k)}\}_{i=1}^n$ of $\mathcal{H}_k$ such that $d_1 \geq d_2 \geq \cdots \geq d_n$ and $\{e_i^{(k)} : 1 \leq i \leq n, k \geq 0\}$ is an ONB of $\mathcal{H}_+$. From the choices of $e_i^{(k)}$ and $w_i^{(k)}$, each w_k has the form (4.6.1) and

$$\varliminf\limits_{k \to \infty} (|\prod_{j=1}^{k} w_i^{(j)}|)^{\frac{1}{k}} = \varliminf\limits_{k \to \infty} \|P_{i-1}^{(k)} M_k e_i^{(0)}\|^{\frac{1}{k}} = d_i (i = 1, 2, \cdots, n).$$

By Proposition 4.6.1, the following lemma can be easily proved.

Lemma 4.6.7 *Let $S \sim \{W_k\}_{k=0}^{\infty}$ be an operator weighted shift, then*

$$
S \cong \begin{bmatrix} S_1 & S_{12} & \cdots & S_{1n} \\ & S_2 & \ddots & \vdots \\ & & \ddots & S_{n-1,n} \\ 0 & & & S_n \end{bmatrix} \begin{matrix} \mathcal{H}_1 \\ \mathcal{H}_2 \\ \vdots \mathcal{H}_n \end{matrix},
$$

where $\mathcal{H}_+ = \mathcal{H}_1^{(n)}$, $S_i \sim \{\omega_i^{(k)}\}_{k=1}^{\infty}$ is an injective unilateral weighted shift and S_{ij} is a scalar-valued weighted shift (not necessarily injective). Without loss of generality, we may assume that $\mathcal{H}_1 = l^2$.

Lemma 4.6.8 *Let $S \sim \{W_k\}$ be an operator weighted shift of the form given in Lemma 4.6.7. Then $\sigma_e(S) = \bigcup_{i=1}^{n} \sigma_e(S_i)$.*

Proof   We only prove the lemma when $n = 2$, and let

$$
S = \begin{bmatrix} S_1 & S_{12} \\ 0 & S_2 \end{bmatrix}.
$$

Let π be the canonical map from $\mathcal{L}(\mathcal{H}_+)$ onto $\mathcal{L}(\mathcal{H}_+)/\mathcal{K}(\mathcal{H}_+)$. Then

$$
\pi(S) = \begin{bmatrix} \pi_1(S_1) & \pi_1(S_{12}) \\ 0 & \pi_1(S_2) \end{bmatrix},
$$

where $\pi_1 : \mathcal{L}(\mathcal{H}_1) \longrightarrow \mathcal{L}(\mathcal{H}_1)/\mathcal{K}(\mathcal{H}_1)$.

We need only to show that $\sigma(\pi(S)) = \sigma(\pi_1(S_1)) \cup \sigma(\pi_1(S_2))$.

By Gelfand-Naimark-Segal Theorem, we need only to prove that

$$
\sigma(S) = \sigma(S_1) \cup \sigma(S_2).
$$

If $\lambda \in \rho(S)$, then $(S - \lambda)^{-1} = (A_{ij})_{2 \times 2}$ satisfying

$$
\begin{bmatrix} S_1 - \lambda & S_{12} \\ 0 & S_2 - \lambda \end{bmatrix} \begin{bmatrix} A_{11} & A_{12} \\ A_{21} & A_{22} \end{bmatrix} = I.
$$

Thus $(S_2 - \lambda)A_{22} = I$. Since $dimker(S_2 - \lambda) \le 1$, by Atkinson Theorem, $(S_2 - \lambda)$ has a left inverse and $\lambda \in \rho(S_2)$. Since $(S_2 - \lambda)A_{21} = 0$, $A_{21} = 0$. Therefore

$$
(S_1 - \lambda)A_{11} = A_{11}(S_1 - \lambda) = I.
$$

This implies that $\lambda \in \rho(S_1)$. Thus $\lambda \in \rho(S_1) \cap \rho(S_2)$.

Conversely, if $\lambda \in \rho(S_1) \cap \rho(S_2)$, set

$$X = \begin{bmatrix} (S_1 - \lambda)^{-1} & -(S_1 - \lambda)^{-1} B(S_2 - \lambda)^{-1} \\ 0 & (S_2 - \lambda)^{-1} \end{bmatrix}.$$

Then $X(S - \lambda) = (S - \lambda)X = I$. This implies $\lambda \in \rho(S)$ and $\sigma(S) = \sigma(S_1) \cup \sigma(S_2)$.

Now we are in a position to prove Theorem 4.6.2.

Proof of Theorem 4.6.2 By Lemma 4.6.8, $\sigma_e(S) = \bigcup_{i=1}^{n} \sigma_e(S_i)$. Since

$$\sigma_e(S) = \sigma_\pi(S) \subset \{\lambda \in \mathbf{C} : r_1(S) \le |\lambda| \le r(S)\} \quad [\text{Lambert, A (1971)}],$$

there exist ε and δ, $0 \le \varepsilon < \delta$, such that $\Omega := \{\lambda : \varepsilon < |\lambda| < \delta\}$ is a bounded component of $\rho_F(S)$. Therefore, for each S_i, at least one of the following holds:

(a) $\sigma(S_i) \subset U_\varepsilon := \{\lambda : |\lambda| \le \varepsilon\}$;
(b) $\sigma_e(S_i) \subset V_\delta := \{\lambda : |\lambda| \ge \delta\}$.

Denote $i_0 = max\{i : \sigma_e(S_i) \subset V_\delta\}$, then by the definitions of $r(S)$ and $r_1(S)$ and Lemma 4.6.6, $r_1(S_{i_0}) \le d_{i_0} \le d_i \le r(S_i)$, if $i \le i_0$. Therefore S_i does not satisfy (a) and $\sigma_e(S_i) \subset V_\delta$ if $i \le i_0$; but if $i > i_0$, S_i does not satisfy (b), thus $\sigma(S_i) \subset U_\varepsilon$. Note that $\{\lambda : |\lambda| = \varepsilon \text{ or } \delta\} \subset \sigma_e(S)$. We have $1 \le i_0 < n$.

Now denote $\mathcal{M}_1 = \bigoplus_{i=1}^{i_0} \mathcal{H}_i$ and $\mathcal{M}_2 = \mathcal{M}_1^\perp = \bigoplus_{i=i_0+1}^{n} \mathcal{H}_i$. Then

$$S = \begin{bmatrix} A & C \\ 0 & B \end{bmatrix} \begin{matrix} \mathcal{M}_1 \\ \mathcal{M}_2 \end{matrix},$$

where A and B are operator weighted shifts with multiplicities i_0 and $n - i_0$ respectively, and C is of the form given in Lemma 4.6.5. By Lemma 4.6.8,

$$\sigma_e(A) = \bigcup_{i=1}^{i_0} \sigma_e(S_i) \subset V_\delta.$$

By Lemma 4.6.8 and its proof $\sigma_r(A) = \sigma(B) = \bigcup_{i=i_0+1}^{n} \sigma(S_i) \subset U_\varepsilon$. Therefore

$$\sigma_e(A) \cap \sigma_r(B) = \emptyset.$$

By Lemma 4.6.5, there exists an operator X such that

$$AX - XB = C.$$

Set

$$Y = \begin{bmatrix} I & X \\ 0 & I \end{bmatrix}.$$

Then

$$YSY^{-1} = \begin{bmatrix} A & 0 \\ 0 & B \end{bmatrix}$$

and $S \notin (SI)$.

Next we will discuss when the adjoint of a unilateral operator weighted shift is a Cowen-Douglas operator.

Proposition 4.6.9 *Let* $S \sim \{W_k\}_{k=1}^{\infty}$ *be a unilateral operator weighted shift. If* $0 \in \sigma_e(S)$, *then* S^* *can not be a Cowen-Douglas operator.*
Proof If $\sigma_e(S) = \sigma(S)$, then S^* is not a Cowen-Douglas operator. Thus we may assume that $\sigma_e(S) \neq \sigma(S)$. Since $0 \in \sigma_e(S)$, $\sigma_e(S)$ is disconnected. Let Ω be a connected component of $\rho_F(S) \cap \sigma(S)$. Repeating the proof of Theorem 4.6.2, we can choose Ω satisfying $\Omega = \{\lambda : \varepsilon < |\lambda| < \delta\}$, where $0 \leq \varepsilon < \delta$. Therefore, there exists an invertible operator Y such that

$$YSY^{-1} = \begin{bmatrix} A & 0 \\ 0 & B \end{bmatrix} \begin{matrix} \mathcal{M}_1 \\ \mathcal{M}_2 \end{matrix} = D.$$

So $D^* = (Y^*)^{-1} S^* Y^* = \begin{bmatrix} A^* & 0 \\ 0 & B^* \end{bmatrix} \begin{matrix} \mathcal{M}_1 \\ \mathcal{M}_2 \end{matrix}$. Since $\sigma(B^*) = \sigma(B) \subset \{\lambda : |\lambda| \leq \varepsilon\}$
(see the proof of Theorem 4.6.2), $B^* - \lambda$ is invertible for all $\lambda \in \Omega$. Thus

$$\ker(D^* - \lambda) \subset \mathcal{M}_1.$$

This implies that $\bigvee \{\ker(S^* - \lambda) : \lambda \in \Omega\} \neq \mathcal{H}_+$.

Theorem 4.6.10 *Let* $S \sim \{W_k\}_{k=1}^{\infty}$ *be a unilateral operator weighted shift. Then the following are equivalent:*
  (i) $M := \sup_k \|W_k^{-1}\| < \infty$;
  (ii) $r_1(S) > 0$;
  (iii) $0 \in \rho_F(S)$;
  (iv) *There exists a connected open set* Ω *containing* 0 *such that* $S^* \in \mathcal{B}_n(\Omega)$.
Proof (i)$\Rightarrow$(ii) and (ii)$\Rightarrow$(iii) are obvious from the definitions.

(iii)$\Rightarrow$(iv). It is easy to see that there exists a connected open set Ω containing 0 and $\Omega \subset \rho_F(S)$. Since $\sigma_p(S) = \emptyset$, $S^* - \lambda$ is surjective and

$$dimker(S^* - \lambda) = ind(S^* - \lambda) = indS^* = n$$

for all $\lambda \in \Omega$. Note that $\bigvee \{ker(S^*)^k : k \geq 1\} = \mathcal{H}_+$, thus $S^* \in \mathcal{B}_n(\Omega)$.

(iv)$\Rightarrow$(i).   Since S^* is surjective, it has a right inverse $T = (T_{k,l})_{k,l=0}^{\infty} \in \mathcal{L}(\mathcal{H}_+)$. Thus $W_k^* T_{k,k-1} = I$ and $\|W_k^{-1}\| = \|T_{k,k-1}\| \leq \|T\|$ for each k.

By Proposition 4.6.9 and Theorem 4.6.10, the following corollary is obtained.

Corollary 4.6.11 *Let $S \sim \{W_k\}_{k=1}^{\infty}$ be a unilateral operator weighted shift. Then B^* is a Cowen-Douglas operator if and only if $\sup_k \|W_k^{-1}\| < +\infty$.*

In what follows we consider the backward operator weighted shift $S \sim \{W_k\}$ with the form

$$S = \begin{bmatrix} 0 & W_1 & & 0 \\ & 0 & W_2 & \\ & & 0 & \ddots \\ 0 & & & \ddots \end{bmatrix} \begin{matrix} \mathbf{C}^n \\ \mathbf{C}^n \\ \mathbf{C}^n \\ \vdots \end{matrix} \quad \text{on} \quad \mathcal{H} = \bigoplus_{n=1}^{\infty} \mathbf{C}^n,$$

and denote $\mathcal{A}_0(S) = \mathcal{A}'(S)|_{kerS}$.

Theorem 4.6.12 *Let $S \sim \{W_k\}$ be a backward operator weighted shift, then the following are equivalent:*
(i) There exists $A_0 \in \mathcal{A}_0(S)$ such that $card(\sigma(A_0)) = m$, $(m \leq n)$;
(ii) $S = S_1 \dot{+} S_2 \dot{+} \cdots \dot{+} S_m$ and for each i, $1 \leq i \leq m$, there exists a unitary operator U_i such that $U_i S_i U_i^* = \overline{S}_i$ is an operator weighted shift with multiplicity n_i and $\sum_{i=1}^{m} n_i = n$;
(iii) S has a spectral family.
Proof   (i)$\Rightarrow$(ii).   For convenience, we assume that $m = 2$. Since $A_0 \in \mathcal{A}_0(S)$, there is $A \in \mathcal{A}'(S)$ such that $A|_{kerS} = A_0$. Let $A_\alpha = diag(A_0, A_1, \cdots, A_k, \cdots)$ be the diagonal part of A and let $p(z)$ be the characteristic polynomial of A_0. Since $A_k = E_k^{-1} A_0 E_k$, $p(A_\alpha) = 0$, where $E_k = W_1 \cdots W_k$ $(k \geq 1)$. Thus $p(\sigma(A_\alpha)) = \{0\}$. This implies that $\sigma(A_0) = \{\lambda_1, \lambda_2\}$. Thus there are Jordan domains Ω_1 and Ω_2 such that $\lambda_1 \in \Omega_1, \lambda_2 \in \Omega_2$ and $\overline{\Omega_1} \cap \overline{\Omega_2} = \emptyset$. Denote $\Gamma_i = \partial \Omega_i$ $(i = 1, 2)$. Clearly,

$A_d \in \mathcal{A}'(S)$. Thus the Riesz idempotents

$$P_i = \frac{1}{2\pi i} \int_{\Gamma_i} (\lambda - A_d)^{-1} d\lambda \in \mathcal{A}'(S) \ (i = 1, 2).$$

Note that

$$(\lambda - A_d)^{-1} = diag((\lambda - A_0)^{-1}, \cdots, E_k^{-1}(\lambda - A_0)^{-1}E_k, \cdots).$$

Thus $P_i = diag(P_0^{(i)}, \cdots, P_k^{(i)}, \cdots)$, where

$$P_k^{(i)} = E_k^{-1}(\frac{1}{2\pi i} \int_{\Gamma_i} (\lambda - A_0)^{-1} d\lambda)E_k = E_k^{-1}P_0^{(i)}E_k, (i = 1, 2)$$

and

$$P_k^{(1)}P_k^{(2)} = 0, \quad P_k^{(1)} + P_k^{(2)} = I_{\mathbf{C}^n}, \ (k = 0, 1, 2, \cdots).$$

Denote $\mathcal{M}_k = P_k^{(1)}\mathbf{C}^n$, $\mathcal{N}_k = P_k^{(2)}\mathbf{C}^n$, $\mathcal{H}_i = P_i\mathcal{H}$, $(i = 1, 2)$. Then

$$\mathcal{H} = \bigoplus_{k=0}^{\infty} (\mathcal{M}_k \dotplus \mathcal{N}_k) = \mathcal{H}_1 \dotplus \mathcal{H}_2,$$

and $S = S_1 \dotplus S_2$, where $S_1 = S|_{\mathcal{H}_1}, S_2 = S|_{\mathcal{H}_2}$.

Since

$$P_k^{(i)} = W_k^{-1}P_{k-1}^{(i)}W_k,$$

$$dim\mathcal{M}_k = dim\mathcal{M}_{k-1} = n_1$$

and

$$dim\mathcal{N}_k = dim\mathcal{N}_{k-1} = n_2 \ (k \geq 1).$$

Therefore

$$W_k = W_k^{(1)} \dotplus W_k^{(2)},$$

where $W_k^{(1)} \in \mathcal{L}(\mathcal{M}_k, \mathcal{M}_{k-1}), W_k^{(2)} \in \mathcal{L}(\mathcal{N}_k, \mathcal{N}_{k-1})$.

Define the unitary operator $U_1 \in \mathcal{L}(\mathcal{H}_1, \bigoplus_{k=0}^{\infty} \mathcal{M}_k)$ as follows

$$U_1 : ((y_0, 0), (y_1, 0), \cdots, (y_k, 0), \cdots) \mapsto (y_0, y_1, \cdots, y_k, \cdots).$$

Then $U_1 S_1 U_1^* = \overline{S_1}$ is an operator weighted shift with weight sequence $\{W_k^{(1)}\}_{k=1}^{\infty}$ and multiplicity n_1. Similarly, we can define a unitary operator

$U_2 \in \mathcal{L}(\mathcal{H}_2, \bigoplus_{k=0}^{\infty} \mathcal{N}_k)$ such that $U_2 S_2 U_2^* = \overline{S_2}$ satisfying the requirements of the theorem.

(ii)$\Rightarrow$(iii). Obvious.

(iii)$\Rightarrow$(i). We still assume that $m = 2$. Since $\{Q_1, Q_2\}$ is a spectral family of S, $Q_1, Q_2 \in \mathcal{A}'(S)$. Let

$$Q_d^{(i)} = diag(Q_0^{(i)}, \cdots, Q_k^{(i)}, \cdots)$$

be the diagonal of Q_i ($i = 1, 2$). Clearly, $\{Q_d^{(1)}, Q_d^{(2)}\}$ is also a spectral family of S. Set $S_i = S|_{ran Q_d^{(i)}}$ ($i = 1, 2$). Then $S = S_1 \dot{+} S_2$. Repeating the argument of (i)$\Rightarrow$(ii), we can prove that $W_k = W_k^{(1)} \dot{+} W_k^{(2)}$, where $W_k^{(1)} \in \mathcal{L}(\mathcal{M}_k, \mathcal{M}_{k-1})$, $W_k^{(2)} \in \mathcal{L}(\mathcal{N}_k, \mathcal{N}_{k-1})$ and $dim \mathcal{M}_k = dim \mathcal{M}_{k-1}, dim \mathcal{N}_k = dim \mathcal{N}_{k-1}$ for all $k \geq 1$. Choose $\mu_1, \mu_2 \in \mathbf{C}$ such that $|\mu_1| < |\mu_2|$.

Set $A_0 = \mu_1 I_{\mathcal{M}_0} \dot{+} \mu_2 I_{\mathcal{N}_0}$ and $A_k = E_k^{-1} A_0 E_k = \mu_1 I_{\mathcal{M}_k} \dot{+} \mu_2 I_{\mathcal{N}_k}$, $k \geq 1$. Then

$$\|A_k\| \leq |\mu_2|(\|Q_1\| + \|Q_2\|).$$

Thus

$$A := diag(A_0, A_1, \cdots, A_k, \cdots) \in \mathcal{A}'(S)$$

and

$$A_0 = A|_{ker S}, \sigma(A_0) = \{\mu_1, \mu_2\},$$

i.e., $card(\sigma(A_0)) = 2$.

Theorem 4.6.13 *Let $S \sim \{W_k\}$ be a backward operator weighted shift and let $r = \max\{card(\sigma(A_0)) : A_0 \in \mathcal{A}_0(S)\}$. Then $S \sim \overline{S} = \bigoplus_{i=1}^{r} \overline{S_i}$, where each $\overline{S_i} \in (SI)$ is an operator weighted shift with multiplicity n_i and $\sum_{i=1}^{r} n_i = n$.*

Corollary 4.6.14 *Let $S \sim \{W_k\}$ be a backward operator weighted shift and $A_0 \in \mathcal{A}_0(S)$, then $\sigma(A_0) = \sigma_{\mathcal{A}_0(S)}(A_0)$.*

Proof Clearly, $\sigma(A_0) \subset \sigma_{\mathcal{A}_0(S)}(A_0)$.

Assume that $A_d = diag(A_0, A_1, \cdots, A_k, \cdots)$, where $A_k = E_k^{-1} A_0 E_k$ ($k \geq 1$). By the preceding discussion, $\sigma(A_0) = \sigma(A_d)$ and $A_d \in \mathcal{A}'(S)$. If $\lambda \notin \sigma(A_0)$, $A_d B = B A_d = I$ for some $B \in \mathcal{L}(\mathcal{H})$ and $B \in \mathcal{A}'(S)$. This implies that $B_0 = B|_{ker S} \in \mathcal{A}_0(S)$ and $A_0 B_0 = B_0 A_0 = I_{\mathbf{C}^n}$. Therefore $\sigma_{\mathcal{A}_0(S)}(A_0) \subset \sigma(A_0)$.

Theorem 4.6.15 *Let $S \sim \{W_k\}$ be an operator weighted shift, then the following are equivalent:*

  (i) $S \in (SI)$;
  (ii) $\sigma(A_0)$ *contains only one point for each* $A_0 \in \mathcal{A}_0(S)$;
  (iii) $\mathcal{A}_0(S)/rad\mathcal{A}_0(S) \cong \mathbf{C}$;
  (iv) *There is no nontrivial idempotent in* $\mathcal{A}_0(S)$.

Proof (i)$\Rightarrow$(ii).   It is a straightforward conclusion of Theorem 4.6.12.

(ii)$\Rightarrow$(iii).   Let $\mathcal{J} = \{A_0 \in \mathcal{A}_0(S) : \sigma(A_0) = \{0\}\}$. Clearly $\mathcal{J} = rad\mathcal{A}_0(S)$. Define a linear mapping $\varphi : \mathcal{A}_0(S)/rad\mathcal{A}_0(S) \longrightarrow \mathbf{C}$ by

$$\varphi[A_0] = \lambda_{A_0}, [A_0] \in \mathcal{A}_0(S)/rad\mathcal{A}_0(S), \text{ where } \lambda_{A_0} \in \sigma(A_0).$$

Note that $\sigma(A_0)$ consists of one point and for each $B_0 \in [A_0]$, $A_0 - B_0 \in rad\mathcal{A}_0(S)$. Thus $\sigma(A_0 - B_0) = \{0\}$. Assume that $\sigma(A_0) = \{\lambda\}$, $\sigma(B_0) = \{\mu\}$, then

$$0 = tr(A_0 - B_0) = trA_0 - trB_0 = n\lambda - n\mu$$

and $\lambda = \nu$. This implies that φ is well defined.

If $[A_0] \neq [B_0]$, then $\sigma(A_0 - B_0) = \{\gamma\}, \gamma \neq 0$, and

$$n\gamma = tr(A_0 - B_0) = trA_0 - trB_0 = n\gamma \neq 0.$$

This means that φ is injective. Clearly φ is surjective. Since

$$[A][B] - \lambda_A\lambda_B = [AB - \lambda_A B + \lambda_A B - \lambda_A\lambda_B] = [A - \lambda_A][B] + \lambda_A[B - \lambda_B],$$

it follows from $\sigma((A - \lambda_A)B) = \sigma(\lambda_A(B - \lambda_B)) = \{0\}$ that $\varphi([AB] - \lambda_A\lambda_B) = 0$. This implies that $\varphi([AB]) = \varphi([A])\varphi([B]) = \lambda_A\lambda_B$ and φ is an isomorphism.

(iii)$\Rightarrow$(iv).   Let φ be an isomorphism from $\mathcal{A}_0(S)/rad\mathcal{A}_0(S)$ to $\mathbf{C}$. Then

$$\varphi([P_0]) = \lambda_0$$

for all $P_0 \in \mathcal{A}_0(S)$, where $\lambda_0 = 0$ or $\lambda_0 = 1$. Since

$$\varphi([P_0 - \lambda I_{\mathbf{C}^n}]) = 0,$$

$$\sigma(P_0 - \lambda) = \{0\}.$$

Therefore, $0 = tr(P_0 - \lambda) = trP_0 - n\lambda$. This implies that $P_0 = 0$ or $P_0 = I_{\mathbf{C}^n}$.

(iv)$\Rightarrow$(i).   Assume that $P \in \mathcal{A}'(S)$ is an idempotent, then $P_0 = P|_{kerS} \in \mathcal{A}_0(S)$ is also an idempotent. Thus $P_0 = 0$ or $P_0 = I_{\mathbf{C}^n}$. Without loss of generality, we assume that $P_0 = 0$. From $PS = SP$,

$$
P = \begin{bmatrix} 0 & P_{01} & & \star \\ & 0 & P_{12} & \\ & & 0 & \ddots \\ & & & \ddots \end{bmatrix} \begin{array}{l} \mathbf{C}^n \\ \mathbf{C}^n \\ \mathbf{C}^n \\ \vdots \end{array} .
$$

Since $P^2 = P, P = 0$. Thus $S \in (SI)$.

Theorem 4.6.16 *Let $S \sim \{W_k\}$ be an operator weighted shift, then $S \in (SI)$ if and only if $\mathcal{J} := \{A \in \mathcal{A}'(S) : \sigma(A|_{kerS}) = \{0\}\}$ is a maximal two sided ideal of $\mathcal{A}'(S)$ and $\mathcal{A}'(S)/\mathcal{J}$ is abelian.*
Proof Suppose that $S \in (SI)$.
{Claim a} $\mathcal{J}$ is a linear space. For $A, B \in \mathcal{J}$,

$$
(A + B)|_{kerS} = A|_{kerS} + B|_{kerS} = A_0 + B_0.
$$

By Theorem 4.6.15, $\sigma(A_0 + B_0) = \{\lambda\}$. Thus

$$
n\lambda = tr(A_0 + B_0) = trA_0 + trB_0 = 0.
$$

This implies that $\lambda = 0$ and $A + B \in \mathcal{J}$.
{Claim b} $\mathcal{J}$ is a closed two-sided ideal. Note that for $A \in \mathcal{J}$ and $B \in \mathcal{A}'(S)$, $AB|_{kerS} = (A|_{kerS})(B|_{kerS}) \in \mathcal{A}_0(S)$. Thus $\sigma(AB|_{kerS}) = \{0\}$ and Claim b holds.
{Claim c} $\mathcal{J}$ is a maximal ideal. Suppose that $\mathcal{J}'$ is a two-sided ideal of $\mathcal{A}'(S)$ satisfying $\mathcal{J}' \supset \mathcal{J}$ and $\mathcal{J}' \neq \mathcal{J}$. Choose $A \in \mathcal{J}' \backslash \mathcal{J}$, then A_d, the diagonal of A with respect to $\mathcal{H} = \bigoplus_{k=0}^{\infty} \mathbf{C}^n$, is in $\mathcal{A}'(S)$. Since $A \notin \mathcal{J}$, $\sigma(A_\alpha) = \sigma(A|_{kerS}) = \{\lambda\}, \lambda \neq 0$. This means that A_d is invertible. Denote $A_r = A - A_d$. Then $A_r \in \mathcal{J}$. Thus $A_d = A - A_r \in \mathcal{J}'$. This implies that $\mathcal{J}' = \mathcal{A}'(S)$. Therefore $\mathcal{J}$ is maximal.
{Claim d} $\mathcal{A}'(S)/\mathcal{J}$ is commutative.
    For $A, B \in \mathcal{A}'(S)$, Set $C = AB - BA$. Then

$$
C_0 = C|_{kerS} = (AB - BA)|_{kerS} = A_0 B_0 - B_0 A_0.
$$

By Theorem 4.6.15, $\sigma(C_0) = \{\mu\}$. Thus

$$
n\mu = trC_0 = tr(A_0 B_0 - B_0 A_0) = tr(A_0 B_0) - tr(B_0 A_0) = 0
$$

and $\mu = 0$. Therefore $C \in \mathcal{J}$ and $\mathcal{A}'(S)/\mathcal{J}$ is commutative.

Conversely, suppose that $\mathcal{J}$ is a maximal two-sided ideal of $\mathcal{A}'(S)$ and $\mathcal{A}'(S)/\mathcal{J}$ is abelian. If P is a nontrivial idempotent in $\mathcal{A}'(S)$.

Set $\mathcal{J}' = \mathcal{A}'(S)P\mathcal{A}'(S) + \mathcal{J}$. It is not difficult to see that $\mathcal{J}'$ is a two-sided ideal of $\mathcal{A}'(S)$ and $\mathcal{J}' \supset \mathcal{J}$. Since $P_0 = P|_{kerS} \neq 0, P \notin \mathcal{J}$ and $\mathcal{J}' \neq \mathcal{J}'$. We assert that the identity $I \notin \mathcal{J}'$. Otherwise $I = APB + C$, where $A, B \in \mathcal{A}'(S)$ and $C \in \mathcal{J}$. Thus $I_{\mathbf{C}^n} = I|_{kerS} = A_0 P_0 B_0 + C_0$, here $A_0 = A|_{kerS}, B_0 = B|_{kerS}$ and $C_0 = C|_{kerS}$. Since $C_0 \in \mathcal{J}|_{kerS}, \sigma(C_0) = \{0\}$. This means that

$$I_{\mathbf{C}^n} - C_0 = A_0 P_0 B_0$$

is invertible. But the determinant $det P_0$ of P_0 is zero ($P_0 \neq 0$). Thus

$$det A_0 P_0 B_0 = (det A_0)(det P_0)(det B_0) = 0.$$

A contradiction. Therefore $\mathcal{J}' \supset \mathcal{J}$ (but $\mathcal{J}' \neq \mathcal{J}$) is also a maximal proper ideal. This contradicts the assumption that $\mathcal{J}$ is a maximal two-sided ideal.

Corollary 4.6.17 *Let* $S \sim \{W_k\}$ *be an operator weighted shift satisfying that* $W_k = W$ *for all* $k \geq 1$. *Then* $S \in (SI)$ *if and only if* $W \in (SI)$.
Proof By Jordan Theorem, there is an invertible matrix $X \in M_n(\mathbf{C})$ such that

$$J = XWX^{-1} = \begin{bmatrix} J_{n_1}(\lambda_1) & & 0 \\ & \ddots & \\ 0 & & J_{n_l}(\lambda_l) \end{bmatrix}$$

and $\sum\limits_{i=1}^{l} n_i = n$.

Set $Y = diag(X, X, \cdots)$, then Y is invertible, and $\overline{S} = YSY^{-1} \sim \{\overline{W_k}\}$ is an operator weighted shift, where $\overline{W_k} = J$. By Theorem 4.6.15, $S \in (SI)$ if and only if $l = 1$ or if and only if $W \in (SI)$.

The following example indicates that the condition $W_k = W$ ($k \geq 1$) can not be omitted.

Example 4.6.18 *Let*

$$W_{2k+1} = \begin{bmatrix} 1 & 1 \\ 0 & 1 \end{bmatrix}, \quad W_{2k} = \begin{bmatrix} 1 & -1 \\ 0 & 1 \end{bmatrix},$$

then

$$W_k \sim \begin{bmatrix} 1 & 1 \\ 0 & 1 \end{bmatrix} \in (SI).$$

Let S be the operator weighted shift with weighted $\{W_k\}$. Set

$$P_{2k+1} = \begin{bmatrix} 1 & 0 \\ 0 & 0 \end{bmatrix}, \quad P_{2k} = \begin{bmatrix} 1 & 1 \\ 0 & 0 \end{bmatrix}$$

and $P = diag(P_1, P_2, P_3, \cdots)$. A simple computation shows that $PS = SP$ and $P^2 = P$, i.e., $S \notin (SI)$.

Now we will discuss the similarity of two operator weighted shifts.

Proposition 4.6.19 *Let $S \sim \{W_k\}$ and $T \sim \{V_k\}$ be two operator weighted shifts. Then $S \sim T$ if and only if there is a sequence of invertible matrices $\{X_i\}$ such that $\sup_i \{\|X_i\|, \|X_i^{-1}\|\} < \infty$ and $W_i = X_i V_i X_{i+1}^{-1}, (i \geq 1)$.*

Proof If there are invertible operators $\{X_i\}_{i=1}^{\infty}$ on $\mathbf{C}^n$ such that

$$\sup_i \{\|X_i\|, \|X_i^{-1}\|\} < \infty$$

and

$$W_i = X_i V_i X_{i+1}^{-1}, (i \geq 1).$$

Set $X = diag(X_1, X_2, \cdots)$. Then X is invertible and $SX = XT$, i.e., $S \sim T$. On the other hand, if $S \sim T$ or $SX = XT$ and $X^{-1}S = TX^{-1}$ for some operator X. Note that X can be expressed as

$$X = \begin{bmatrix} X_1 & X_{12} & X_{13} & \cdots \\ & X_2 & X_{23} & \cdots \\ & & X_3 & \ddots \\ 0 & & & \ddots \end{bmatrix},$$

where $W_i X_{i+1} = X_i V_i$ and $\sup_i \{\|X_i\|\} < \infty$. Since X^{-1} has the form

$$X^{-1} = \begin{bmatrix} Y_1 & Y_{12} & Y_{13} & \cdots \\ & X_2 & Y_{23} & \cdots \\ & & Y_3 & \ddots \\ 0 & & & \ddots \end{bmatrix},$$

where $Y_i W_i = V_i Y_{i+1}$ and $\sup_i \{\|Y_i\|\} < \infty$. Since

$$XX^{-1} = X^{-1}X = I,$$

$$X_i Y_i = Y_i X_i = I_{\mathbf{C}^n}, \ (i = 1, 2, \cdots).$$

Corollary 4.6.20 *Let* $S \sim \{W_k\}_{k=1}^{\infty}, T \sim \{W_k\}_{k=2}^{\infty}$ *and* $S \in \mathcal{B}_n(\Omega)$, *then* $S \sim T$.

Proof By Corollary 4.6.11, $S \in \mathcal{B}_n(\Omega)$ implies that

$$\sup_i \{\|W_i\|, \|W_i^{-1}\|\} < \infty.$$

Set $X_i = W_i$, then $W_i = X_i W_{i+1} X_{i+1}^{-1}$. By Proposition 4.6.19, $S \sim T$.

Proposition 4.6.21 *Let* $S \sim \{W_k\}$ *and* $\overline{S} \sim \{V_k\}$ *be two operator weighted shifts, then* $S \sim \overline{S}$ *if and only if* $S \sim_{q.s} \overline{S}$, *where "$\sim_{q.s}$" means quasisimilar.*
Proof We need only to verify that $S \sim_{q.s} \overline{S}$ implies $S \sim \overline{S}$. By the quasisimilarity, we can assume that S and $\overline{S}$ have the same multiplicity n and there exist $X, Y \in \mathcal{L}(\mathcal{H})$ with trivial kernels and dense ranges such that $SX = X\overline{S}$ and $YS = \overline{S}Y$. Then X and Y have the form

$$X = \begin{bmatrix} X_1 & & \star \\ & X_2 & \\ 0 & & \ddots \end{bmatrix}, \quad Y = \begin{bmatrix} Y_1 & & \star \\ & Y_2 & \\ 0 & & \ddots \end{bmatrix}.$$

Since $ker X_1 \subset ker X = \{0\}, X_1$ is invertible. If follows from $SXY = X\overline{S}Y + XYS$ that $XY \in \mathcal{A}'(S)$ and

$$XY = \begin{bmatrix} X_1 Y_1 & & \star \\ & X_2 Y_2 & \\ 0 & & \ddots \end{bmatrix}.$$

Thus $diag(X_1 Y_1, X_2 Y_2, \cdots) \in \mathcal{A}'(A)$. From $SX = X\overline{S}$, we have

$$X_2 = W_1^{-1} X_1 V_1, \cdots, X_{k+1} = W_k^{-1} X_k V_k, \ (k \geq 1).$$

This implies that $X_k \ (k \geq 1)$ is invertible. Similarly, $Y_k \ (k \geq 1)$ is invertible. Thus $X_k Y_k \ (k \geq 1)$ and $diag(X_1 Y_1, X_2 Y_2, \cdots)$ are invertible, and

$$[diag(X_1 Y_1, X_2 Y_2, \cdots)]^{-1} = diag(Y_1^{-1} X_1^{-1}, Y_2^{-1} X_2^{-1}, \cdots).$$

Therefore, there exists $M > 0$ such that

$$\sup_{k}\{\|Y_k^{-1}X_k^{-1}\|\} < M.$$

Since $diag(X_1, X_2, \cdots) \in ker\tau_{S,\overline{S}}$, there exists $N > 0$ such that $\sup_{k}\{\|X_k\|\} < N$. Furthermore,

$$\sup_{k}\{\|Y_k^{-1}\|\} \le \sup_{k}\{\|Y_k^{-1}X_k^{-1}\| \cdot \|X_k\|\} \le \sup_{k}\{\|Y_k^{-1}X_k^{-1}\|\} \cdot \sup_{k}\{\|X_k\|\} \le MN.$$

This implies that $diag(Y_1, Y_2, \cdots)$ is invertible and $diag(Y_1, Y_2, \cdots) \in ker\tau_{\overline{S},S}$. Thus $S \sim \overline{S}$.

Proposition 4.6.22 *Let $S \sim \{W_k\}$ and $\overline{S} \sim \{V_k\}$ be two operators weighted shifts with multiplicity n. $S, \overline{S} \in (SI)$ and S is not similar to $\overline{S}$. $T = S \oplus \overline{S}$. Then $\mathcal{A}_0(T)/rad\mathcal{A}_0(T) \cong \mathbf{C} \oplus \mathbf{C}$.*

Proof Note that

$$\mathcal{A}'(T) = \left\{ \begin{bmatrix} X_S & X_{12} \\ X_{21} & X_{\overline{S}} \end{bmatrix} : \begin{array}{cc} X_S \in \mathcal{A}'(S) & X_{\overline{S}} \in \mathcal{A}'(\overline{S}) \\ X_{12} \in ker\tau_{S,\overline{S}} & X_{21} \in ker\tau_{\overline{S},S} \end{array} \right\}.$$

For arbitrary $X_{12} \in ker\tau_{S,\overline{S}}$ and $X_{21} \in ker\tau_{\overline{S},S}$, we have

$$X_{12} = \begin{bmatrix} X_1 & & \star \\ & X_2 & \\ 0 & & \ddots \end{bmatrix}, \quad X_{21} = \begin{bmatrix} Y_1 & & \star \\ & Y_2 & \\ 0 & & \ddots \end{bmatrix}.$$

Computations indicate that

$$X_{12}X_{21} \in \mathcal{A}'(S), X_{21}X_{12} \in \mathcal{A}'(\overline{S})$$

and

$$X_{12}X_{21}|_{kerS} = X_1Y_1, X_{21}X_{12}|_{ker\overline{S}} = Y_1X_1.$$

Since $S, \overline{S} \in (SI)$, it follows from Theorem 4.6.15 that

$$\sigma(X_1Y_1) = \sigma(Y_1X_1) = \{\lambda\}.$$

{Claim} $\lambda = 0$.

We assume that X_1 and X_2 are invertible. Set $X' = diag(X_1, X_2, \cdots)$, then $kerX' = \{0\}$ and $ranX' = \mathcal{H}$. Similarly, if Y' denotes $Y' = diag(Y_1, Y_2, \cdots)$ then $kerY' = \{0\}$ and $ranY' = \mathcal{H}$. Since $X' \in ker\tau_{S,\overline{S}}$ and $Y' \in ker\tau_{\overline{S},S}$, $S \sim \overline{S}$. A contradiction.

For each

$$\begin{bmatrix} E_{11} & E_{12} \\ E_{21} & E_{22} \end{bmatrix} \in \mathcal{A}_0(T)$$

and $X \in ker\tau_{S,\overline{S}}$, let

$$D = \begin{bmatrix} E_{11} & E_{12} \\ E_{21} & E_{22} \end{bmatrix} \begin{bmatrix} 0 & X|_{ker\overline{S}} \\ 0 & 0 \end{bmatrix} = \begin{bmatrix} 0 & E_{11}X|_{ker\overline{S}} \\ 0 & E_{21}X|_{ker\overline{S}} \end{bmatrix}.$$

By the claim above, $\sigma(E_{21}X|_{ker\overline{S}}) = \{0\}$. Thus $\sigma(D) = \{0\}$. This implies that

$$\begin{bmatrix} 0 & X|_{ker\overline{S}} \\ 0 & 0 \end{bmatrix} \in rad\mathcal{A}_0(T).$$

Repeating the argument above, if $Y \in ker\tau_{\overline{S},S}$, then

$$\begin{bmatrix} 0 & 0 \\ Y|_{kerS} & 0 \end{bmatrix} \in rad\mathcal{A}_0(T).$$

Thus

$$\mathcal{A}_0(T)/rad\mathcal{A}_0(T) = \left\{ \begin{bmatrix} X_s + rad\mathcal{A}_0(S) & 0 \\ 0 & X_{\overline{S}} + rad\mathcal{A}_0(\overline{S}) \end{bmatrix} : \begin{array}{l} X_S \in \mathcal{A}'(S) \\ X_{\overline{S}} \in \mathcal{A}'(\overline{S}) \end{array} \right\}.$$

By Theorem 4.6.15, $\mathcal{A}_0(T)/rad\mathcal{A}_0(T) \cong \mathbf{C} \oplus \mathbf{C}$.

Proposition 4.6.23　　*Let S and $\overline{S}$ be two operator weighted shifts, $S, \overline{S} \in (SI)$ and $S \sim \overline{S}$. $T = S \oplus \overline{S}$, then*

$$\mathcal{A}_0(T)/rad\mathcal{A}_0(T) \cong M_2(\mathbf{C}).$$

Proof　　Without loss of generality, we assume that $S = \overline{S}$. Then

$$\mathcal{A}_0(T) = \left\{ \begin{bmatrix} X_{11}|_{kerS} & X_{12}|_{kerS} \\ X_{21}|_{kerS} & X_{22}|_{kerS} \end{bmatrix} : X_{ij} \in \mathcal{A}'(S), i,j = 1,2. \right\}.$$

Since $S \in (SI)$, by Theorem 4.6.15, $\mathcal{A}_0(T)/rad\mathcal{A}_0(T) \cong \mathbf{C}$.
Thus we have $\mathcal{A}_0(T)/rad\mathcal{A}_0(T) \cong M_2(\mathbf{C})$.

Proposition 4.6.24　　*Let $S \sim \{W_k\}$ and $\overline{S} \sim \{V_k\}$ be two operators weighted shifts. $S, \overline{S} \in (SI)$ and S is not similar to $\overline{S}$. $T = S \oplus \overline{S}$. Then*

$$\mathcal{A}_0(T)/rad\mathcal{A}_0(T) \cong \mathbf{C} \oplus \mathbf{C}.$$

Proof   Assume that S and $\overline{S}$ have multiplicities m and n respectively. By Theorem 4.6.15, we need only to prove the proposition in the case of $m < n$.

For

$$\begin{bmatrix} E_{11} & E_{12} \\ E_{21} & E_{22} \end{bmatrix} \begin{matrix} ker S \\ ker \overline{S} \end{matrix} \in \mathcal{A}_0(T),$$

since $E_{12}E_{21} \in \mathcal{A}_0(S)$ is not invertible, by Theorem 4.6.15, $\sigma(E_{12}E_{21}) = \{0\}$.

Denote

$$\mathcal{J} = \left\{ \begin{bmatrix} rad\mathcal{A}'(S) & E_{12} \\ E_{21} & rad\mathcal{A}'(\overline{S}) \end{bmatrix} : \begin{matrix} E_{12} \in ker \tau_{S,\overline{S}} \\ E_{21} \in ker \tau_{\overline{S},S} \end{matrix} \right\}.$$

By the arguments used in the proof of Proposition 4.6.21 we have $\mathcal{J} = rad\mathcal{A}_0(T)$. Thus $\mathcal{A}_0(T)/rad\mathcal{A}_0(T) \cong \mathbf{C} \oplus \mathbf{C}$.

Summarizing the discussion above, we have the following theorem.

Theorem 4.6.25   *Let S and $\overline{S}$ be two operator weighted shifts. $S, \overline{S} \in (SI)$ and $T = S \oplus \overline{S}$. Then*

1. $S \sim \overline{S}$ if and only if $\mathcal{A}_0(T)/rad\mathcal{A}_0(T) \cong M_2(\mathbf{C})$;

2. S is not similar to $\overline{S}$ if and only if $\mathcal{A}_0(T)/rad\mathcal{A}_0(T) \cong \mathbf{C} \oplus \mathbf{C}$.

As a matter of fact, we have a more general result by Proposition 4.6.21, Proposition 4.6.22 and Proposition 4.6.23.

Theorem 4.6.26   *Let S and $\overline{S}$ be two operator weighted shifts and $S \sim \bigoplus_{i=1}^{m} S_i^{(k_i)}$, where $S_i \in (SI)$ is an operator weighted shift and $S_i \nsim S_j$ for $i \neq j$ (Corollary 4.6.13). Let $T = S \oplus \overline{S}$, then $S \sim \overline{S}$ if and only if $\mathcal{A}_0(T)/rad\mathcal{A}_0(T) \cong \bigoplus_{i=1}^{m} M_{2k_i}(\mathbf{C})$.*

Proposition 4.6.27   *Let $S \sim \{W_k\}$ be an operator weighted shift and*

$$W_k = \begin{bmatrix} 1 & \lambda_k \\ 0 & 1 \end{bmatrix},$$

then $\mathcal{A}'(S)$ is commutative if and only if $\sup_{k} | \sum_{i=1}^{k} \lambda_i | = +\infty$.

Proof   If $T \in \mathcal{A}'(S)$, then T is of the form

$$T = \begin{bmatrix} T_{11} & T_{12} & T_{13} & \cdots \\ & T_{22} & T_{23} & \cdots \\ & & & \\ 0 & & T_{33} & \ddots \\ & & & \ddots \end{bmatrix}$$

and $T_{k+1,k+1} = E_{1k}^{-1} T_{11} E_{1,k+l-1}$ $(k \geq 1, l \geq 1)$, where

$$E_{lk} = W_l W_{l+1} \cdots W_k = \begin{bmatrix} 1 & \sum\limits_{i=l}^{k} \lambda_i \\ 0 & 1 \end{bmatrix}.$$

Set

$$T_{1l} = \begin{bmatrix} t_{11} & t_{12} \\ t_{21} & t_{22} \end{bmatrix},$$

then

$$T_{k+1,k+l} = E_{1k}^{-1} T_{1l} E_{l,k+l-1} = \begin{bmatrix} t_{11}' & t_{12}' \\ t_{21}' & t_{22}' \end{bmatrix},$$

where

$$t_{11}' = t_{11} - \left(\sum_{i=1}^{k} \lambda_i \right) t_{21}$$

$$t_{12}' = t_{12} + \left(\sum_{i=l}^{k+l-1} \lambda_i \right) t_{11} - \left(\sum_{i=1}^{k} \lambda_i \right) t_{22} - \left(\sum_{i=1}^{k} \lambda_i \right) \left(\sum_{i=l}^{k+l-1} \lambda_i \right) t_{21}$$

$$= t_{12} + \left(\sum_{i=1}^{k} \lambda_i \right) (t_{11} - t_{22}) + \left(-\sum_{i=1}^{l-1} \lambda_i + \sum_{i=k+1}^{k+l-1} \lambda_i \right) t_{11} - \left(\sum_{i=1}^{k} \lambda_i \right) \left(\sum_{i=l}^{k+l-1} \lambda_i \right) t_{21}$$

$$t_{22}' = t_{22} + \left(\sum_{i=l}^{k+l-1} \lambda_i \right) t_{21}.$$

If $\sup\limits_{k} \left| \sum\limits_{i=1}^{k} \lambda_i \right| = +\infty$, since t_{11}' are uniformly bounded, $t_{21} = 0$. Since

$$\sup_{k} \{ \| W_k \| \} < +\infty, \ |\lambda_i| < M$$

for some number M. Thus $|-\sum\limits_{i=1}^{l-1}\lambda_i + \sum\limits_{i=k+1}^{k+l-1}\lambda_i| < 2lM$. Since t'_{12} are uniformly bounded, $t_{11} = t_{12}$. Thus

$$T_{k+1,k+l} = E_{1k}^{-1}T_{11}E_{l,k+l-1} = t_{11}I + t_{12}J_2(0).$$

Therefore, $\mathcal{A}'(S)$ is commutative.

If $\sup\limits_{k}|\sum\limits_{i=1}^{k}\lambda_i| = N < +\infty$, set $A = diag(A_1, A_2, \cdots)$ and $B = diag(B_1, B_2, \cdots)$, where

$$A_{k+1} = \begin{bmatrix} -\sum\limits_{i=1}^{k}\lambda_i & (\sum\limits_{i=1}^{k}\lambda_i)^2 \\ 1 & \sum\limits_{i=1}^{k}\lambda_i \end{bmatrix},$$

$B_1 = \begin{bmatrix} 0 & 1 \\ 0 & 0 \end{bmatrix}$. Then $A, B \in \mathcal{A}'(S)$, but $AB - BA \neq 0$.

4.7 Open Problem

1. Is every operator in $\mathcal{L}(\mathcal{H})$ is a direct integral of strongly irreducible operators?

2. What is the necessary and sufficient conditions for an operator $T \in \mathcal{L}(\mathcal{H})$ to have only finitely many Banach reducing subspaces?

3. If $T \in \mathcal{L}(\mathcal{H}) \cap (SI)$, is $\mathcal{A}'(T)/rad\mathcal{A}'(T)$ commutative?

4. Does the following statement holds for arbitrary $T_1, T_2 \in \mathcal{L}(\mathcal{H}) \cap (SI)$?
    $T_1 \sim T_2$ if and only if $K_0(\mathcal{A}'(T_1 \oplus T_2)) \cong K_0(\mathcal{A}'(T_1)) \cong K_0(\mathcal{A}'(T_2))$.

5. Let $T \in \mathcal{L}(\mathcal{H}) \cap (SI)$, then for each natural number n, does $T^{(n)}$ have a unique (SI) decomposition up to similarity?

6. If $T \in \mathcal{L}(\mathcal{H})$ is a direct sum of finitely many (SI) operators, does T have a unique (SI) decomposition up to similarity?

7. What is the "Beurling" Theorem for M_z in Sobolev disk algebra?

8. Given a necessary and sufficient condition for an injective unilateral operator weighted shift $S \sim \{W_k\}$ to be strongly irreducible.

4.8 Remark

Theorems 4.1.3–4.1.20 are given by [Fang, J.S., Jiang, C.L. and Wu, P.Y. (2003)]. Theorem 4.2.1, Proposition 4.2.13 and Theorem 4.2.14 are due

to [Cao, Y., Fang, J.S. and Jiang, C.L.(2002)]. Theorem 4.3.1 belongs
to [Fang, J.S. and Jiang, C.L. (1999)]. Proposition 4.3.5, Theorem 4.3.9,
Proposition 4.3.11 and Proposition 4.3.12 are proved by [Ji, Y.Q. and Yang,
Y.H. (2003)]. Example 4.3.8 is due to [Fang, J.S. and Jiang, C.L. (1999)].
Theorem 4.3.6 is given by [Jiang, C.L. (2004)]. Theorem 4.3.13 and Corol-
lary 4.3.14 are proved by [Wang, Z.Y. and Xue, Y.F. (2000)]. Example
4.4.1 and Proposition 4.4.2 belong to [Jiang, C.L. and Li, J.X. (2000)].
C.L. Jiang also proved Theorem 4.4.3 [Jiang, C.L. (2004)]. The all re-
sults in Section 4.5 are proved by [Wang, Z.Y. (1993)], [Wang, Z.Y. and
Liu, Y.Q.], [Liu, Y.Q. and Wang, Z.Y. (2004)], [Liu, Y.Q. and Wang,
Z.Y.], [Jin, Y.F. and Wang, Z.Y.(1)]]. Theorem 4.6.2, Proposition 4.6.9
and Theorem 4.6.10 are given by [Ji, Y.Q., Li, J.X. and Sun, S.L. (2003)].
Theorem 4.6.12–Theorem 4.6.26 are proved by [Jiang, C.L. and Li, J.X.
(2000)]. Example 4.6.18, Proposition 4.6.19–Proposition 4.6.27 are given
by jia-jin-wan. The reader can refer to [Davidson, K.R. and Herrero, D.A.
(1990)] about the (SI) decomposition of some special operator classes and
[Behncke, H.], [Yan, C.Q. (1993)] about to the unique (SI) decomposition
up to unitary equivalence.

Chapter 5

The Similarity Invariant of Cowen-Douglas Operators

5.1 The Cowen-Douglas Operators with Index 1

The backward unilateral shift is a typical Cowen-Douglas operator with index 1. Denote $\mathcal{H} = l^2 = \{(x_1, x_2, \cdots) : \sum |x_i|^2 < \infty\}$. For $(\alpha_0, \alpha_1, \cdots) \in l^2$, define $T_z^*(\alpha_0, \alpha_1, \cdots) = (\alpha_1, \alpha_2, \cdots)$. For $|\lambda| < 1$, we have

$$T_z^*(1, \lambda, \lambda^2, \cdots) = \lambda(1, \lambda, \lambda^2, \cdots).$$

This implies that $D \subset \sigma_p(T_z^*)$. Clearly, $T_z^* \in \mathcal{B}_1(D)$ and $\mathcal{A}'(T_z) \cong H^\infty$. Note that T_z, the adjoint of T_z^*, is an analytic Teolitz operator and also a pure isometry operator.

An operator $S \in \mathcal{L}(\mathcal{H})$ is called a pure isometry operator if $\bigcap_{n=1}^{\infty} S^n \mathcal{H} = \{0\}$.

von-Neumann-Wold Theorem *Let $S \in \mathcal{L}(\mathcal{H})$ be a pure isometry operator. Then $S \cong \bigoplus_{k=1}^{l} T_z$, where $l = dim ker S^*$.*

It is easily seen that $T_z^{(n)}$ is also a pure isometry operator for each natural number n. The following is a well-known result.

Lemma 5.1.1 *Let $S \in \mathcal{L}(\mathcal{H})$ be a pure isometry operator, then*
 (i) $S \cong T_z^{(l)}$ and $S^* \in \mathcal{B}_l(D)$, where $l = dim ker S^*$;
 (ii) $S \in (SI)$ if and only if $S^* \in \mathcal{B}_1(D)$.

Lemma 5.1.2 *Let P be an idempotent in $\mathcal{A}'(T_z^{(n)})$, let $S = T_z^{(n)}|_{P\mathcal{H}^{(n)}}$ and $m = dim ker S^*$. Then there is a unitary operator U such that*

(i) $U(P\mathcal{H}^{(n)}) = \mathcal{H}^{(m)} \oplus 0^{(n-m)}$, i.e.,

$$UPU^* = \begin{bmatrix} I_{\mathcal{H}^{(m)}} & \star \\ 0 & 0 \end{bmatrix} \begin{matrix} \mathcal{H}^{(m)} \\ \mathcal{H}^{(n-m)} \end{matrix} ;$$

(ii) Let $V = U|_{P\mathcal{H}^{(n)}}$, then $VSV^* = T_z^{(m)}$, i.e., S is unitarily equivalent to $T_z^{(m)}$.

Proof It is obvious that S is a pure isometry. By von-Neumann-Wold Theorem, S is unitarily equivalent to $T_z^{(m)}$. Thus there is a unitary operator

$$V : P\mathcal{H}^{(n)} \longrightarrow \mathcal{H}^{(n)}$$

such that

$$VSV^* = T_z^{(m)}.$$

Note that if $m < n$, $\mathcal{H}^{(n)} \ominus P\mathcal{H}^{(m)}$ is infinite dimensional. Therefore, there exists a unitary operator

$$W : \mathcal{H}^{(n)} \ominus P\mathcal{H}^{(n)} \longrightarrow \mathcal{H}^{(n-m)}.$$

Set $U = V \oplus W$, then U satisfies the requirements of the lemma.

Lemma 5.1.3 Let P be an idempotent in $\mathcal{A}'((T_z^*)^{(n)})$ and let $S = (T_z^*)^{(n)}|_{P\mathcal{H}^{(n)}}$. If $m = dimker S$, then there exists a unitary operator U such that:

(i) $U(P\mathcal{H}^{(n)}) = \mathcal{H}^{(m)} \oplus 0^{(n-m)}$, i.e., $UPU^* = \begin{bmatrix} I_{\mathcal{H}^{(m)}} & \star \\ 0 & 0 \end{bmatrix} \begin{matrix} \mathcal{H}^{(m)} \\ \mathcal{H}^{(n-m)} \end{matrix} ;$

(ii) Let $V = U|_{P\mathcal{H}^{(n)}}$, then $VSV^* = (T_z^*)^{(m)}$, i.e., S is unitarily equivalent to $(T_z^*)^{(m)}$.

Proof Let $Q = (I_{\mathcal{H}^{(n)}} - P)^*$, then $Q \in \mathcal{A}'((T_z)^{(n)})$ is an idempotent and

$$(Q(T_z)^{(n)}Q)^* = (I_{\mathcal{H}^{(n)}} - P)(T_z^*)^{(n)}(I_{\mathcal{H}^{(n)}} - P).$$

Thus

$$((T_z)^{(n)}|_{Q\mathcal{H}^{(n)}})^* = (T_z^*)^{(n)}|_{(I_{\mathcal{H}^{(n)}} - P)\mathcal{H}^{(n)}}.$$

Since $P \in \mathcal{A}'((T_z^*)^{(n)})$,

$$dimker((T_z)^{(n)}|_{Q\mathcal{H}^{(n)}})^* = dimker((T_z^*)^{(n)})|_{(I_{\mathcal{H}^{(n)}} - P)\mathcal{H}^{(n)}}$$

$$= n - dimker((T_z^*)^{(n)}) = n - m.$$

By Lemma 5.1.2, there exists a unitary operator U_1 such that

$$U_1 Q U_1^* = \begin{bmatrix} I_{\mathcal{H}^{(n-m)}} & \star \\ 0 & 0 \end{bmatrix} \begin{matrix} \mathcal{H}^{(n-m)} \\ \mathcal{H}^{(m)} \end{matrix}$$

and

$$U_1 Q T_z^{(n)} Q U_1^* = \begin{bmatrix} T_z^{(n-m)} & \star \\ 0 & 0 \end{bmatrix} \begin{matrix} \mathcal{H}^{(n-m)} \\ \mathcal{H}^{(m)} \end{matrix} \ .$$

Thus

$$U_1 P U_1^* = \begin{bmatrix} 0 & 0 \\ \star & I_{\mathcal{H}^{(m)}} \end{bmatrix} \begin{matrix} \mathcal{H}^{(n-m)} \\ \mathcal{H}^{(m)} \end{matrix}$$

and

$$U_1 P (T_z^*)^{(n)} P U_1^* = \begin{bmatrix} 0 & 0 \\ \star & (T_z^*)^{(m)} \end{bmatrix} \begin{matrix} \mathcal{H}^{(n-m)} \\ \mathcal{H}^{(m)} \end{matrix} \ .$$

Define $U_2 : \mathcal{H}^{(n)} = \mathcal{H}^{(n-m)} \oplus \mathcal{H}^{(m)} \overset{\text{onto}}{\longrightarrow} \mathcal{H}^{(m)} \oplus \mathcal{H}^{(m)}$ by $U_2(x \oplus y) = y \oplus x$ for $x \in \mathcal{H}^{(n-m)}$ and $y \in \mathcal{H}^{(m)}$. Then U_2 is a unitary operator. Let $U = U_2 U_1$, then U satisfies the requirements of the lemma.

For $T \in \mathcal{B}_n(\Omega)$ and $z \in \Omega$, $S(T - zI) = (T - zI)S$ for all $S \in \mathcal{A}'(T)$. Thus

$$S \ker(T - zI) \subset \ker(T - zI).$$

Define $(\Gamma_T S)(z) = S|_{\ker(T-zI)}$. In general, we substitute $S(z)$ for $(\Gamma_T S)(z)$ (see Chapter 3). Clearly, Γ_T is an injective contraction. But for $T = (T_z^*)^{(n)} \in \mathcal{B}_n(D)$, we can easily obtain the following lemma.

Lemma 5.1.4 *Let* $T = (T_z^*)^{(n)}$, *then* Γ_T *is an isometry isomorphism from* $\mathcal{A}'(T)$ *onto* $M_n(H^\infty)$.

Lemma 5.1.5 *Let* $\mathcal{H} = H^2, A \in \mathcal{B}_1(\Omega) \cap \mathcal{L}(\mathcal{H})$ *and* $T = A^{(n)}$. *For each idempotent* $P \in \mathcal{A}'(T)$, *denote* $T_1 = T|_{P\mathcal{H}^{(n)}}$. *If* $T_1 \in \mathcal{B}_m(\Omega)$, *then* $T_1 \cong A^{(m)}$.
Proof Without loss of generality, we can assume that $\overline{D} \subset \Omega$. Then we can find $\mathcal{H}$-valued holomorphic functions $v(z)$ and $e(z)$ on D such that

$$(A - z)v(z) = 0, \quad (T_z^* - z)e(z) = 0,$$

and $v(z)$ and $e(z)$ can be chosen to be the holomorphic frames of $\ker(A - z)$

and $ker(T_z^* - z)$ respectively. Set

$$v_k(z) = (0, \cdots, 0, v(z), 0, \cdots, 0),$$
$$k = 1, 2, \cdots, n, z \in D.$$
$$e_k(z) = (0, \cdots, 0, e(z), 0, \cdots, 0),$$

Let $P(z) = (\Gamma_T P)(z), z \in D$, then $P(z) = (P_{ij}(z))_{n \times n} \in M_n(H^\infty)$ is an idempotent. By Lemma 5.1.4, $P(z)$ is an idempotent in $\mathcal{A}'((T_z^*)^{(n)})$. Set

$$Q = P(z)$$

and

$$S = (T_z^*)^{(n)}|_{Q\mathcal{H}^{(n)}}.$$

Since $T_1 \in \mathcal{B}_m(\Omega)$, $dimkerS = ranP(0) = dimkerT_1 = m$.

By Lemma 5.1.3, there exists a unitary operator U such that

$$U(P\mathcal{H}^{(n)}) = \mathcal{H}^{(m)} \oplus 0^{(n-m)},$$

i.e.,

$$UPU^* = \begin{bmatrix} I_{\mathcal{H}^{(m)}} & \star \\ 0 & 0 \end{bmatrix} \begin{matrix} \mathcal{H}^{(m)} \\ \mathcal{H}^{(n-m)} \end{matrix} \qquad (5.1.1)$$

and if $V = U|_{P\mathcal{H}^{(n)}}$, then $VSV^* = (T_z^*)^{(m)}$.

Since $V^*(\mathcal{H}^{(m)} \oplus 0^{(n-m)}) = U^*(\mathcal{H}^{(m)} \oplus 0^{(n-m)}) = Q\mathcal{H}^{(n)}$ and $VSV^* = (T_z^*)^{(m)}$, $U^*e_i(z) \in ker(S - z) \subset ker((T_z^*)^{(n)} - z), 1 \leq i \leq m$. Note that $(e_1(z), \cdots, e_n(z))$ is a holomorphic frame of $ker((T_z^*)^{(n)} - z)$. Thus

$$U^*e_i(z) = \lambda_{i1}(z)e_1(z) + \cdots + \lambda_{in}(z)e_n(z), 1 \leq i \leq m,$$

where $\lambda_{ij} \in \mathbf{C}$. Since $< e_i(z), e_j(z) >= \delta_{ij} < e(z), e(z) >, 1 \leq i, j \leq n$ and U^* is a unitary operator, we have

$$\lambda_{i1}(z)\overline{\lambda_{j1}(z)} + \cdots + \lambda_{in}(z)\overline{\lambda_{jn}(z)} = \delta_{ij}, 1 \leq i, j \leq m, z \in D. \qquad (5.1.2)$$

From (5.1.1), $UP(z)U^*e_i(z) = I_{\mathcal{H}^{(m)}}e_i(z) = e_i(z), 1 \leq i \leq m, z \in D$. Therefore,

$$P(z)U^*e_i(z) = U^*e_i(z),$$

i.e.,

$$(P_{ij}(z))_{n \times n}(\lambda_{i1}(z), \cdots, \lambda_{in}(z)) = (\lambda_{i1}(z), \cdots, \lambda_{in}(z)), \qquad (5.1.3)$$

where $1 \leq i \leq m$ and $z \in D$.

Set $w_i(z) = \lambda_{i1}(z)v_1(z) + \cdots + \lambda_{in}(z)v_n(z), 1 \leq i \leq m$. Since

$$< v_i(z), v_j(z) >= \delta_{ij} < v(z), v(z) >, 1 \leq i \neq j \leq n \qquad (5.1.4)$$

by (5.1.2), $< w_i(z), w_j(z) >= \delta_{ij} < v(z), v(z) >, 1 \leq i \neq j \leq m$. It follows from (5.1.3) that $P(z)w_i(z) = w_i(z), 1 \leq i \leq m$ and $z \in D$. Note that

$$P(z)ker(T - z) = ker(T_1 - z),$$

thus $w_i(z) \in ker(T_1 - z), 1 \leq i \leq m$ and $(w_1(z), \cdots, w_m(z))$ forms a holomorphic frame of $ker(T_1 - z)$ for each $z \in D$.

For $z \in D$, define $U(z) : ker(A^{(m)} - z) \longrightarrow ker(T_1 - z)$ as follows

$$U(z)v_i(z) = w_i(z), \quad 1 \leq i \leq m.$$

It follows from (5.1.4) and (5.1.5) that

$$< U(z)v_i(z), U(z)v_j(z) >=< v_i(z), v_j(z) >= \delta_{ij} < v(z), v(z) >, 1 \leq i, j \leq m.$$

Since $U(z)$ is a holomorphic isometric bundle map, using Rigidity Theorem we have $T_1 \cong A^{(m)}$.

Theorem 5.1.6 *Let* $A \in \mathcal{B}_1(\Omega) \cap \mathcal{L}(\mathcal{H})$, *then* $\bigvee(\mathcal{A}'(A)) \cong \mathbf{N}$ *and* $K_0(\mathcal{A}'(A)) \cong \mathbf{Z}$.

Proof By Theorem 4.2.1, for every natural number n and idempotent $P \in (\mathcal{A}'(A^{(n)}))$, if $A_1 = A^{(n)}|_{P\mathcal{H}^{(n)}}$, then $A_1 \sim A$. This is a straightforward corollary of Lemma 5.1.5.

Proposition 5.1.7 *Let* $A, B \in \mathcal{B}_1(\Omega)$, *then the following are equivalent.*
  (i) $A \sim B$;
  (ii) $K_0(\mathcal{A}'(A \oplus B)) \cong \mathbf{Z}$.

Proof (i)$\Rightarrow$(ii). It is a straightforward conclusion of Theorem 5.1.6.

(ii)$\Rightarrow$(i). We need only to show that if $A \nsim B$, then $K_0(\mathcal{A}'(A \oplus B)) \neq \mathbf{Z}$. Otherwise, we assume that $K_0(\mathcal{A}'(A \oplus B)) \cong \mathbf{Z}$. Since $A \nsim B$, there exists a maximal ideal $\mathcal{J}$ in $\mathcal{A}'(A \oplus B)$ such that $\mathcal{A}'(A \oplus B)/\mathcal{J} \cong \mathbf{C}$, where

$$\mathcal{J} = \begin{bmatrix} \mathcal{J}' & ker\tau_{A,B} \\ ker\tau_{B,A} & \mathcal{A}'(B) \end{bmatrix}$$

and $\mathcal{J}'$ is a maximal ideal of $\mathcal{A}'(A)$ (see Section 5.2) . Thus we have the following separating exact sequence:

$$0 \longrightarrow \mathcal{J} \overset{l}{\longrightarrow} \mathcal{A}'(A \oplus B) \underset{\lambda}{\overset{\pi}{\rightleftharpoons}} \mathcal{A}'(A \oplus B)/\mathcal{J} \longrightarrow 0$$

and the semi-exact sequence:

$$K_0(\mathcal{J}) \xrightarrow{l_*} K_0(\mathcal{A}'(A\oplus B)) \underset{\lambda_*}{\overset{\pi_*}{\rightleftarrows}} K_0(\mathcal{A}/\mathcal{J}).$$

Observing the six-term exact sequence:

$$\begin{array}{ccc}
K_0(\mathcal{J}) & \xrightarrow{l_*} K_0(\mathcal{A}'(A\oplus B)) \xrightarrow{\pi_*} K_0(\mathcal{A}'(A\oplus B)/\mathcal{J}) \\
\partial\uparrow & \partial\downarrow \\
K_1(\mathcal{A}'(A\oplus B)/\mathcal{J}) \longleftarrow K_1(\mathcal{A}'(A\oplus B)) \longleftarrow & K_1(\mathcal{J})
\end{array}$$

It is easy to see that $\partial = 0$. Thus we get the exact sequence:

$$0\longrightarrow K_0(\mathcal{J}) \xrightarrow{l_*} K_0(\mathcal{A}'(A\oplus B)) \xrightarrow{\pi_*} K_0(\mathcal{A}/\mathcal{J})\longrightarrow 0.$$

Note that $K_0(\mathcal{A}'(A\oplus B))\cong K_0(\mathcal{A}/\mathcal{J})\cong\mathbf{Z}$, therefore $K_0(\mathcal{J}) = 0$. This contradicts that $K_0(\mathcal{J}) \neq 0$.

5.2   Cowen-Douglas Operators with Index n

Lemma 5.2.1 Let $A\in\mathcal{B}_n(\Omega)\cap(SI)$, $T = A^{(l)}\in\mathcal{L}(\mathcal{H}^{(l)})$ and let P be an idempotent in $\mathcal{A}'(T)$ satisfying $T|_{P\mathcal{H}^{(l)}}\in(SI)$. Then $A_1 := T|_{P\mathcal{H}^{(l)}}$ is similar to A.

Proof Without loss of generality, we may assume that $\Omega = D$ an n is the minimal index of A. For convenience we will prove the lemma only in the case of $n = 2$. Then $T = A\oplus A$. Note that P is an idempotent in $\mathcal{A}'(T)$, by Theorem 4.4.3 we can find an idempotent $P_1\in\mathcal{A}'(T)$ and $B\in rad\mathcal{A}'(T)$ such that $P(z) = P_1(z) + B(z)$, where

$$P_1(z) = \begin{bmatrix} f_{11}(z) & f_{12}(z) \\ f_{21}(z) & f_{22}(z) \end{bmatrix} \begin{matrix} ker(A - z) \\ ker(A - z) \end{matrix}$$

and

$$B(z) = \begin{bmatrix} B_{11}(z) & B_{12}(z) \\ B_{21}(z) & B_{22}(z) \end{bmatrix} \begin{matrix} ker(A - z) \\ ker(A - z) \end{matrix},$$

$f_{ij}\in H^\infty$ and $B_{ij}\in rad\mathcal{A}'(A)$. Set $G = -I_{\mathcal{H}^{(2)}} + (2P_1 + B)$. Since $B\in rad\mathcal{A}'(T)$, G is invertible in $\mathcal{A}'(T)$ and $PG = GP_1$. This implies that $G^{-1}PG = P_1\in\mathcal{A}'(T)$.

Without loss of generality, we may assume that $P = P_1$, i.e.,

$$P(z) = \begin{bmatrix} f_{11}(z) & f_{12}(z) \\ f_{21}(z) & f_{22}(z) \end{bmatrix} \begin{array}{l} ker(A - z) \\ ker(A - z) \end{array}.$$

Set

$$P'(z) = \begin{bmatrix} f_{11}(z) & f_{12}(z) \\ f_{21}(z) & f_{22}(z) \end{bmatrix} \begin{array}{l} ker(T_{z^*} - z) \\ ker(T_{z^*} - z) \end{array}.$$

Since $T|_{P\mathcal{H}^{(2)}} \in (SI)$, $tr(P'(z)) = 1$ for all $z \in D$. By the arguments similar to that used in the proof of Lemma 5.1.5, there exists an invertible element $X(z)$ in $M_2(H^\infty)$ such that

$$X(z)P'(z)X^{-1}(z) = \begin{bmatrix} I_\mathbf{C} & 0 \\ 0 & 0 \end{bmatrix}$$

and

$$X(z)(I - P'(z))X^{-1}(z) = \begin{bmatrix} 0 & 0 \\ 0 & I_\mathbf{C} \end{bmatrix}.$$

Furthermore, $X(z)|_{ranP'(z)}$ and $X(z)|_{ran(I-P'(z))}$ are isomorphic bundle maps from $ranP'(z)$ and $ran(I - P'(z))$ respectively onto $ker(T_{z^*} - z)$.

Set

$$X(z) = \begin{bmatrix} u_{11}(z) & u_{12}(z) \\ u_{21}(z) & u_{22}(z) \end{bmatrix}$$

and

$$\widehat{X}(z) = \begin{bmatrix} u_{11}(z)I_{ker(A-z)} & u_{12}(z)I_{ker(A-z)} \\ u_{21}(z)I_{ker(A-z)} & u_{22}(z)I_{ker(A-z)} \end{bmatrix}.$$

Then

$$\widehat{X}(z)P(z)\widehat{X}(z) = \begin{bmatrix} I_{ker(A-z)} & 0 \\ 0 & 0 \end{bmatrix}.$$

Note that $\widehat{X}(z)ker(T - z) = ker(T - z)$.

{**Claim**} $\widehat{G}(z) = \widehat{X}(z)|_{ranP(z)}$ is an isomorphic bundle map from $ranP(z)$ onto $ker(A - z)$.

Note that $G(z) = X(z)|_{ranP'(z)}$ is an isomorphic bundle map from $ranP'(z)$ onto $ker(T_{z^*} - z)$.

Let $e(z)$ be a holomorphic frame of $ker(T_{z^*} - z)$ and

$$t_1(z) = e(z) \oplus 0, \quad t_2(z) = 0 \oplus e(z).$$

Then $(t_1(z), t_2(z))$ is a holomorphic frame of $ker(T_{z^*}^{(2)} - z)$.

Set

$$\overline{A}_1 = T_{z^*}^{(2)}|_{P'(z)\mathcal{H}^{(2)}}$$

and $l(z)$ is a holomorphic frame of $ker(\overline{A}_1 - z)$, then

$$l(z) = \alpha(z)t_1(z) + \beta(z)t_2(z),$$

where $\alpha(z), \beta(z)$ are analytic functions in D.

Since $G(z)$ is a holomorphic isomorphic map, we can find a function $c(z)$ holomorphic in D such that

$$G(z)l(z) = c(z)e(z)$$

and

$$\|l(z)\|^2 = (|\alpha(z)|^2 + |\beta(z)|^2)\|e(z)\| = |c(z)|^2\|e(z)\|$$

for $z \in D$.

Let $(S_1(z), \cdots, S_n(z))$ be a holomorphic frame of $ker(A - z)$. Set

$$v_j(z) = S_j(z) \oplus 0, u_j(z) = 0 \oplus S_j(z), (j = 1, 2, \cdots, n).$$

Then $(v_1(z), \cdots, v_n(z), u_1(z), \cdots, u_n(z))$ is a holomorphic frame of $ker(T - z)$.

Let $f_j(z) = \alpha(z)v_j(z) + \beta(z)u_j(z), (j = 1, 2, \cdots, n)$. Then $(f_1(z), \cdots, f_n(z))$ is a holomorphic frame of $ker(A_1 - z)$. Set

$$\overline{G}(z)f_j(z) = c(z)v_j(z).$$

Let $K_1(z), \cdots, K_n(z)$ be analytic functions in D and

$$g(z) = K_1(z)f_1(z) + \cdots + K_n(z)f_n(z)$$

$$= K_1(z)(\alpha(z)v_1(z) + \beta(z)u_1(z)) + \cdots + K_n(z)(\alpha(z)v_n(z) + \beta(z)u_n(z)).$$

Then $\overline{G}(z)g(z) = c(z)(K_1(z)v_1(z) + \cdots + K_n(z)v_n(z)) := g'(z)$.

Since $< v_i(z), v_j(z) >=< u_i(z), u_j(z) >=< S_i(z), S_j(z) >$ for $z \in D$,

$$< g(z), g(z) > = \sum_{i=1}^{n} |K_i(z)|^2(|\alpha(z)|^2 + |\beta(z)|^2)\|S_i(z)\|^2$$

$$+ \sum_{i,j=1}^{n} K_i(z)\overline{K}_j(z)(|\alpha(z)|^2 + |\beta(z)|^2) < S_i(z), S_j(z) > .$$

Furthermore,

$$< g'(z), g'(z) > = \sum_{i=1}^{n} |K_i(z)|^2 |c(z)|^2 \|S_i(z)\|^2$$

$$+ \sum_{i,j=1}^{n} K_i(z)\overline{K}_j(z)|c(z)|^2 < S_i(z), S_j(z) > .$$

This implies that $\|\widehat{G}(z)g(z)\| = \|g(z)\|$ and it verifies our claim.

Similarly, we can prove that $\overline{X}(z)|_{ran(I-P(z))}$ is also a holomorphic isomorphism from $ran(I - P(z))$ onto $ker(A - z)$. By Rigidity Theorem, we can find two isomorphisms

$$U_1 \in \mathcal{L}(P\mathcal{H}^{(2)}, \mathcal{H} \oplus 0) \quad \text{and} \quad U_2 \in \mathcal{L}((I - P)\mathcal{H}^{(2)}, 0 \oplus \mathcal{H})$$

such that $X = U_1 + U_2 \in \mathcal{A}'(T)$ and

$$XPX^{-1} = \begin{bmatrix} I_\mathcal{H} & 0 \\ 0 & 0 \end{bmatrix}.$$

This indicates that $A_1 \sim A$ and ends the proof of the lemma.

By Theorem 4.2.1 and Lemma 5.2.1, we get the following theorem.

Theorem 5.2.2 *Let* $A \in \mathcal{B}_n(\Omega) \cap (SI)$, *then* $\bigvee(\mathcal{A}'(A)) \cong \mathbf{N}$ *and* $K_0(\mathcal{A}'(A)) \cong \mathbf{Z}$.

Similar to the proof of Proposition 5.1.7, we have the following result.

Proposition 5.2.3 *Let* $A, B \in \mathcal{B}_n(\Omega) \cap (SI)$, *then the following two statements are equivalent:*

    (i) $A \sim B$;

    (ii) $K_0(\mathcal{A}'(A \oplus B)) \cong \mathbf{Z}$.

5.3 The Commutant of Cowen-Douglas Operators

In this section, we always assume that $T = \bigoplus_{k=1}^{n} T_k$, where $T_k \in \mathcal{B}_{n_k}(\Omega_k) \cap (SI)$ and $\bigvee_{z \in \Omega_k} ker(T_k - z) = \mathcal{H}_k$. By the basic operator theory, we have the following properties:

    (5.3.1) $\mathcal{A}'(T) = \{(S_{ij})_{n \times n} | S_{ij} \in ker\tau_{T_i, T_j}, 1 \leq i, j \leq n\}$ is a unital Banach algebra.

(5.3.2) $ker\tau_{T_i,T_j}$ is a linear space and $ker\tau_{T_i,T_i} = \mathcal{A}'(T_i)$ is a unital Banach algebra.

(5.3.3) Denote $e_{\mathcal{A}'(T)}$ the identity in $\mathcal{A}'(T)$. Then

$$e_{\mathcal{A}'(T)} = e_{\mathcal{A}'(T_1)} \oplus \cdots \oplus e_{\mathcal{A}'(T_n)}.$$

(5.3.4) If $S_{ij} \in ker\tau_{T_i,T_j}$ and $S_{jk} \in ker\tau_{T_j,T_k}$, then $S_{ij}S_{jk} \in ker\tau_{T_i,T_k}$.

(5.3.5) If $(S_{ij})_{n \times n} \in \mathcal{A}'(T)$, then

$$S(i,j) \triangleq \begin{bmatrix} 0 \cdots & 0 & \cdots 0 \\ \vdots & \vdots & \vdots \\ 0 \cdots & S_{ij} & \cdots 0 \\ \vdots & \vdots & \vdots \\ 0 \cdots & 0 & \cdots 0 \end{bmatrix} \in \mathcal{A}'(T).$$

By (5.3.5), we can define a canonical map $\Phi_{ij} : \mathcal{A}'(T) \longrightarrow ker\tau_{T_i,T_j}$ as follows:

$$\Phi_{ij}(S) = S_{ij}, \quad S = (S_{ij})_{n \times n} \in \mathcal{A}'(T).$$

(5.3.6) Φ_{ij} is a linear map and $\Phi_{ii}(S) \in \mathcal{A}'(T_i)$ for all $S \in \mathcal{A}'(T)$.
In this section, $\mathcal{J}$ denotes a proper two-sided ideal.

(5.3.7) Let $\mathcal{J}$ be an ideal of $\mathcal{A}'(T)$. Define

$$J_{ij} = \{S_{ij} : S_{ij} \in ker\tau_{T_i,T_j} \text{ and } \begin{bmatrix} 0 \cdots & 0 & \cdots 0 \\ \vdots & \vdots & \vdots \\ 0 \cdots & S_{ij} & \cdots 0 \\ \vdots & \vdots & \vdots \\ 0 \cdots & 0 & \cdots 0 \end{bmatrix} \in \mathcal{J}\}.$$

Then

(5.3.7.1) J_{ii} is an ideal of $\mathcal{A}'(T_i)$ or $J_{ii} = \mathcal{A}'(T_i)$;

(5.3.7.2) J_{ij} is a subspace of $ker\tau_{T_i,T_j}$;

(5.3.7.3) $S(i,j) \in \mathcal{J}$ for $S = (S_{ij})_{n \times n} \in \mathcal{J}$.

By (5.3.7), we can define a canonical map from $ker\tau_{T_i,T_j}$ onto $ker\tau_{T_i,T_j}/\Phi_{ij}(\mathcal{J})$ as follows: $S_{ij} \longrightarrow [S_{ij}]_{\mathcal{J}}$, where $ker\tau_{T_i,T_j}/\Phi_{ij}(\mathcal{J})$ is the quotient of $ker\tau_{T_i,T_j}$ modulo $\Phi_{ij}(\mathcal{J})$. If $\mathcal{J}$ is closed, then

$$\mathcal{A}'(T)/\mathcal{J} = \{([S_{ij}]_{\mathcal{J}})_{n \times n} : S_{ij} \in ker\tau_{T_i,T_j}\}$$

is a unital Banach algebra. Thus we have a canonical map

$$\Phi_{\mathcal{J}} : \mathcal{A}'(T) \longrightarrow \mathcal{A}'(T)/\mathcal{J}$$

as follows:

$$\Phi_{\mathcal{J}}((S_{ij})_{n \times n}) = ([S_{ij}]_{\mathcal{J}})_{n \times n}.$$

(5.3.8) Let $\mathcal{J}$ be a closed ideal of $\mathcal{A}'(T)$. If

$$([S_{ij}]_{\mathcal{J}})_{n \times n} = \Phi_{\mathcal{J}}(S) \in \mathcal{A}'(T)/\mathcal{J},$$

then

$$\begin{bmatrix} 0 \cdots & 0 & \cdots 0 \\ \vdots & \vdots & \vdots \\ 0 \cdots & [S_{ij}]_{\mathcal{J}} & \cdots 0 \\ \vdots & \vdots & \vdots \\ 0 \cdots & 0 & \cdots 0 \end{bmatrix} = \Phi_{\mathcal{J}}(S(i,j)) \in \mathcal{A}'(T)/\mathcal{J}.$$

Lemma 5.3.1(Lifting Lemma)   *Let $T = \bigoplus\limits_{k=1}^{2} T_k$ and $\mathcal{J}_1$ be an ideal of $\mathcal{A}'(T_1)$, then there exists an ideal $\mathcal{J}$ of $\mathcal{A}'(T)$ such that $\Phi_{11}(\mathcal{J}) = \mathcal{J}_1$. Furthermore, if there is another ideal $\mathcal{J}'$ of $\mathcal{A}'(T)$ such that $\Phi_{11}(\mathcal{J}') = \mathcal{J}_1$, then $\mathcal{J} \subset \mathcal{J}'$, where Φ_{11} is given in (5.3.5).*
Proof   Set

$$\chi = \left\{ \begin{bmatrix} R_1 & R_2 A_{12} \\ A_{21} R_3 & B_{21} R_4 B_{12} \end{bmatrix} : R_i \in \mathcal{J}_1, i = 1, 2, 3, 4; B_{ij}, A_{ij} \in ker\tau_{T_i, T_j}, i = 1, 2 \right\},$$

and

$$\mathcal{J} = \{ x_1 + x_2 + \cdots + x_n : 1 \leq n < \infty, x_i \in \chi \}.$$

{Claim} $\mathcal{J}$ is an ideal of $\mathcal{A}'(T)$.

Clearly, $\mathcal{J}$ is an additive group. Set

$$W = \begin{bmatrix} W_{11} & W_{12} \\ W_{21} & W_{22} \end{bmatrix} \in \mathcal{A}'(T) \quad \text{and} \quad X = \begin{bmatrix} R_1 & R_2 A_{12} \\ A_{21} R_3 & B_{21} R_4 B_{12} \end{bmatrix} \in \chi.$$

Then

$$WX = \begin{bmatrix} W_{11} R_1 & (W_{11} R_2) A_{12} \\ W_{21} R_1 & W_{21} R_2 A_{12} \end{bmatrix} + \begin{bmatrix} (W_{12} A_{21}) R_3 & (W_{12} B_{21} R_4) B_{12} \\ (W_{22} A_{21}) R_3 & (W_{22} B_{21}) R_4 B_{12} \end{bmatrix}.$$

Since

$$\begin{bmatrix} W_{11}R_1 & (W_{11}R_2)A_{12} \\ W_{21}R_1 & W_{21}R_2A_{12} \end{bmatrix}, \begin{bmatrix} (W_{12}A_{21})R_3 & (W_{12}B_{21}R_4)B_{12} \\ (W_{22}A_{21})R_3 & (W_{22}B_{21})R_4B_{12} \end{bmatrix} \in \chi,$$

$WX \in \mathcal{J}$. Similarly, $XW \in \mathcal{J}$. Thus $\mathcal{J}$ is an ideal of $\mathcal{A}'(T)$. Since $\Phi_{11}(X) \in \mathcal{J}_1$ for all $X \in \mathcal{J}$, $e_{\mathcal{A}'(T)} \notin \mathcal{J}$. Therefore $\mathcal{J}$ is a proper ideal of $\mathcal{A}'(T)$ and $\Phi_{11}(\mathcal{J}) = \mathcal{J}_1$. If $\Phi_{11}(\mathcal{J}') = \mathcal{J}_1$ for another ideal $\mathcal{J}'$ of $\mathcal{A}'(T)$, by (5.3.4) and (5.3.7), $\mathcal{J} \subset \mathcal{J}'$.

Corollary 5.3.2 *Let* $T = \bigoplus_{k=1}^{n} T_k$ *and* $\mathcal{J}_1$ *be an ideal* $\mathcal{J}$ *of* $\mathcal{A}'(T_1)$, *then there exists an ideal* $\mathcal{J}$ *of* $\mathcal{A}'(T)$ *such that* $\Phi_{11}(\mathcal{J}) = \mathcal{J}_1$. *Furthermore, if* $\Phi_{11}(\mathcal{J}') = \mathcal{J}_1$ *for another ideal* $\mathcal{J}'$ *of* $\mathcal{A}'(T)$, *then* $\mathcal{J} \subset \mathcal{J}'$.

Corollary 5.3.3 *Let* $T = \bigoplus_{k=1}^{n} T_k$ *and* $\mathcal{J} \in \mathcal{M}(\mathcal{A}'(T))$, *then* $\Phi_{kk}(\mathcal{J}) = \mathcal{A}'(T_k)$ *or* $\Phi_{kk}(\mathcal{J}) \in \mathcal{M}(\mathcal{A}'(T_k)), k = 1, 2, \cdots, n$.

Lemma 5.3.4 *Let* $T = \bigoplus_{k=1}^{n} T_k$ *and* $S = (S_{ij})_{n \times n} \in \mathcal{A}'(T)$. *If for each* $R_{ji} \in \ker \tau_{T_j, T_i}$, $R_{ji}S_{ij} = 0$, *then* $S(i,j) \in rad\mathcal{A}'(T)$.
Proof For each

$$R = \begin{bmatrix} R_{11} & \cdots & R_{1n} \\ \vdots & & \vdots \\ R_{n1} & \cdots & R_{nn} \end{bmatrix},$$

$$RS(i,j) = \begin{bmatrix} 0 & \cdots & R_{1i}S_{ij} & \cdots & 0 \\ \vdots & & \vdots & & \vdots \\ 0 & \cdots & R_{ji}S_{ij} & \cdots & 0 \\ \vdots & & \vdots & & \vdots \\ 0 & \cdots & R_{ni}S_{ij} & \cdots & 0 \end{bmatrix}.$$

Since $R_{ji}S_{ij} = 0$, $(RS(i,j))^n = 0$. This implies that $S(i,j) \in rad\mathcal{A}'(T)$.

Corollary 5.3.5 *Let* $T = \bigoplus_{k=1}^{n} T_k$, *then*

$$\Phi_{kk}(rad\mathcal{A}'(T)) = rad\mathcal{A}'(T_k), k = 1, 2, \cdots, n.$$

Corollary 5.3.6 *Let* $T = \bigoplus_{k=1}^{n} T_k$, $\mathcal{J} \in \mathcal{M}(\mathcal{A}'(T))$ *and* $S_{ij} \in ker\tau_{T_i,T_j}/$
$\Phi_{ij}(\mathcal{J})$. *If* $S_{ij}r_{ji} = 0$ *for all* $r_{ji} \in ker\tau_{T_j,T_i}/\Phi_{ji}(\mathcal{J})$, *then* $S_{ij} = 0$.

Theorem 5.3.7 *Let* $T = \bigoplus_{k=1}^{n} T_k$, *then for each* $\mathcal{J} \in \mathcal{M}(\mathcal{A}'(T))$, *there is a positive integer* $l_{\mathcal{J}} \leq n$ *such that* $\mathcal{A}'(T)/\mathcal{J} \cong M_{l_{\mathcal{J}}}(\mathbf{C})$. *Furthermore, if* $T_k \sim T_1$ *for* $k = 1, 2, \cdots, n$, *then* $\mathcal{A}'(T)/\mathcal{J} \cong M_n(\mathbf{C})$ *for all* $\mathcal{J} \in \mathcal{M}(\mathcal{A}'(T))$.
Proof By Corollary 5.3.3,

$$\Phi_{kk}(\mathcal{J}) = \mathcal{A}'(T_k)$$

or

$$\Phi_{kk}(\mathcal{J}) \in \mathcal{M}(\mathcal{A}'(T_k)), k = 1, 2, \cdots, n.$$

By Theorem 4.4.3, $\mathcal{A}'(T_k)/rad\mathcal{A}'(T_k)$ is commutative for $k = 1, 2, \cdots, n$. Thus $\mathcal{A}'(T_k)/\Phi_{kk}(\mathcal{J}) \cong \mathbf{C}$ or $\mathcal{A}'(T_k)/\Phi_{kk}(\mathcal{J}) = \{0\}, k = 1, 2, \cdots, n$. Without loss of generality, we may assume that there exists an integer $l_{\mathcal{J}} \leq n$ such that

$$\mathcal{A}'(T_k)/\Phi_{kk}(\mathcal{J}) \cong \mathbf{C}, k = 1, 2, \cdots, l_{\mathcal{J}},$$

and

$$\mathcal{A}'(T_k)/\Phi_{kk}(\mathcal{J}) = \{0\}, k = l_{\mathcal{J}} + 1, \cdots, n.$$

Thus,

$$\mathcal{A}'(T)/\mathcal{J} = \{([S_{ij}]_{\mathcal{J}})_{n \times n} : S_{ij} \in ker\tau_{T_i,T_j}$$

and

$$[S_{kk}] = 0, l_{\mathcal{J}} \leq k \leq n\}.$$

By (5.3.4), $0 = [S_{ij}R_{ji}]_{\mathcal{J}} = [S_{ij}]_{\mathcal{J}} \in \mathcal{A}'(T)/\Phi_{ij}(\mathcal{J})$, where $S_{ij} \in ker\tau_{T_i,T_j}, R_{ji} \in ker\tau_{T_j,T_i}$ and $l_{\mathcal{J}} < i \leq n$. By Corollary 5.3.6, $[S_{ij}]_{\mathcal{J}} = [R_{ij}]_{\mathcal{J}} = 0$ for $l_{\mathcal{J}} < i \leq n$. Therefore $l_{\mathcal{J}} \geq 1$ and

$$\mathcal{A}'(T)/\mathcal{J} = \left\{ \begin{bmatrix} ([S_{ij}])_{l_{\mathcal{J}} \times l_{\mathcal{J}}} & 0 \\ 0 & 0 \end{bmatrix} : S_{ij} \in ker\tau_{T_i,T_j} \right\}.$$

{Claim 1} For $1 \leq i, j, k \leq l_{\mathcal{J}}$, if

$$ker \tau_{T_i, T_j} / \Phi_{ij}(\mathcal{J}) \neq \{0\}$$

and

$$ker \tau_{T_j, T_k} / \Phi_{jk}(\mathcal{J}) \neq \{0\},$$

then

$$ker \tau_{T_i, T_k} / \Phi_{ik}(\mathcal{J}) \neq \{0\}.$$

Note that $\mathcal{A}'(T_i)/\Phi_{ii}(\mathcal{J}) \cong \mathbf{C}$ and $\mathcal{A}'(T_j)/\Phi_{jj}(\mathcal{J}) \cong \mathbf{C}, 1 \leq i, j \leq l_{\mathcal{J}}$. If

$$ker \tau_{T_i, T_j} / \Phi_{ij}(\mathcal{J}) \neq \{0\},$$

by Corollary 5.3.6, there exist $S_{ij} \in ker \tau_{T_i, T_j} / \Phi_{ij}(\mathcal{J})$ and $S_{ji} \in ker \tau_{T_j, T_i} / \Phi_{ji}(\mathcal{J})$ such that $S_{ij} S_{ji} = [e_{\mathcal{A}'(T_i)}]_{\mathcal{J}} = 1$.

Similarly, there exist $S_{jk} \in ker \tau_{T_j, T_k} / \Phi_{jk}(\mathcal{J})$ and $S_{kj} \in ker \tau_{T_k, T_j} / \Phi_{kj}(\mathcal{J})$ such that $S_{jk} S_{kj} = [e_{\mathcal{A}'(T_k)}]_{\mathcal{J}} = 1$. Thus, $S_{ij} S_{jk} \neq 0$ and $ker \tau_{T_i, T_k} / \Phi_{ik}(\mathcal{J}) \neq \{0\}$.

{Claim 2} $\Phi_{1i}(\mathcal{J}) \neq ker \tau_{T_1, T_i}$ for $1 \leq i \leq l_{\mathcal{J}}$.

Otherwise we may assume that there is a $j_0 : 1 \leq j_0 < l_{\mathcal{J}}$ such that

$$ker \tau_{T_1, T_i} / \Phi_{1i}(\mathcal{J}) \neq \{0\}, 1 \leq i \leq j_0$$

and

$$ker \tau_{T_1, T_j} / \Phi_{1j}(\mathcal{J}) = \{0\}, j_0 < j \leq l_{\mathcal{J}}.$$

By Claim 1, we have

$$ker \tau_{T_j, T_i} / \Phi_{ji}(\mathcal{J}) = \{0\}, 1 \leq i \leq j_0, j_0 < j \leq l_{\mathcal{J}}.$$

By Corollary 5.3.6,

$$ker \tau_{T_j, T_i} / \Phi_{ji}(\mathcal{J}) = \{0\}, 1 \leq i \leq j_0, j_0 < j \leq l_{\mathcal{J}}.$$

Thus

$$\mathcal{A}'(T)/\mathcal{J} = \left\{ \begin{bmatrix} diag(([S_{ij}]_{\mathcal{J}})_{j_0 \times j_0}, ([S_{ij}]_{\mathcal{J}})_{(l_{\mathcal{J}} - j_0)(l_{\mathcal{J}} - j_0)}) & 0 \\ 0 & 0 \end{bmatrix}_{n \times n} : S_{ij} \in ker \tau_{T_i, T_j} \right\}.$$

This contradicts $\mathcal{J} \in \mathcal{M}(\mathcal{A}'(T))$.

{Claim 3} $\mathcal{A}'(T)/\mathcal{J} \cong M_{l_{\mathcal{J}}}(\mathbf{C})$.

For $1 \leq i \leq l_{\mathcal{J}}$, denote $e_{ii} = [e_{\mathcal{A}'(T_i)}] = 1$. By Claim 1 and Claim 2, there are $e_{1i} \in ker \tau_{T_1, T_i} / \Phi_{1i}(\mathcal{J})$ and $e_{i1} \in ker \tau_{T_i, T_1} / \Phi_{i1}(\mathcal{J})$ such that

$e_{1i}e_{i1} = e_{11}, 1 \leq i \leq l_{\mathcal{J}}$. Since $\mathcal{A}'(T_i)/\Phi_{ii}(\mathcal{J}) \cong \mathbf{C}, e_{i1}e_{1i} = e_{ii}$. Assume that $e_{ij} = e_{i1}e_{1j}$, then

$$e_{ij}e_{ji} = e_{i1}e_{1j}e_{j1}e_{1i} = e_{ii}$$

and

$$e_{ji}e_{ij} = e_{j1}e_{1i}e_{i1}e_{1j} = e_{jj}.$$

Since for each $S_{ij} \in ker\tau_{T_i,T_j}/\Phi_{ij}(\mathcal{J})$ $(1 \leq i, j \leq l_{\mathcal{J}})$ there exists $\lambda_{ij} \in \mathbf{C}$ such that

$$\lambda_{ij} = S_{ij}e_{ij}$$

and since

$$S_{ij} - \lambda_{ij}e_{ij} = (S_{ij} - \lambda_{ij}e_{ij})e_{jj} = (S_{ij} - \lambda_{ij}e_{ij})e_{ji}e_{ij}$$

$$= (S_{ij}e_{ji} - \lambda_{ij}e_{ii})e_{ij}$$

$$= (\lambda_{ij} - \lambda_{ij})e_{ii}e_{ij} = 0,$$

$S_{ij} = \lambda_{ij}e_{ij}$. Thus the first part of the proof of Theorem 5.3.7 is now complete.

If $T_k \sim T_1, k = 1, 2, \cdots, n$, then $\mathcal{A}'(T_k)/\Phi_{kk}(\mathcal{J}) \cong \mathbf{C}$ for all $\mathcal{J} \in \mathcal{M}(\mathcal{A}'(T))$. This completes the proof of the theorem.

Theorem 5.3.7 implies the following properties.

(5.3.9) If $\mathcal{A}'(T_i)/\Phi_{ii}(\mathcal{J}) \neq \{0\}$ and $\mathcal{A}'(T_j)/\Phi_{jj}(\mathcal{J}) \neq \{0\}$ for $\mathcal{J} \in \mathcal{M}(\mathcal{A}'(T))$, then

$$ker\tau_{T_i,T_j}/\Phi_{ij}(\mathcal{J}) \cong ker\tau_{T_j,T_i}/\Phi_{ji}(\mathcal{J}) \cong \mathbf{C}.$$

(5.3.10) If $\mathcal{A}'(T_i)/\Phi_{ii}(\mathcal{J}) = \{0\}$ for some $i : 1 \leq i \leq n$, and $\mathcal{J} \in \mathcal{M}(\mathcal{A}'(T))$, then

$$ker\tau_{T_i,T_k}/\Phi_{ik}(\mathcal{J}) = ker\tau_{T_k,T_i}/\Phi_{ki}(\mathcal{J}) = \{0\}, \ k = 1, 2, \cdots, n.$$

Theorem 5.3.8 *Let* $T = \bigoplus_{k=1}^{n} T_k$, *then for each* $\mathcal{J}_1 \in \mathcal{M}(\mathcal{A}'(T_1))$, *there is a unique* $\mathcal{J} \in \mathcal{M}(\mathcal{A}'(T))$ *such that* $\Phi_{11}(\mathcal{J}) = \mathcal{J}_1$.

Proof   We only prove the theorem when $n = 2$ and $T = T_1 \oplus T_2$. The general case can be proved similarly. By Lemma 5.3.1, there is an ideal $\mathcal{J}_0$ of $\mathcal{A}'(T)$ such that $\Phi_{11}(\mathcal{J}_0) = \mathcal{J}_1$. Set $\mathcal{J}' = \mathcal{J}_0 + rad\mathcal{A}'(T)$.

Then $\mathcal{J}'$ is still an ideal of $\mathcal{A}'(T)$ and $\Phi_{11}(\mathcal{J}') = \mathcal{J}_1$. By Corollary 5.3.6, $rad\mathcal{A}'(T_2) \subset \Phi_{22}(\mathcal{J}')$. Therefore we may assume that $\mathcal{J}_0 = \mathcal{J}', \mathcal{A}'(T_1)/\Phi_{11}(\mathcal{J}_0) \cong \mathbf{C}$ and $\mathcal{A}'(T_2)/\Phi_{22}(\mathcal{J}_0)$ is semisimple. By Theorem 4.4.3, $\mathcal{A}'(T_k)/rad\mathcal{A}'(T_k)$ is commutative, $k = 1, 2$.

Note that

$$\mathcal{A}'(T)/\mathcal{J}_0 = \{\begin{bmatrix} S_{11} & S_{12} \\ S_{21} & S_{22} \end{bmatrix} : S_{ij} \in ker\tau_{T_i,T_j}/\Phi_{ij}(\mathcal{J}_0), \ 1 \leq i, j \leq 2\}.$$

Denote $e_{kk} = [e_{\mathcal{A}'(T_k)}]_{\mathcal{J}}, k = 1, 2$. Since $\mathcal{A}'(T_1)/\Phi_{11}(\mathcal{J}_0) \cong \mathbf{C}$, $e_{11} = 1$.

{Case 1}   Assume that there are

$$e_{12} \in ker\tau_{T_1,T_2}/\Phi_{12}(\mathcal{J}_0), \quad e_{21} \in ker\tau_{T_2,T_1}/\Phi_{21}(\mathcal{J}_0)$$

such that $e_{12}e_{21} = 1$. Set $Q_1 = e_{21}e_{12}$ and $Q_2 = e_{22} - Q_1$, then Q_1 and Q_2 are idempotents in $\mathcal{A}'(T_2)/\Phi_{22}(\mathcal{J}_0)$ and $Q_1Q_2 = Q_2Q_1 = 0$. Let

$$\mathcal{A}' = \{\begin{bmatrix} S_{11} & S_{12} \\ S_{21} & Q_1S_{22} \end{bmatrix} : S_{ij} \in ker\tau_{T_i,T_j}/\Phi_{ij}(\mathcal{J}_0), \ 1 \leq i, j \leq 2\}$$

and

$$\mathcal{A}'' = \{\begin{bmatrix} 0 & 0 \\ 0 & Q_2S_{22} \end{bmatrix} : S_{22} \in \mathcal{A}'(T_2)/\Phi_{22}(\mathcal{J}_0)\}.$$

{Claim 1} $\mathcal{A}'(T)/\mathcal{J}_0 = \mathcal{A}' \oplus \mathcal{A}''$.

It is obvious that for each $S = (S_{ij})_{2 \times 2} \in \mathcal{A}'(T)/\mathcal{J}_0$,

$$S = \begin{bmatrix} S_{11} & S_{12} \\ S_{21} & Q_1S_{22} \end{bmatrix} + \begin{bmatrix} 0 & 0 \\ 0 & Q_2S_{22} \end{bmatrix},$$

where

$$\begin{bmatrix} S_{11} & S_{12} \\ S_{21} & Q_1S_{22} \end{bmatrix} \in \mathcal{A}'$$

and

$$\begin{bmatrix} 0 & 0 \\ 0 & Q_2S_{22} \end{bmatrix} \in \mathcal{A}''.$$

For

$$t = \begin{bmatrix} S_{11} & S_{12} \\ S_{21} & Q_1S_{22} \end{bmatrix} \in \mathcal{A}'$$

and

$$r = \begin{bmatrix} 0 & 0 \\ 0 & Q_2 S_{22} \end{bmatrix} \in \mathcal{A}'',$$

we have that

$$tr = \begin{bmatrix} 0 & S_{12} Q_2 S_{22} \\ 0 & 0 \end{bmatrix}, \quad rt = \begin{bmatrix} 0 & 0 \\ Q_2 S_{22} S_{12} & 0 \end{bmatrix}.$$

To verify Claim 1, we need only to show that $S_{12} Q_2 = Q_2 S_{21} = 0$.

For arbitrary $S_{12} \in ker \tau_{T_1, T_2} / \Phi_{12}(\mathcal{J}_0)$ and $S_{21} \in ker \tau_{T_2, T_1} / \Phi_{21}(\mathcal{J}_0)$, we can find a $\lambda \in \mathbf{C}$ such that $S_{12} Q_2 e_{11} = \lambda e_{12} e_{21}$. By $(5.3.4)$, we have $(S_{12} Q_2 - \lambda e_{12}) e_{12} = 0$ and $\sigma(e_{21}(S_{12} Q_2 - \lambda e_{12})) = \{0\}$.

Since $\mathcal{A}'(T_2) / \Phi_{22}(\mathcal{J}_0)$ is semisimple and commutative,

$$e_{21}(S_{12} Q_2 - \lambda e_{12}) = 0.$$

This implies that $(e_{21} S_{12}) Q_2 = \lambda Q_1$ and therefore $\lambda = 0$ and $e_{21} S_{12} = 0$. Thus,

$$S_{12} Q_2 = e_{11}(S_{12} Q_2) = e_{12} e_{21}(S_{12} Q_2) = e_{12}(e_{21} S_{12} Q_2) = 0.$$

Similarly, we can show that $Q_2 S_{12} = 0$. Thus $rt = tr = 0$ and this proves Claim 1.

{Claim 2} $\mathcal{A}' \cong M_2(\mathbf{C})$.

Let

$$\mathcal{A}_1 = \{S_{11} : S_{11} \in \mathcal{A}'(T_1) / \Phi_{11}(\mathcal{J}_0)\}$$

and

$$\mathcal{A}_2' = \{Q_1 S_{22} : S_{22} \in \mathcal{A}'(T_2) / \Phi_{22}(\mathcal{J}_0)\}.$$

Note that $\mathcal{A}_1 \cong \mathbf{C}$. We define a map $\phi : \mathcal{A}_2 \longrightarrow \mathcal{A}_1$ as follows:

$$\phi(b) = e_{12} b e_{21} \quad \text{for all} \quad b \in \mathcal{A}_2.$$

Clearly, ϕ is a homomorphism. Since $\phi(Q_1) = e_{11} = 1$, ϕ is surjective. If $\phi(b) = \phi(b')$, then $e_{21} \phi(b) e_{12} = e_{21} \phi(b') e_{12}$. Thus $Q_1 b Q_1 = Q_1 b' Q_1$ and $\mathcal{A}_2 \cong \mathcal{A}_1 \cong \mathbf{C}$. Similar to the proof of Theorem 5.3.7, we can get $\mathcal{A}' \cong M_2(\mathbf{C})$. Now we define a surjective map $\pi : \mathcal{A}'(T) \longrightarrow \mathcal{A}'$ as follows:

$$\pi((S_{ij})_{2 \times 2}) = \begin{bmatrix} ([S_{11}])_{\mathcal{J}_0} & ([S_{12}])_{\mathcal{J}_0} \\ ([S_{21}])_{\mathcal{J}_0} & ([S_{22}])_{\mathcal{J}_0} \end{bmatrix}.$$

Then π is a homomorphism. Since $\mathcal{A}' \cong M_2(\mathbf{C})$, $\mathcal{J} = ker\pi \in \mathcal{M}(\mathcal{A}'(T))$ and

$$\Phi_{11}(\mathcal{J}) = \mathcal{J}_1.$$

{**Case 2**}

If there is no $e_{12} \in ker\tau_{T_1,T_2}/\Phi_{12}(\mathcal{J}_0)$ and $e_{21} \in ker\tau_{T_2,T_1}/\Phi_{21}(\mathcal{J}_0)$ such that $e_{12}e_{21} = e_{11} = 1$. Since $\mathcal{A}'(T_1)/\Phi_{11}(\mathcal{J}_0) \cong \mathbf{C}$, for each $S_{12} \in ker\tau_{T_1,T_2}/\Phi_{12}(\mathcal{J}_0)$ and each $S_{21} \in ker\tau_{T_2,T_1}/\Phi_{21}(\mathcal{J}_0)$ we have $S_{12}S_{21} = 0$. By Lemma 5.3.4,

$$ker\tau_{T_1,T_2}/\Phi_{12}(\mathcal{J}_0) = \{0\}$$

and

$$ker\tau_{T_2,T_1}/\Phi_{21}(\mathcal{J}_0) = \{0\}.$$

Thus

$$\mathcal{A}'(T)/\mathcal{J}_0 \cong \mathcal{A}'(T_1)/\Phi_{11}(\mathcal{J}_0) \oplus \mathcal{A}'(T_2)/\Phi_{22}(\mathcal{J}_0) \cong \mathbf{C} \oplus \mathcal{A}'(T_2)/\Phi_{22}(\mathcal{J}_0).$$

Similar to the proof in Case 1, we can find a $\mathcal{J} \in \mathcal{M}(\mathcal{A}'(T))$ such that

$$\Phi_{11}(\mathcal{J}) = \mathcal{J}_1.$$

Now we prove the uniqueness. Suppose that there are $\mathcal{J}$ and $\mathcal{J}' \in \mathcal{M}(\mathcal{A}'(T))$ such that $\Phi_{11}(\mathcal{J}) = \Phi_{11}(\mathcal{J}') = \mathcal{J}_1$. Let

$$\overline{\mathcal{J}} = \mathcal{J} + \mathcal{J}' = \{S + S' : S \in \mathcal{J}, \ S' \in \mathcal{J}'\}.$$

Then $\overline{\mathcal{J}}$ is an ideal of $\mathcal{A}'(T)$. Since $\Phi_{11}(\overline{\mathcal{J}}) = \mathcal{J}_1$, $\mathcal{J} = \mathcal{J}' = \overline{\mathcal{J}}$.

Theorem 5.3.9 *Let* $T = \displaystyle\bigoplus_{k=1}^{n} T_k$, *then for each* $S \in \mathcal{A}'(T)$,

$$\sigma(S) = \bigcup_{\mathcal{J} \in \mathcal{M}(T)} \sigma(\Phi_{\mathcal{J}}(S)),$$

where $\Phi_{\mathcal{J}}$ *is a canonical map:* $\mathcal{A}'(T) \longrightarrow \mathcal{A}'(T)/\mathcal{J}$, $\sigma(\Phi_{\mathcal{J}}(S))$ *is the spectrum of* $\Phi_{\mathcal{J}}(S)$ *in* $\mathcal{A}'(T)/\mathcal{J}$.

Proof We only prove the theorem when $n = 2$ and $T = T_1 \oplus T_2$. The general case can be proved similarly.

If $\mathcal{A}'(T_1)/\Phi_{11}(\mathcal{J}) \cong \mathbf{C}$ and $\mathcal{A}'(T_2)/\Phi_{22}(\mathcal{J}) = \{0\}$ for every $\mathcal{J} \in \mathcal{M}(\mathcal{A}'(T))$, then

$$\Phi_{\mathcal{J}}(S) = \begin{bmatrix} \lambda e_{11} & 0 \\ 0 & 0 \end{bmatrix}, \quad \lambda \in \mathbf{C} \qquad \text{(see Theorem 5.3.8)}$$

If $\mathcal{A}'(T_1)/\Phi_{11}(\mathcal{J})\cong\mathcal{A}'(T_2)/\Phi_{22}(\mathcal{J})\cong\mathbf{C}$, then

$$\Phi_\mathcal{J}(S) = \begin{bmatrix} \lambda_{11}e_{11} & \lambda_{12}e_{12} \\ \lambda_{21}e_{21} & \lambda_{22}e_{22} \end{bmatrix},$$

where $e_{12}e_{21} = e_{11}, e_{21}e_{12} = e_{22}$. This means that

$$\bigcup_{\mathcal{J}\in\mathcal{M}(T)} \sigma(\Phi_\mathcal{J}(S))\subset\sigma(S).$$

If $\lambda\in\sigma(S)$, consider the maximal two-sided ideal $\mathcal{J}$ generated by $(\lambda I_{\mathcal{H}\oplus\mathcal{H}} - S)$ in $\mathcal{A}'(T)$, clearly $\lambda\in\sigma(\Phi_\mathcal{J}(S))$ and this completes the proof of the theorem.

Lemma 5.3.10 *Let $T = \bigoplus_{k=1}^{2} T_k$, then the following are equivalent:*

(i) There exist a positive integer n and $x_i\in ker\tau_{T_1,T_2}$, $y_i\in ker\tau_{T_2,T_1}$, where $i = 1, 2, \cdots, n$, such that $\sum_{i=1}^{n} x_iy_i = I_{\mathcal{A}'(T)}$;

(ii) There exists an idempotent $e\in M_n(\mathcal{A}'(T_2))$ such that

$$I_{\mathcal{A}'(T_1)}\oplus 0\sim_a 0\oplus e \quad in \quad \begin{bmatrix} \mathcal{A}'(T_1) & \star \\ \star & M_n(\mathcal{A}'(T_2)) \end{bmatrix}.$$

Proof (i)$\Rightarrow$(ii). Let

$$e = \begin{bmatrix} y_1 \\ \vdots \\ y_n \end{bmatrix} \begin{bmatrix} x_1 & \cdots & x_n \end{bmatrix} = \begin{bmatrix} y_1x_1 & \cdots & y_nx_1 \\ \vdots & \ddots & \vdots \\ y_nx_1 & \cdots & y_nx_n \end{bmatrix} \in M_n(\mathcal{A}'(T_2)).$$

By (i),

$$e^2 = \begin{bmatrix} y_1 \\ \vdots \\ y_n \end{bmatrix} \begin{bmatrix} x_1 & \cdots & x_n \end{bmatrix} \begin{bmatrix} y_1 \\ \vdots \\ y_n \end{bmatrix} \begin{bmatrix} x_1 & \cdots & x_n \end{bmatrix} = e.$$

Now set

$$u = \begin{bmatrix} 0 & \begin{bmatrix} x_1 & \cdots & x_n \end{bmatrix} \\ 0 & 0 \end{bmatrix}_{(n+1)\times(n+1)}, \quad v = \begin{bmatrix} 0 & 0 \\ \begin{bmatrix} y_1 \\ \vdots \\ y_n \end{bmatrix} & 0 \end{bmatrix}_{(n+1)\times(n+1)}.$$

Then

$$uv = \begin{bmatrix} \mathcal{A}'(T_1) & \star \\ \star & M_n(\mathcal{A}'(T_2)) \end{bmatrix}$$

and

$$I_{\mathcal{A}'(T_1)} \oplus 0 = uv, \ 0 \oplus e = vu.$$

This implies that $I_{\mathcal{A}_1} \oplus 0 \sim_a 0 \oplus e$ in

$$\begin{bmatrix} \mathcal{A}'(T_1) & \star \\ \star & M_n(\mathcal{A}'(T_2)) \end{bmatrix}.$$

(ii)$\Rightarrow$(i). If $I_{\mathcal{A}_1} \oplus 0 \sim_a 0 \oplus e$ in

$$\begin{bmatrix} \mathcal{A}'(T_1) & \star \\ \star & M_n(\mathcal{A}'(T_2)) \end{bmatrix},$$

then by the basic properties of K-theory, we find

$$u, v \in \begin{bmatrix} \mathcal{A}'(T_1) & \star \\ \star & M_2(\mathcal{A}'(T_2)) \end{bmatrix}$$

such that $I_{\mathcal{A}'(T_1)} \oplus 0 = uv$ and $0 \oplus e = vu$, and

$$u = (I_{\mathcal{A}'(T_1)}) u (0 \oplus e), v = (0 \oplus e) v (I_{\mathcal{A}'(T_1)}).$$

Since $I_{\mathcal{A}'(T_1)} \oplus 0 = uv$, $\sum_{i=1}^{n} x_i y_i = I_{\mathcal{A}_1}$.

Proposition 5.3.11 *Let $T = A^{(l)}$ and $\{P_1, \cdots, P_m\}$ be an (SI) decomposition, then $m = l$ and $A_i = A^{(l)}|_{P_i \mathcal{H}^{(l)}} \in \mathcal{B}_n(\Omega)$.*
Proof We first show that $m \le l$. By Theorem 4.4.3, $\mathcal{A}'(A)/rad\mathcal{A}'(A)$ is commutative. It follows from Gelfand Theorem that there exists a continuous natural homomorphism $\varphi : \mathcal{A}'(A) \longrightarrow C(\mathcal{M}(\mathcal{A}'(A)))$ and φ induces a continuous homomorphism $\psi : \mathcal{A}'(T) \longrightarrow M_l(\mathcal{M}(\mathcal{A}'(A)))$ as follows:

$$\psi(s)(\mathcal{J}) = (\varphi(s_{ij})(\mathcal{J}))_{l \times l},$$

where $S = (S_{ij})_{l \times l} \in \mathcal{A}'(A), \mathcal{J} \in \mathcal{M}(\mathcal{A}'(A))$. Set

$$P_k = (P_{ij}^k)_{l \times l}, k = 1, 2, \cdots, m.$$

Then

$$\psi(P_k)(\mathcal{J}) = (\varphi(P_{ij}^k)(\mathcal{J}))_{l \times l}.$$

Denote $tr(\psi(P_k)(\mathcal{J})) := \sum\limits_{i=1}^{l} \varphi(P_{ii}^k)(\mathcal{J})$. Thus $tr(\cdot)$ defines a continuous function on $\mathcal{M}(\mathcal{A}'(A))$. Since $\mathcal{A}'(A)/rad\mathcal{A}'(A)$ is commutative and $A\in(SI), \mathcal{M}(\mathcal{A}'(A))$ is connected to Proposition 1.17 of [Jiang, C.L. and Wang, Z.Y. (1996a)]. Since $\psi(P_k)(\mathcal{J})$ is an idempotent, $tr(\psi(P_k)(\mathcal{J})) = n_k\geq 1$. Note that $\sum\limits_{k=1}^{m} P_k = I$ and $P_k P_{k'} = \delta_{kk'} P_k$, thus $\sum\limits_{k=1}^{m} tr(\psi(P_k)(\mathcal{J})) = l$ or $\sum\limits_{k=1}^{m} tr(\psi(P_k)(\mathcal{J})) = \sum\limits_{k=1}^{m} n_k = l$. Therefore $m\leq l$.

Now we prove that $A_i\in\mathcal{B}_n(\Omega)$. Otherwise, assume that $A_1\in\mathcal{B}_k(\Omega)$, and $k < n$. Let $S = A\oplus A_1$. Since $k < n$, a simple calculation indicates that

$$\begin{bmatrix} 0 & ker\tau_{A_1,A_2} \\ ker\tau_{A_2,A_1} & 0 \end{bmatrix} \in\mathcal{A}'(S).$$

By the arguments similar to that used in the proof of Theorem 5.3.7, we can find $\mathcal{J}_1\in\mathcal{M}(\mathcal{A}'(S))$ such that $\mathcal{A}'(S)/\mathcal{J}_1\cong\mathbf{C}$. Set

$$T_1 = A\oplus T = A^{(l+1)}.$$

By Theorem 5.3.7 $\mathcal{A}'(T)/\mathcal{J}\cong M_{l+1}(\mathbf{C})$ for all $\mathcal{J}\in\mathcal{M}(\mathcal{A}'(T_1))$. Note that $T_1\cong A\oplus A_1\oplus\cdots\oplus A_m$ and $m\leq l$. Repeating the proof of Theorem 5.3.7 and using Lemma 5.3.10 we can find $\mathcal{J}_2\in\mathcal{M}(\mathcal{A}'(T_1))$ such that $\mathcal{A}'(T_1)/\mathcal{J}_2\cong M_d(\mathbf{C})$, where $d < l+1$. This contradicts $\mathcal{A}'(T_1)/\mathcal{J}\cong M_{l+1}(\mathbf{C})$ for each $\mathcal{J}\in\mathcal{M}(\mathcal{A}'(T_1))$. Since every $A_i\in\mathcal{B}_n(\Omega), m = l$.

5.4   The Commutant of a Classes of Operators

FIR algebra. Let $\mathcal{A}$ be a Banach algebra. π is a representation of $\mathcal{A}$ on a Banach space $\mathcal{X}$, $dim\mathcal{X}\geq 1$, if π is a nontrivial continuous homomorphism from $\mathcal{A}$ onto $\mathcal{L}(\mathcal{H})$. If a linear subspace $\mathcal{Y}$ of $\mathcal{X}$ satisfies $\pi(a)\mathcal{Y}\subset\mathcal{Y}$ for all $a\in\mathcal{A}$, $\mathcal{Y}$ is said to be an invariant subspace of $\pi(\mathcal{A})$. A representation π is said to be irreducible if a subspace $\mathcal{Y}$ of $\mathcal{X}$ satisfying $\pi(a)\mathcal{Y}\subset\mathcal{Y}$ for all $a\in\mathcal{A}$ is either $\mathcal{Y} = \{0\}$ or $\mathcal{Y} = \mathcal{X}$. An ideal $\mathcal{J}\subset\mathcal{A}$ is called a Primitive ideal of $\mathcal{A}$ if $\mathcal{J} = ker\pi$, when π is an irreducible representation of $\mathcal{A}$.

Definition 5.4.1 A Banach algebra $\mathcal{A}$ is called an FIR algebra if for every irreducible representation $\pi : \mathcal{A}\longrightarrow\mathcal{L}(\mathcal{X}), \pi(\mathcal{A})$ is finite dimensional, i.e., $dim\mathcal{X} < \infty$. A Banach algebra $\mathcal{A}$ is said to be n-homogeneous if there is a natural number n such that $\pi(\mathcal{A}) \cong M_n(\mathbf{C})$, where π is a irreducible representation of $\mathcal{A}$.

Definition 5.4.2   Let $\mathcal{A}$ be a unital Banach algebra, $\mathcal{A}$ is said to be essentially commutative if $\mathcal{A}/rad\mathcal{A}$ is commutative.

Proposition 5.4.3   *The commutant of every strongly irreducible Cowen-Douglas operator is essentially commutative.*

Definition 5.4.4   Let $\mathcal{AL}(\mathcal{H})$. A is called typical strongly irreducible operator $A \in (SI)$ and A is essentially commutative. A is called a type-1 operator if A has a finite (SI) decomposition $(P_1, P_2, \cdots, P_n)$ and each $A|_{P_i \mathcal{H}}$ is a typical (SI) operator.

Proposition 5.4.5   *Every Cowen-Douglas operator is a type-1 operator.*

By Gelfand Theorem, we know that an essentially commutative Banach algebra is 1-homogeneous. On the other hand, every 1-homogeneous algebra is essentially commutative. In fact, we have the following stronger result.

Proposition 5.4.6   *Let $\mathcal{A}$ be a unital Banach algebra and let $\chi(\mathcal{A})$ be the set of all nonzero multiplicative linear functionals. For $x \in \mathcal{A}$, $r(x)$ denotes the spectral radius of x. Then the following are equivalent:*
   (i) $\mathcal{A}$ is essentially commutative;
   (ii) $\mathcal{A}$ is 1-homogeneous;
   (iii) $\sigma(x) = \{\varphi(x) : \varphi \in \chi(\mathcal{A})\}$ for each $x \in \mathcal{A}$;
   (iv) $\sigma(xyz) = \sigma(xzy)$ for $x, y, z \in \mathcal{A}$.

Proof   (i)$\Rightarrow$(ii).   It follows from Gelfand Theory.

(ii)$\Rightarrow$(iii).   First we consider $\overline{\mathcal{A}} = \mathcal{A}/rad\mathcal{A}$. For $x \in rad\mathcal{A}, \varphi(x) = 0$ for all $\varphi \in \chi(\mathcal{A})$. Thus $\chi(\overline{\mathcal{A}}) = \chi(\mathcal{A})$. Since $\sigma(x + rad\mathcal{A}) = \sigma(x)$, using Gelfand Theory again, (ii)$\Rightarrow$(iii).

(iii)$\Rightarrow$(iv).   For arbitrary $x, y, z \in \mathcal{A}$,

$$\sigma(xyz) = \{\varphi(xyz) : \varphi \in \chi(\mathcal{A})\}$$

$$= \{\varphi(x)\varphi(y)\varphi(z) : \varphi \in \chi(\mathcal{A})\}$$

$$= \{\varphi(x)\varphi(z)\varphi(y) : \varphi \in \chi(\mathcal{A})\}$$

$$= \{\varphi(xzy) : \varphi \in \chi(\mathcal{A})\} = \sigma(xzy).$$

(iv)$\Rightarrow$(i).   Given $x, y \in \mathcal{A}$, $\sigma(xy) = \sigma(yx)$.

Note that for each $n \geq 1$ and $0 \leq k \leq n-1$,

$$\sigma((xy)^{n-k}x^k y^k) = \sigma((xy)^{n-k-1}xyx^k y^k) = \sigma(y^k(xy)^{n-k-1}xyx^k)$$

$$= \sigma(y^k(xy)^{n-k-1}x^{k+1}y) = \sigma((xy)^{n-k-1}x^{k+1}y^{k+1}).$$

This implies that $\sigma((xy)^n) = \sigma(x^n y^n)$. Thus

$$r(x,y) = (r((xy)^n))^{\frac{1}{n}} = (r(x^n y^n))^{\frac{1}{n}} = \|x^n\|^{\frac{1}{n}}\|y^n\|^{\frac{1}{n}}.$$

Let $n \longrightarrow \infty$, we have $r(xy) \leq r(x)r(y)$. It follows from the Corollary 5.2.3 of [Aupetit, B. (1991)] that $\mathcal{A}$ is essentially commutative.

Free matrix algebra. Let $\mathcal{X}_1, \cdots, \mathcal{X}_n$ be Banach spaces and $\mathcal{X} = \mathcal{X}_1 \oplus \cdots \oplus \mathcal{X}_n$ be the product Banach space. For each i, let $\|\cdot\|_i$ be the norm of $\mathcal{X}_i$. Then the norm $\|\cdot\|$ of $\mathcal{X}$ is defined by $\|x\| = \max\limits_{1 \leq i \leq n}\{\|x_i\|_i\}$, where $x = \{x_1, \cdots, x_n\} \in \mathcal{X}$. Let $S \in \mathcal{L}(\mathcal{X})$, then $S = (S_{ij})_{n \times n}$, where $S_{ij} \in \mathcal{L}(\mathcal{X}_i, \mathcal{X}_j)$.

Definition 5.4.7 Let $\mathcal{X} = \mathcal{X}_1 \oplus \cdots \oplus \mathcal{X}_n$ be the product Banach space of $\{\mathcal{X}_i\}_{i=1}^n$. An algebra $\mathcal{A} \subset \mathcal{L}(\mathcal{X})$ is called a free matrix algebra of order n if $S = (S_{ij})_{n \times n} \in \mathcal{A}$ implies that

$$S(i,j) := \begin{bmatrix} 0 & \cdots & 0 & \cdots & 0 \\ \vdots & & \vdots & & \vdots \\ 0 & \cdots & S_{ij} & \cdots & 0 \\ \vdots & & \vdots & & \vdots \\ 0 & \cdots & 0 & \cdots & 0 \end{bmatrix} \in \mathcal{A};$$

$\mathcal{A}$ is called a unital free matrix algebra of order n if I, the identity on $\mathcal{X}$, is in $\mathcal{A}$.

Let $\mathcal{A} \subset \mathcal{B}(\mathcal{X})$ be a free matrix algebra of order n. Denote

$$\mathcal{A}_{ij} = \left\{ S_{ij} : \begin{bmatrix} 0 & \cdots & 0 & \cdots & 0 \\ \vdots & & \vdots & & \vdots \\ 0 & \cdots & S_{ij} & \cdots & 0 \\ \vdots & & \vdots & & \vdots \\ 0 & \cdots & 0 & \cdots & 0 \end{bmatrix} \in \mathcal{A} \right\}$$

and $\mathcal{A}_i = \mathcal{A}_{ii}$, then it has the following properties:

(5.4.1) $\mathcal{A}_i$ is a Banach algebra and $\mathcal{A}$ is unital if and only if each $\mathcal{A}_i$ is unital, $1 \leq i \leq n$.

(5.4.2) $\mathcal{A}_{ij}$ is a commutative additive group for $1 \leq i, j \leq n$.

(5.4.3) If $S_{ij} \in \mathcal{A}_{ij}$ and $S_{jk} \in \mathcal{A}_{jk}$, then $S_{ij}S_{jk} \in \mathcal{A}_{ik}$. Particularly, if $S_{ij} \in \mathcal{A}_{ij}$ and $S_{ji} \in \mathcal{A}_{ji}$, then $S_{ij}S_{ji} \in \mathcal{A}_i$.

Definition 5.4.8   The map $\pi_{ij} : \mathcal{A} \xrightarrow{onto} \mathcal{A}_{ij}$ is called an entry map if

$$\pi_{ij}((S_{ij})_{n \times n}) = S_{ij}$$

for $(S_{ij})_{n \times n} \in \mathcal{A}$.

In most cases, we take great care of the diagonal elements $\mathcal{A}_i$'s in a free matrix algebra. So we denote the free matrix algebra $\mathcal{A}$ of order n with the diagonal elements $\mathcal{A}_1, \cdots, \mathcal{A}_n$ by $\mathcal{A} \sim diag(\mathcal{A}_1, \cdots, \mathcal{A}_n)$.

Example 5.4.9   *Let $\mathcal{H}_i$ be a separable Hilbert space, $1 \leq i \leq n$, and $\mathcal{H} = \bigoplus_{i=1}^{n} \mathcal{H}_i$. Let $T_i \in \mathcal{L}(\mathcal{H}_i)$ and $T = \bigoplus_{i=1}^{n} T_i \in \mathcal{L}(\mathcal{H})$. Then $\mathcal{A} = \mathcal{A}'(T)$ is a typical free matrix algebra, where $\mathcal{A}_i = \mathcal{A}'(T_i), \mathcal{A}_{ij} = ker\tau_{T_i, T_j}$.*

Definition 5.4.10   Let $\mathcal{A}$ be a unital Banach algebra. Two idempotents $e, f \in \mathcal{A}$ are called orthogonal, denoted by $e \perp f$, if $ef = fe = 0$. A set of elements $\{e_i\}_{i=1}^{m} \subset \mathcal{A}$; $n < +\infty$ is called a finite decomposition of $\mathcal{A}$ if $e_i \perp e_j$ for $1 \leq i \neq j \leq n$ and

$$e_1 + \cdots + e_n = 1.$$

Example 5.4.11   *Let $\{e_1, \cdots, e_m\} \subset \mathcal{A}$ be a finite decomposition of $\mathcal{A}$. Define*

$$\mathcal{M} := \{(S_{ij})_{n \times n} : S_{ij} = e_i S e_j, S \in \mathcal{A}\},$$

$$\chi_i = \{e_i S e_i : S \in \mathcal{A}\} \quad 1 \leq i \leq n.$$

Then χ_i is a Banach space and $\mathcal{M}$ is a free matrix algebra of order n, $\mathcal{M} \in \mathcal{L}(\mathcal{X})$, acting on $\mathcal{X} = \mathcal{X}_1 \oplus \cdots \oplus \mathcal{X}_n$.
    Define a map $\phi : \mathcal{A} \longrightarrow \mathcal{M}$ by

$$\phi(S) = (S_{ij})_{n \times n}, \quad S \in \mathcal{A},$$

then ϕ is a continuous isomorphism.

Lemma 5.4.12   *Let $\mathcal{A} \sim diag(\mathcal{A}_1, \mathcal{A}_2, \cdots, \mathcal{A}_n)$ and $\mathcal{J} \subset \mathcal{A}$ be an ideal. Then $\mathcal{J}$ is a free matrix algebra and $\mathcal{J} \sim diag(\mathcal{J}_1, \mathcal{J}_2, \cdots, \mathcal{J}_n)$. Furthermore, either $\mathcal{J}_i = \mathcal{A}_i$ or $\mathcal{J}_i$ is an ideal of $\mathcal{A}_i$, $1 \leq i \leq n$.*
Proof   For $S = (S_{ij})_{n \times n} \in \mathcal{J}, S(i,j) = I_{\mathcal{A}}(i,j) S I_{\mathcal{A}}(i,j) \in \mathcal{J}_i$, where $S(i,j)$ is defined in Definition 5.4.7. Thus $\mathcal{J}$ is a free matrix algebra of order n.

Since $\mathcal{A}$ is a free matrix algebra, a simple computation shows that either $\mathcal{J}_i = \mathcal{A}_i$ or $\mathcal{J}_i$ is an ideal of $\mathcal{A}_i$, $1 \leq i \leq n$.

Lemma 5.4.13 *Let $\mathcal{A} \sim diag(\mathcal{A}_1, \mathcal{A}_2, \cdots, \mathcal{A}_n)$ be a unital free matrix algebra of order n acting on $\mathcal{X} = \mathcal{X}_1 \oplus \cdots \oplus \mathcal{X}_n$. If $\mathcal{A}$ is an irreducible subalgebra of $\mathcal{L}(\mathcal{X})$, then $\mathcal{A}_i$ is an irreducible subalgebra of $\mathcal{L}(\mathcal{X}_i)$, $1 \leq i \leq n$.*

Proof For $1 \leq i \leq n$, since $\mathcal{A}$ is an irreducible subalgebra of $\mathcal{L}(\mathcal{X})$, for arbitrary $x, y \in \mathcal{X}$ there exists $(S_{ij})_{n \times n} \in \mathcal{A}$ such that $(S_{ij})_{n \times n} x = y$. A simple computation shows that

$$S_{ii} x_i = y_i, \ \forall \ x_i, y_i \in \mathcal{X}_i.$$

Thus $\mathcal{A}_i$ is an irreducible subalgebra of $\mathcal{L}(\mathcal{X}_i)$, $1 \leq i \leq n$.

Definition 5.4.14 Let

$$\mathcal{A} \sim diag(\mathcal{A}_1, \mathcal{A}_2, \cdots, \mathcal{A}_n)$$

and

$$\mathcal{B} \sim diag(\mathcal{B}_1, \mathcal{B}_2, \cdots, \mathcal{B}_n)$$

be two free matrix algebras of order n. The map $\phi : \mathcal{A} \longrightarrow \mathcal{B}$ is called a *freely matrical morphism* if
  (i) $\phi(\mathcal{A}) \subseteq \mathcal{B}$;
  (ii) $\phi(\alpha S_1 + \beta S_2) = \alpha \phi(S_1) + \beta \phi(S_2)$ for $S_1, S_2 \in \mathcal{A}$ and $\alpha, \beta \in \mathbf{C}$;
  (iii) $\phi(S_1 S_2) = \phi(S_1) \phi(S_2)$ for $S_1, S_2 \in \mathcal{A}$;
  (iv) If $\phi(S) = T$, then $\phi(S(i,j)) = T(i,j)$.

If in addition $\phi(I_\mathcal{A}) = I_\mathcal{B}$, then ϕ is called a *freely matrical homomorphism*.

If ϕ is a freely matrical morphism from $\mathcal{A}$ to $\mathcal{B}$, then ϕ induces a morphism from $\mathcal{A}_{ij}$ to $\mathcal{B}_{ij}$. In general, we will not distinguish ϕ with the morphism induced by it from $\mathcal{A}_{ij}$ to $\mathcal{B}_{ij}$.

Lemma 5.4.15 *Let $\mathcal{A}$ be a unital free matrix algebra of order n and $\mathcal{J} \subset \mathcal{A}$ be a closed ideal of $\mathcal{A}$. Then $\mathcal{A}/\mathcal{J}$ is a unital free matrix algebra of order n. Furthermore, there is a freely matrical morphism $\phi_\mathcal{J}$ from $\mathcal{A}$ to $\mathcal{A}/\mathcal{J}$.*

Proof Let $\phi_\mathcal{J}$ be the canonical map from $\mathcal{A}$ to $\mathcal{A}/\mathcal{J}$. Denote $e_i = I_\mathcal{A}(i,i)$ and $f_i = \phi_\mathcal{J}(e_i)$ for $1 \leq i \leq n$. Then $\{f_1, \cdots, f_n\} \subset \mathcal{A}/\mathcal{J}$ is a decomposition of $\mathcal{A}/\mathcal{J}$. By Example 5.4.11, $\mathcal{A}/\mathcal{J}$ is a free matrix algebra of order n. It is easily seen that $\phi_\mathcal{J}$ is a freely matrical morphism from $\mathcal{A}$ to $\mathcal{A}/\mathcal{J}$.

Theorem 5.4.16   *Let $\mathcal{A} \sim diag(\mathcal{A}_1, \mathcal{A}_2, \cdots, \mathcal{A}_n)$, then $\mathcal{A}$ is an FIR algebra if and only if each $\mathcal{A}_i$ is an FIR algebra.*

Proof   "$\Leftarrow$"   Suppose that π is a continuous irreducible representation of $\mathcal{A}$ acting on a Banach space $\mathcal{X}$. Let $P_i = \pi(e_i)$, then $\pi(\mathcal{A})$ is a unital free matrix algebra and $\pi(\mathcal{A}) \sim diag(\pi(\mathcal{A}_1), \pi(\mathcal{A}_2), \cdots, \pi(\mathcal{A}_n))$ acts on $\mathcal{X} = P_1 \mathcal{X} \oplus \cdots \oplus P_n \mathcal{X}$. By Lemma 5.4.13, every $\pi(\mathcal{A}_i)$ is a unital irreducible algebra, $\pi(\mathcal{A}_i) \subset \mathcal{L}(P_i \mathcal{X})$ or $P_i \mathcal{X} = 0, 1 \leq i \leq n$. Since $\mathcal{A}_i$ is FIR for $1 \leq i \leq n, dim P_i \mathcal{X} < \infty$. Thus

$$dim\mathcal{X} = \sum_{i=1}^{n} dim P_i \mathcal{X} < \infty.$$

"$\Rightarrow$"   Suppose that π is a continuous irreducible representation of $\mathcal{A}_1$ and $\mathcal{J}_1 = ker\pi$. By Kaplansky [Bonsall, F.F. and Duncan, J. (1973)], there exists a unique primitive ideal $\mathcal{J} \subset \mathcal{A}$ such that $\mathcal{J}_1 = \mathcal{J} \cap \mathcal{A}_1$. Since $\mathcal{A}$ is FIR, $\mathcal{A}/\mathcal{J}$ is a finite dimensional algebra and $\mathcal{A}_1/\mathcal{J}_1$ is a subalgebra of $\mathcal{A}/\mathcal{J}$. Thus $\mathcal{A}_1/\mathcal{J}_1$ is finite dimensional and $\mathcal{A}_1$ is FIR. Similarly, we can prove that $\mathcal{A}_i$ is FIR for all i.

By Example 5.4.11, we can restate Theorem 5.4.16 as follows.

Theorem 5.4.16′   *Let $\mathcal{A}$ be a unital Banach algebra and*

$$\{e_1, e_2, \cdots, e_n\} \subset \mathcal{A}$$

be a decomposition of $\mathcal{A}$. Denote $\mathcal{A}_i = e_i \mathcal{A} e_i, 1 \leq i \leq n$. Then $\mathcal{A}$ is FIR if and only if $\mathcal{A}_i$ is FIR for all i.

An FIR algebra $\mathcal{A}$ is said to be stable finite if $M_n(\mathcal{A})$ is FIR for each natural number n.

Corollary 5.4.17   *Let $\mathcal{A}$ be a unital FIR algebra, then for each natural number, $M_n(\mathcal{A})$ is FIR.*

Proof   This is a straightforward corollary of Theorem 5.4.16′.

Similar to the discussion in Section 5.3, we can get the following result about the type-1 operator $T = \bigoplus_{k=1}^{n} T_k$.

Theorem   5.4.18   *Let $\mathcal{J}_1 \in \mathcal{M}(\mathcal{A}'(T_1))$,   then   there   exists   a   unique $\mathcal{J} \in \mathcal{M}(\mathcal{A}'(T))$ such that $\pi_{11}(\mathcal{J}) = \mathcal{J}_1$, where π_{11} is defined in Section 5.3.*

Theorem 5.4.19   *Let $T = T_1 \oplus T_2$ be a type-1 operator. If $\overline{\mathcal{J}}_1$ is a subalgebra generated by $ker\tau_{T_1, T_2}$ and $ker\tau_{T_2, T_1}$, then*

(i) $\mathcal{A}'(T)$ *is 1-homogeneous if and only if* $\overline{\mathcal{J}}_1 \subset rad\mathcal{A}'(T_1)$. *Moreover, if* $\mathcal{A}'(T)$ *is semisimple, i.e.,* $rad\mathcal{A} = \{0\}$, *then* $\mathcal{A}'(T)$ *is 1-homogeneous if and only if*

$$\mathcal{A}'(T) \cong \mathcal{A}'(T_1) \oplus \mathcal{A}'(T_2).$$

(ii) $\mathcal{A}'(T)$ *is 2-homogeneous if and only if there exist a positive integer* n *and* $x_i \in ker\tau_{T_1,T_2}$, $y_i \in ker\tau_{T_2,T_1}$, $1 \leq i \leq n$, *such that*

$$\sum_{i=1}^{n} x_i y_i = I_{\mathcal{A}'(T_1)}$$

and

$$\sum_{i=1}^{n} y_i x_i = I_{\mathcal{A}'(T_2)} + R,$$

where $R \in rad\mathcal{A}'(T_2)$. *Furthermore, if* $\mathcal{A}'(T)$ *is semisimple, then* $\sum_{i=1}^{n} y_i x_i = I_{\mathcal{A}'(T_2)}$.

Proof The first part of the theorem is obvious. We need only to show the second part.

Suppose that $\mathcal{A}'(T)$ is 2-homogeneous.

{Claim} There exist a positive integer n and $x_i \in ker\tau_{T_1,T_2}$, $y_i \in ker\tau_{T_2,T_1}$, $1 \leq i \leq n$, such that $\sum_{i=1}^{n} x_i y_i = I_{\mathcal{A}'(T_1)}$. Otherwise, $\overline{\mathcal{J}}_1$ generated by $ker\tau_{T_1,T_2}$ and $ker\tau_{T_2,T_1}$ is an ideal of $\mathcal{A}'(T_1)$. By Kaplansky Theorem, there is a $\mathcal{J} \in \mathcal{M}(\mathcal{A}'(T))$ such that $\pi_{11}(\mathcal{J}) = \mathcal{J}_1 \supset \overline{\mathcal{J}}_1$. Thus for arbitrary $x \in ker\tau_{T_1,T_2}$ and $y \in ker\tau_{T_2,T_1}$, $xy \in \pi_{11}(\mathcal{J})$. Then it follows from Corollary 5.3.6, $ker\tau_{T_1,T_2}/\pi_{12}(\mathcal{J}) = ker\tau_{T_2,T_1}/\pi_{21}(\mathcal{J}) = \{0\}$. This implies that $\mathcal{A}'(T)/\mathcal{J} \cong \mathbf{C}$. A contradiction.

Therefore, there exist $x_i \in ker\tau_{T_1,T_2}$ and $y_i \in ker\tau_{T_2,T_1}$ such that $\sum_{i=1}^{n} x_i y_i = I_{\mathcal{A}'(T_1)}$. Note that $\sum_{i=1}^{n} y_i x_i - I_{\mathcal{A}'(T_2)}$ is a nilpotent and idempotent, thus

$$\sum_{i=1}^{n} y_i x_i - I_{\mathcal{A}'(T_2)} \in rad\mathcal{A}'(T_2) \quad [\text{Antonevich, A. and Krupmk, N. (2000)}].$$

Conversely, if there exist a positive integer n and $x_i \in ker\tau_{T_1,T_2}$, $y_i \in ker\tau_{T_2,T_1}$, $1 \leq i \leq n$, such that $\sum_{i=1}^{n} x_i y_i = I_{\mathcal{A}'(T_1)}$ and $\sum_{i=1}^{n} y_i x_i = I_{\mathcal{A}'(T_2)} +$

R, $R{\in}rad\mathcal{A}'(T_2)$). Similar to the Proof of Theorem 5.3.7, we can get $\mathcal{A}'(T)/\mathcal{J}{\cong}M_2(\mathbf{C})$ for all $\mathcal{J}{\in}\mathcal{M}(\mathcal{A}'(T))$. Thus $\mathcal{A}'(T)$ is 2-homogeneous.

5.5 The (SI) Representation Theorem of Cowen-Douglas Operators

In this section, we always assume that $T{\in}\mathcal{L}(\mathcal{H})$ is a Cowen-Douglas operator with index n.

Lemma 5.5.1 *Let* $T{\in}\mathcal{B}_n(\Omega)$ *and* $T = A{\oplus}B$, $T = A\dot{+}D$ *be two decompositions of* T, *then* $D{\sim}B$.

Proof We may find three idempotents $P_A, P_B, P_D{\in}\mathcal{A}'(T)$ such that

$$T|_{ranP_A} = A, \; T|_{ranP_B} = B, \; T|_{ranP_D} = D,$$

then $P_A + P_B = I$, $P_A + P_D = I$. Thus P_D is can be regarded as an invertible operator from $ranP_B$ to $ranP_D$.

Suppose $A{\in}\mathcal{B}_m(\Omega)$, then $B, D{\in}\mathcal{B}_{n-m}(\Omega)$. Let $(e_1^A(\lambda), \cdots, e_m^A(\lambda))$ be the holomorphic frame of $ker(A - \lambda)$ and $(f_1^B(\lambda), \cdots, f_{n-m}^B(\lambda))$ be the holomorphic frame of $ker(B - \lambda)$. Then $(P_D f_1^B(\lambda), \cdots, P_D f_{n-m}^B(\lambda))$ is the holomorphic frame of $ker(D - \lambda)$. Denote $(P_D|_{ranP_B})^{-1} = X_D$, then

$$X_D D P_D f_i^B(\lambda) = \lambda X_D P_D f_i^B(\lambda) = \lambda f_i^B(\lambda) = B f_i^B(\lambda).$$

So $B{\sim}D$.

Lemma 5.5.2 *Let*

$$T = A_1^{(m_1)}{\oplus}A_2^{(m_2)}{\oplus}\cdots{\oplus}A_k^{(m_k)}, \; A_i{\in}(SI){\cap}\mathcal{B}_{n_i}(\Omega), \; i = 1, 2, \cdots, k$$

and $A_i{\not\sim}A_j$ *for* $i \neq j$, *then* $\bigvee(\mathcal{A}'(T)){\cong}\mathbf{N}^{(k)}$ *if and only if* $\bigvee(\mathcal{A}'(\bigoplus\limits_{i=1}^{k} A_i^{(n_i)})){\cong}\mathbf{N}^{(k)}$, *where* $\{m_1, \cdots, m_k\}$ *and* $\{n_1, \cdots, n_k\}$ *are two ordered sets of positive integers.*

Proof We need only prove the necessity of the lemma. By Theorem 4.2.1, $\bigvee(\mathcal{A}'(T)){\cong}\mathbf{N}^{(k)}$ implies that $\bigoplus\limits_{i=1}^{k} A_i^{(mm_i)}$ has a unique finite (SI) decomposition up to similarity, where $m = \sum\limits_{i=1}^{k} nn_i$. Set $T_1 = \bigoplus\limits_{i=1}^{k} A_i^{(n_i)}$, then $T_1^{(n)} = \bigoplus\limits_{i=1}^{k} A_i^{(nn_i)}$. Since $mm_i{\geq}nn_i, 1{\leq}i{\leq}k$ and $\bigoplus\limits_{i=1}^{k} A_i^{(mm_i)} =$

$T_1^{(n)} \oplus \bigoplus_{i=1}^{k} A_i^{(mm_i - nn_i)}$. By Theorem 4.2.1 again, $T^{(n)}$ has a unique (SI) decomposition up to similarity and

$$\bigvee (\mathcal{A}'(\bigoplus_{i=1}^{k} A_i^{(n_i)})) \cong \mathbf{N}^{(k)}.$$

Remark 4.5.3 By Theorem 4.2.1 and Lemma 5.5.2, if $A_i \in (SI), 1 \leq i \leq k$ and $A_i \not\sim A_j$ $(1 \leq i \neq j \leq k)$, then $(\bigoplus_{i=1}^{k} A_i^{(n_i)})^{(n)}$ has a unique (SI) decomposition up to similarity if and only if $(\bigoplus_{i=1}^{k} A_i)^{(n)}$ has a unique (SI) decomposition up to similarity.

Lemma 5.5.4 *Let $\mathcal{A}$ be a finite irreducible algebra and $\mathcal{J} \subseteq \mathcal{A}$ be a closed ideal and $0 \rightarrow \mathcal{J} \xrightarrow{i} \mathcal{A} \xrightarrow{\pi} \mathcal{A}/\mathcal{J} \rightarrow 0$ be an exact sequence. If $\bigvee(\mathcal{A}) \cong N$ and $[I_\mathcal{A}] = 1$, then the induced map*

$$\pi_* : K_0(\mathcal{A}) \rightarrow K_0(\mathcal{A}/\mathcal{J})$$

is injective.

Proof Suppose that n is a natural number and $p, q \in M_n(\mathcal{A})$ are two idempotents. Since $\bigvee(\mathcal{A}) \cong N$, $[p] = [e_r]$ and $[q] = [e_s]$, where $e_k = diag(I_\mathcal{A}, \cdots, I_\mathcal{A}, 0, \cdots)$ with k $I_\mathcal{A}$'s in the diagonal for $k = r, s$. If $\pi_*([p]) = \pi_*([q])$, then $[\pi(e_r)] = [\pi(e_s)]$. Since $\mathcal{A}$ is FIR, $\mathcal{A}/\mathcal{J}$ is FIR. Moreover, $\mathcal{A}/\mathcal{J}$ is stably finite by Corollary 5.4.17, thus $r = s$ and $[p] = [q]$. This proves that $\pi_* : K_0(\mathcal{A}) \rightarrow K_0(\mathcal{A}/\mathcal{J})$ is injective.

Lemma 5.5.5 *Let $\mathcal{A}$ be a unital FIR algebra and $\mathcal{J}_1 \neq \mathcal{J}_2$ be two maximal ideals of $\mathcal{A}$. Denote $\mathcal{J} = \mathcal{J}_1 \cap \mathcal{J}_2$, then $\mathcal{A}/\mathcal{J} \cong \mathcal{A}/\mathcal{J}_1 \oplus \mathcal{A}/\mathcal{J}_2$.*
Proof Suppose that ϕ_i be a quotient map from $\mathcal{A}$ to $\mathcal{A}/\mathcal{J}_i$, $i = 1, 2$. Define $\phi : \mathcal{A} \longrightarrow \mathcal{A}/\mathcal{J}_1 \oplus \mathcal{A}/\mathcal{J}_2$ as follows:

$$\phi(a) = \phi_1(a) \oplus \phi_2(a), \quad a \in \mathcal{A}.$$

Then ϕ is a homomorphism. By Chinese Remainder Theorem, ϕ is surjective. Since $ker\phi = ker\phi_1 \cap ker\phi_2 = \mathcal{J}_1 \cap \mathcal{J}_2 = \mathcal{J}$, $\mathcal{A}/\mathcal{J} \cong \mathcal{A}/\mathcal{J}_1 \oplus \mathcal{A}/\mathcal{J}_2$.

Lemma 5.5.6 *Let A_i $(i = 1, 2, \cdots, k)$ be strongly irreducible operators and $A_i \not\sim A_j$ for $1 \leq i \neq j \leq k$. Let $\{n_1, n_2, \cdots, n_k\}$ be an ordered set of positive* *integers and* $T = \bigoplus\limits_{i=1}^{k} A_i^{(n_i)} \oplus B$ *be a Cowen-Douglas operator. Denote*

$$S_1 = A_1^{(n_1)} \oplus B, \quad S_2 = \bigoplus_{i=2}^{k} A_i^{(n_i)},$$

i.e., $T = S_1 \oplus S_2$. *If* $\bigvee(\mathcal{A}'(S_2)) \cong N^{(k-1)}$, *then*

$$\overline{\mathcal{J}}_1 = \{\sum_{i=1}^{n} x_i y_i, \ x_i \in ker\tau_{S_1,S_2}, \ y_i \in ker\tau_{S_2,S_1}, \ 1 \leq i \leq n, \ n = 1, 2, \cdots\}$$

is a proper ideal of $\mathcal{A}'(S_1)$.

Proof If $\overline{\mathcal{J}}_1 = \mathcal{A}'(S_1)$, then there exist

$$x_1, x_2, \cdots, x_n \in ker\tau_{S_1,S_2}, \ y_1, y_2, \cdots, y_n \in ker\tau_{S_2,S_1}$$

such that

$$x_1 y_1 + \cdots + x_n y_n = I_{\mathcal{A}'(S_1)}.$$

So

$$P = \begin{bmatrix} y_1 \\ \vdots \\ y_n \end{bmatrix} \begin{bmatrix} x_1 & \cdots & x_n \end{bmatrix} \in M_n(\mathcal{A}'(S_2))$$

is an idempotent. By the similar argument used in the proof of Lemma 5.3.10, we have $1_{\mathcal{A}'(S_1)} \oplus 0 \sim_a 0 \oplus P$ in $\mathcal{A}'(S_1 \oplus S_2^{(n)})$. Let $S_1 \in \mathcal{L}(\mathcal{K}_1)$, $S_2^{(n)} \in \mathcal{L}(\mathcal{K}_2)$. Then by Lemma 4.2.4,

$$A_1^{(n)} \oplus B = S_1$$
$$= (S_1 \oplus S_2^{(n)})|_{(I_{\mathcal{A}'(S_1)} \oplus 0)(\mathcal{K}_1 \oplus \mathcal{K}_2)} \sim (S_1 \oplus S_2^{(n)})|_{(0 \oplus P)(\mathcal{K}_1 \oplus \mathcal{K}_2)} = S_2^{(n)}|_{P\mathcal{K}_2}.$$

Since $\bigvee(\mathcal{A}'(S_2)) \cong N^{(k-1)}$, it follows from Theorem 4.2.1 that $S_2^{(n)}$ has a unique (SI) decomposition up to similarity. Since $A_1^{(n)} \oplus B \sim S_2^{(n)}|_{P\mathcal{K}_2}$, $A_1 \sim A_j$, where $2 \leq j \leq k$. This contradicts to our assumption that $A_i \nsim A_j$ for $1 \leq i \neq j \leq k$. Thus $\overline{\mathcal{J}}_1$ is a proper ideal of $\mathcal{A}'(S_1)$.

Lemma 5.5.7 *Let* $\mathcal{A} \sim diag\{\mathcal{A}_1, \cdots, \mathcal{A}_n\}$ *be a unital Banach algebra and* $\mathcal{J} \sim diag\{\mathcal{J}_1, \cdots, \mathcal{J}_n\}$ *be a closed ideal of* $\mathcal{A}$. *Suppose that*

$$\pi : \mathcal{A} \longrightarrow \mathcal{A}/\mathcal{J}, \quad \pi_1 : \mathcal{A}_1 \longrightarrow \mathcal{A}_1/\mathcal{J}_1$$

are quotient maps and $\mathcal{A}/\mathcal{J} \cong (\mathcal{A}_1/\mathcal{J}_1, 0, \cdots, 0)$, *then*

$$\pi_{1*}(K_0(\mathcal{A}_1)) \cong \pi_*(K_0(\mathcal{A})).$$

Proof Define $\alpha_* : \pi_{1*}(K_0(\mathcal{A}_1)) \longrightarrow \pi_*(K_0(\mathcal{A}))$ as follows

$$\alpha_*(\pi_{1*}([e])) = \pi_*([e \oplus 0 \oplus \cdots \oplus 0]) \text{ for all } e \in M_k(\mathcal{A}_1).$$

We first show that α_* is injective. If $\pi_*([e \oplus 0 \oplus \cdots \oplus 0]) = 0$, then

$$\pi_*(e \oplus 0 \oplus \cdots \oplus 0) \sim_s 0.$$

Thus there exists r such that $\pi_*(e \oplus 0 \oplus \cdots \oplus 0) \oplus r \sim_a 0 \oplus r$. Then

$$\pi_{1*}(e) \oplus 0 \oplus \cdots \oplus 0 \oplus r \sim_a 0 \oplus r.$$

Note that $0 \oplus r \sim_a r$ for each r. Let $r' = 0 \oplus \cdots \oplus 0 \oplus r$, then

$$\pi_{1*}(e) \oplus r' \sim_a 0 \oplus r'.$$

This shows that $\pi_{1*}(e) \sim_s 0$. Therefore, $[\pi_{1*}(e)] = \pi_{1*}([e]) = 0$.

Remark For unital Banach algebra $\mathcal{A}$, let $p, q \in M_\infty(\mathcal{A})$ be two idempotents, we say $p \sim_s q$ if there exists an idempotent r in $M_\infty(\mathcal{A})$ such that $p \oplus r \sim_a q \oplus r$.

Second, we will prove that α_* is surjective. It follows from $\mathcal{A}/\mathcal{J} \cong \mathcal{A}_1/\mathcal{J}_1$ that for every $\beta \in K_0(\mathcal{A})$, there is an e such that

$$\pi_{1*}([e]) = \pi_*([\beta]).$$

Similarly, for $(\beta_{ij})_{n \times n} \in M_n(K_0(\mathcal{A}))$, we have $[e_{ij}]_{n \times n} \in M_n(K_0(\mathcal{A}_1))$ such that

$$\pi_{1*}([e_{ij}]) = \pi_*([\beta_{ij}]).$$

By the basic K-theory, we have

$$[(\pi_*(e_{ij}) \oplus 0 \cdots \oplus 0)_{n \times n}] = [\pi_*((e_{ij})_{n \times n}) \oplus 0].$$

Thus α_* is surjective and so α_* is an isomorphism.

Lemma 5.5.8 *Let* $T = A_1 \oplus A_2$, *where* A_1 *and* A_2 *are strongly irreducible Cowen-Douglas operators and* $A_1 \not\sim A_2$. *Suppose that* n *is a positive integer and*

$$T^{(n)} \sim A_1^{(m_1)} \oplus A_2^{(m_2)} \oplus B_1 \oplus \cdots \oplus B_m$$

is another finite decomposition of $T^{(n)}$, *where* $m_1, m_2, m \geq 0$, $B_i \in (SI)$ *and* $B_i \not\sim A_j$ *for* $1 \leq i \leq m$ *and* $j = 1, 2$. *Then* $m_i + m = n, i = 1, 2$.

Proof Since $T = A_1 \oplus A_2$, $T^{(n)} = A_1^{(n)} \oplus A_2^{(n)}$.

{Claim 1} $m_i + m \leq n$, $i = 1, 2$.

Otherwise assume that $m_1 + m > n$.

Let $B = B_1 \oplus B_2 \oplus \cdots \oplus B_m$, $R = T^{(n)} \oplus A_2$,   then

$$\mathcal{A}'(R) = \begin{bmatrix} \mathcal{A}'(T^{(n)}) & ker\tau_{T^{(n)}, A_2} \\ ker\tau_{A_2, T^{(n)}} & \mathcal{A}'(A_2) \end{bmatrix}.$$

Suppose that $\hat{\mathcal{J}}$ is a subalgebra generated by $ker\tau_{T^{(n)}, A_2}$ and $ker\tau_{A_2, T^{(n)}}$. By Theorem 5.2.2, $\bigvee \mathcal{A}'(A_2) \cong \mathbf{N}$. By Lemma 5.5.6, $\hat{\mathcal{J}}$ is a proper ideal of $\mathcal{A}'(T^{(n)})$. Let $\mathcal{J}_1$ be the closure of $\hat{\mathcal{J}}$, then $\mathcal{J}_1$ is a closed ideal of $\mathcal{A}'(T^{(n)})$.

Set $\mathcal{J} = \begin{bmatrix} \mathcal{J}_1 & ker\tau_{T^{(n)}, A_2} \\ ker\tau_{A_2, T^{(n)}} & \mathcal{A}'(A_2) \end{bmatrix} \subset \mathcal{A}'(R)$, then $\mathcal{J}$ is a closed ideal of $\mathcal{A}'(R)$.

When $T^{(n)} = A_1^{(n)} \oplus A_2^{(n)}$, we have:

$$ker\tau_{T^{(n)}, A_2} = \begin{bmatrix} ker\tau_{A_1, A_2} \\ \vdots \\ ker\tau_{A_1, A_2} \\ \mathcal{A}'(A_2) \\ \cdots \\ \mathcal{A}'(A_2) \end{bmatrix}_{2n \times 1}$$

$$:= \left\{ \begin{bmatrix} x_1 \\ \vdots \\ x_n \\ y_1 \\ \vdots \\ y_n \end{bmatrix} : x_i \in ker\tau_{A_1, A_2}, \ y_i \in \mathcal{A}'(A_2), i = 1, 2, \cdots, n \right\}$$

and

$$ker\tau_{A_2, T^{(n)}}$$

$$= [ker\tau_{A_2, A_1}, \ \cdots, \ ker\tau_{A_2, A_1}, \ \mathcal{A}'(A_2), \ \cdots, \ \mathcal{A}'(A_2)]_{1 \times 2n}$$

$$= \{(x_1', \cdots, x_n', y_1', \cdots, y_n') : x_i' \in ker\tau_{A_2, A_1}, \ y_i' \in \mathcal{A}'(A_2), i = 1, 2, \cdots, n\}.$$

Thus

$$ker\tau_{T^{(n)},A_2}\cdot ker\tau_{A_2,T^{(n)}} = \begin{bmatrix} [ker\tau_{A_1,A_2}\cdot ker\tau_{A_2,A_1}]_{n\times n} & * \\ * & M_n(\mathcal{A}'(A_2)) \end{bmatrix}$$

(5.5.1)

where

$$ker\tau_{A_1,A_2}\cdot ker\tau_{A_2,A_1} = \{xx' : x\in ker\tau_{A_1,A_2}, x'\in ker\tau_{A_2,A_1}\},$$

$$[ker\tau_{A_1,A_2}\cdot ker\tau_{A_2,A_1}]_{n\times n} = \left\{ \begin{bmatrix} x_1x'_1 & \cdots & x_1x'_n \\ \vdots & & \vdots \\ x_nx'_1 & \cdots & x_nx'_n \\ y_1y'_1 & \cdots & y_1y'_n \\ \vdots & & \vdots \\ y_ny'_1 & \cdots & y_ny'_n \end{bmatrix} : \begin{array}{l} x_i\in ker\tau_{A_1,A_2} \\ \\ x'_i\in ker\tau_{A_2,A_1} \\ \\ y_i, y'_i\in \mathcal{A}'(A_2) \\ \\ i = 1, 2, \cdots, n \end{array} \right\}.$$

Consider another decomposition of $T^{(n)}$:

$$T^{(n)}\sim A_1^{(m_1)}\oplus A_2^{(m_2)}\oplus B_1\oplus\cdots\oplus B_m = A_1^{(m_1)}\oplus A_2^{(m_2)}\oplus B\sim A_1^{(m_1)}\oplus B\oplus A_2^{(m_2)}.$$

Similarly, we have that

$$ker\tau_{T^{(n)},A_2}\cdot ker\tau_{A_2,T^{(n)}}$$

$$= diag([ker\tau_{A_1,A_2}\cdot ker\tau_{A_2,A_1}]_{m_1\times m_1}, ker\tau_{B_1,A_2}\cdot ker\tau_{A_2,B_1},$$

$$\cdots, \ ker\tau_{B_m,A_2}\cdot ker\tau_{A_2,B_m}, \cdots, M_{m_2}(\mathcal{A}'(A_2)))$$

(5.5.2)

Note that $\mathcal{J}$ can not be a maximal ideal of $\mathcal{A}'(R)$. Otherwise, by the construction of $\mathcal{J}$, $\mathcal{J}_1$ is a maximal ideal of $\mathcal{A}'(T^{(n)}$ and

$$\mathcal{A}'(R)/\mathcal{J} = \mathcal{A}'(A_1^{(n)}\oplus A_2^{(n+1)})/\mathcal{J}\cong\mathcal{A}'(T^{(n)}/\mathcal{J}_1\cong M_n(\mathbf{C}).$$

But from (5.5.1) and (5.5.2),

$$\mathcal{A}'(R)/\mathcal{J}\cong\mathcal{A}'(A_1^{(m_1)}\oplus B\oplus A_2^{(m_2+1)})/\mathcal{J}\cong M_{m_1+m}(\mathbf{C}).$$

This contradicts $m_1 + m > n$.

Now consider

$$\mathcal{A}'(T^{(n)}) = diag(\mathcal{A}'(A_1^{(n)}), \mathcal{A}'(A_2^{(n)}))$$

and

$$\mathcal{J}_1\sim diag(\mathcal{J}''_{11}, \mathcal{J}'_{22}).$$

Denote $\mathcal{A} = \mathcal{A}'(T^{(n)})/\mathcal{J}_1$. By Theorem 5.2.2, $\bigvee(\mathcal{A}'(A_2)) \cong \mathbf{N}$. Moreover, by Lemma 5.5.6, $\mathcal{J}_{11}''$ is a closed ideal of $\mathcal{A}'(A_1^{(n)})$, and $\mathcal{J}_{22}'' = \mathcal{A}'(A_2^{(n)}) = M_n(\mathcal{A}'(A_2))$. Therefore,

$$\mathcal{A} = (\mathcal{A}'(A_1^{(n)})/\mathcal{J}_{11}'') \oplus 0. \tag{5.5.3}$$

On the other hand, we consider

$$\mathcal{A}'(T^{(n)}) \sim diag(\mathcal{A}'(A_1^{(m_1)}), \mathcal{A}'(B_1), \cdots, \mathcal{A}'(B_m), \mathcal{A}'(A_2^{(m_2)})).$$

Now,

$$\mathcal{J}_1 = \mathcal{J}_1 \sim diag(\mathcal{J}_{11}, \mathcal{J}_{22}, \cdots, \mathcal{J}_{m+2,m+2}).$$

By Theorem 5.2.2 and Lemma 5.5.6 again, $\mathcal{J}_{11}$ is a closed ideal of $\mathcal{A}'(A_1^{(m_1)})$, $\mathcal{J}_{ii}$ is a closed ideal of $\mathcal{A}'(B_{i-1})$, $2 \leq i \leq m+1$, and $\mathcal{J}_{m+2,m+2} = \mathcal{A}'(A_2^{(m_2)})$. Therefore,

$$\mathcal{A} = \mathcal{A}'(T^{(n)})/\mathcal{J}_1 = (\mathcal{A}'(A_1^{(m_1)} \oplus B)/\mathcal{J}_1') \oplus 0, \tag{5.5.4}$$

where $\mathcal{J}_1' = \mathcal{J}_1' = diag(\mathcal{J}_{11}, \mathcal{J}_{22}, \cdots, \mathcal{J}_{m+1,m+1})$.

Without loss of generality, we can assume $m_1, m_2 > 0$. Otherwise, we can consider that:

$$T^{(2n)} = T^{(n)} \oplus T^{(n)} \sim A_1^{(n+m_1)} \oplus A_2^{(n+m_2)} \oplus B_1 \oplus \cdots \oplus B_m$$

and

$$T^{(2n)} = A_1^{(2n)} \oplus A_2^{(2n)}.$$

By (5.5.4), there exists a surjective homomorphism

$$\phi : \mathcal{A}'(A_1^{(m_1)} \oplus B) \longrightarrow \mathcal{A}.$$

By Theorem 4.4.3, $\mathcal{A}'(A_1)/rad\mathcal{A}'(A_1)$ is commutative. Thus $\mathcal{A}'(A_1^{(n)})$ and $\mathcal{A}$ both are n-homogeneous. Therefore $\mathcal{A}$ is an FIR algebra.

Furthermore,

$$\mathcal{A} = \mathcal{A} \sim diag(\mathcal{A}_1, \mathcal{A}_2, \cdots, \mathcal{A}_{m+1})$$

$$= \mathcal{A} \sim diag(\mathcal{A}'(A_1^{(m_1)})/\mathcal{J}_{11}, \mathcal{A}'(B_1)/\mathcal{J}_{22}, \cdots, \mathcal{A}'(B_m)/\mathcal{J}_{m+1,m+1}).$$

Since each $\mathcal{J}_{ii}$ is a proper ideal, $\mathcal{A}_i \neq 0$, $i = 1, 2, \cdots, m+1$. Suppose $\mathcal{J}_{11}'$ is a maximal ideal of $\mathcal{A}_1$. By Kaplansky theorem [Bonsall, F.F. and

Duncan, J. (1973)], there exists a unique maximal ideal $\mathcal{J}_2 \subset \mathcal{A}$ such that $\Phi_{11}(\mathcal{J}_2) = \mathcal{J}'_{11}$. Then

$$\mathcal{A}/\mathcal{J}_2 = \mathcal{A}/\mathcal{J}_2 \sim diag(\mathcal{A}_1/\mathcal{J}'_{11}, \ \mathcal{A}_2/\Phi_{22}(\mathcal{J}_2), \ \cdots, \ \mathcal{A}_{1+m}/\Phi_{1+m,1+m}(\mathcal{J}_2)).$$

Since $\mathcal{A}$ is n-homogeneous, $\mathcal{A}/\mathcal{J}_2 \cong M_n(C)$. Note that $m_1 + m > n$, by the arguments similar to that used in the proof of Theorem 5.3.7, we can see that there exist $m_1 + m - n$ natural numbers $\{k_1, k_2, \cdots, k_{m+m_1-n}\}$ in $\{1, 2, \cdots, m+1\}$ such that

$$\mathcal{A}_j/\Phi_{jj}(\mathcal{J}_2) = 0, \ j \in \{k_1, k_2, \cdots, k_{m+m_1-n}\}.$$

Without loss of generality, assume that when $j = m+1$,

$$\mathcal{A}_{1+m,1+m}/\Phi_{1+m}(\mathcal{J}_2) = 0.$$

Suppose that $\mathcal{J}'_{1+m,1+m}$ is a maximal ideal of $\mathcal{A}_{1+m}$, using Kaplansky theorem again, we can find a unique maximal ideal $\mathcal{J}_3$ of $\mathcal{A}$ such that

$$\Phi_{1+m,1+m}(\mathcal{J}_3) = \mathcal{J}'_{1+m,1+m}.$$

Thus $\mathcal{J}_2 \neq \mathcal{J}_3$. Since $\mathcal{A}$ is n-homogeneous, $\mathcal{A}/\mathcal{J}_3 \cong M_n(C)$. Denote $\mathcal{J}_4 = \mathcal{J}_2 \cap \mathcal{J}_3$. By Lemma 5.5.5, there exists an isomorphism

$$\Phi_1 : \mathcal{A} \longrightarrow \mathcal{A}/\mathcal{J}_4 \cong M_n(C) \oplus M_n(C)$$

such that

$$\Phi_1(I_{\mathcal{A}_1} \oplus 0 \oplus \cdots \oplus 0) = (1 \oplus 0 \oplus \cdots \oplus 0) \oplus P, \ P \neq 0,$$

$$\Phi_1(0 \oplus \cdots \oplus 0 \oplus I_{\mathcal{A}_{1+m}}) = 0 \oplus (0 \oplus \cdots \oplus 0 \oplus 1).$$

Set $\Phi = \Phi_1 \cdot \phi$. Then Φ is a surjective homomorphism from $\mathcal{A}'(A_1^{(m_1)} \oplus B)$ to $M_n(C) \oplus M_n(C)$ such that

$$\Phi(I_{\mathcal{A}'(A_1^{(m_1)})} \oplus 0 \oplus \cdots \oplus 0) = (1 \oplus 0 \oplus \cdots \oplus 0) \oplus P, \ P \neq 0,$$

$$\Phi(0 \oplus \cdots \oplus 0 \oplus I_{\mathcal{A}'(B_m)}) = 0 \oplus (0 \oplus \cdots \oplus 0 \oplus 1).$$

Since $\mathcal{A}'(R)/\mathcal{J} = \mathcal{A}'(T^{(n)})/\mathcal{J}_1 \oplus 0 = \mathcal{A} \oplus 0$, there exists a closed ideal $E \supset \mathcal{J}$ such that $\mathcal{A}'(R)/E = \mathcal{A}/\mathcal{J}_4 \oplus 0 = \mathcal{A}'(A_1^{(m_1)} \oplus B)/ker\Phi \oplus 0$.

Suppose that $\pi : \mathcal{A}'(R) \rightarrow \mathcal{A}'(R)/J$, $\pi_2 : \mathcal{A}'(A_1^{(m_1)} \oplus B) \longrightarrow \mathcal{A}'(A_1^{(m_1)} \oplus B)/ker\Phi$ and

$\pi_1 : \mathcal{A}'(A_1^{(n)}) \to \mathcal{A}'(A_1^{(n)})/\mathcal{J}_1''$ are canonical maps. Then by Lemma 5.5.7,

$$\pi_{1*}(K_0(\mathcal{A}'(A_1^{(n)}))) \cong \pi_*(K_0(\mathcal{A}'(R))) \cong \pi_{2*}(K_0(\mathcal{A}'(A_1^{(m_1)} \oplus B))).$$

By Lemma 5.5.4, π_{1*} is injective. So

$$\pi_{2*}(K_0(\mathcal{A}'(A_1^{(m_1)} \oplus B))) \cong \pi_*(K_0(\mathcal{A}'(R))) \cong \pi_{1*}(K_0(\mathcal{A}'(A_1^{(n)}))) \cong K_0(\mathcal{A}'(A_1^{(n)})) \cong Z$$

Furthermore, Φ induces a homomorphism

$$\Psi : \mathcal{A}'(A_1^{(m_1)} \oplus B)/ker\Phi \longrightarrow M_n(C) \oplus M_n(C).$$

Therefore,

$$\Phi_* = \Psi_* \cdot \pi_{2*} : K_0(\mathcal{A}'(A_1^{(m_1)} \oplus B)) \longrightarrow K_0(M_n(C) \oplus M_n(C)) \cong Z \oplus Z.$$

Since Ψ_* is an isomorphism,

$$\Phi_*(K_0(\mathcal{A}'(A_1^{(m_1)} \oplus B))) = \Psi_*(\pi_{2*}(K_0(\mathcal{A}'(A_1^{(m_1)} \oplus B)))) \cong Z. \qquad (5.5.5)$$

Since

$$\Phi(I_{\mathcal{A}'(A_1^{(m_1)})} \oplus 0 \oplus \cdots \oplus 0) = (1 \oplus 0 \oplus \cdots \oplus 0) \oplus P$$

and

$$\Phi(0 \oplus \cdots \oplus 0 \oplus I_{\mathcal{A}'(B_m)}) = 0 \oplus (0 \oplus \cdots \oplus 0 \oplus 1)$$

we have

$$\Phi_*([I_{\mathcal{A}'(A_1^{(m_1)})} \oplus 0 \oplus \cdots \oplus 0]) = [1 \oplus 0 \oplus \cdots \oplus 0] \oplus [P] = 1 \oplus [P];$$

$$\Phi_*([0 \oplus \cdots \oplus 0 \oplus I_{\mathcal{A}'(B_m)}]) = [0] \oplus [0 \oplus \cdots \oplus 0 \oplus 1] = 0 \oplus 1.$$

By (5.5.5) again, there exists $n \in Z$ such that

$$\Phi_*([I_{\mathcal{A}'(A_1^{(m_1)})} \oplus 0 \oplus \cdots \oplus 0]) = n\Phi_*([0 \oplus \cdots \oplus 0 \oplus I_{\mathcal{A}'(B_m)}]),$$

i.e. $1 \oplus P = n(0 \oplus 1) = 0 \oplus n \in Z \oplus Z$. But this is impossible and so we verifies out claim that $m_i + m \leq n$, $i = 1, 2$.

{Claim 2} $m_i + m = n$.

By the Claim 1, we need only to prove that: $m_i + m \geq n$, and it suffices to prove that $m_1 + m \geq n$.

By Theorem 4.4.3, $\mathcal{A}'(B_i)/rad\mathcal{A}'(B_i)$, $1 \leq i \leq m$ is commutative. Since

$$\mathcal{A} = \mathcal{A}'(T^{(n)})/\mathcal{J}_1 = \mathcal{A}'(A_1^{(n)})/\mathcal{J}_{11}$$

is n-homogeneous and since

$$\mathcal{A} = \mathcal{A} {\sim} diag(\mathcal{A}'(A_1^{(n)})/\mathcal{J}_{11}, \ \mathcal{A}'(B_1)/\mathcal{J}_{22}, \ \cdots, \ \mathcal{A}'(B_m)/\mathcal{J}_{1+m,1+m}, \ 0).$$

By the similar arguments used in the proof of Theorem 5.3.7, we have

$$\mathcal{A}/\mathcal{J}' \cong M_l(\mathbf{C})$$

for each $\mathcal{J}' \in \mathcal{M}(\mathcal{A})$ and $l \leq m_1 + n$. Since $\mathcal{A}$ is n-homogeneous, $m_1 + m \geq n$. Similarly, we may obtain that $m_2 + m \geq n$.

Therefore, $m_i + m = n$ for $i = 1, 2$.

Lemma 5.5.9 *Let A_1, A_2 be two strongly irreducible Cowen-Douglas operators. Assume that $A_1 \not\sim A_2$ and $T = A_1^{(n_1)} \oplus A_2^{(n_2)}$, where n_1, n_2 are two natural numbers. If $\mathcal{A}'(T)/rad\mathcal{A}'(T)$ is commutative, then $\bigvee(\mathcal{A}'(T)) \cong N^{(2)}$ and $K_0(\mathcal{A}'(T)) \cong Z^{(2)}$.*

Proof Since $\mathcal{A}'(T)/rad\mathcal{A}'(T)$ is commutative, $\mathcal{A}'(T)$ is 1-homogeneous and

$$\mathcal{A}'(T)/rad\mathcal{A}'(T) \cong (\mathcal{A}'(A_1^{(n_1)})/rad\mathcal{A}'(A_1^{(n_1)})) \oplus (\mathcal{A}'(A_2^{(n_2)})/rad\mathcal{A}'(A_2^{(n_2)}))$$
$$\cong \mathbf{N} \oplus \mathbf{N}.$$

Thus, $K_0(\mathcal{A}'(T)) \cong Z^{(2)}$.

Lemma 5.5.10 *Let A_1, A_2 be two strongly irreducible Cowen-Douglas operators. Assume that $A_1 \not\sim A_2$ and $T = A_1^{(n_1)} \oplus A_2^{(n_2)}$, where n_1, n_2 are two natural numbers, then*

$$\bigvee(\mathcal{A}'(T)) \cong N^{(2)} \ and \ K_0(\mathcal{A}'(T)) \cong Z^{(2)}.$$

Proof By Remark 5.5.3, we may assume that $T = A_1 \oplus A_2$. By theorem 4.2.1, we only to prove that for each natural number n, $T^{(n)}$ has a unique (SI) decomposition up to similarity.

Let $T^{(n)} = A_1^{(n)} \oplus A_2^{(n)}$ and assume that

$$T^{(n)} {\sim} A_1^{(m_1)} \oplus A_2^{(m_2)} \oplus B_1 \oplus \cdots \oplus B_m$$

is another finite decomposition of $T^{(n)}$, where $m_1, m_2, m \geq 0$, $B_j \in (SI)$, $B_j \not\sim A_i$ for $i = 1, 2$ and $1 \leq j \leq m$.

By Lemma 5.5.9, $m_i + m = n$, i.e., $m_i = n - m$ for $i = 1, 2$.

{**Claim**} $m = 0$, i.e. $m_i = n$, $i = 1, 2$.

Since $T = A_1 \oplus A_2$, $\mathcal{A}'(T) = \mathcal{A}'(T) {\sim} diag(\mathcal{A}'(A_1), \mathcal{A}'(A_2))$. By Theorem 4.4.3, $\mathcal{A}'(A_i)/rad\mathcal{A}'(A_i)$ is commutative for $i = 1, 2$. From the proof of

Theorem 5.3.7, for every $\mathcal{J}\in\mathcal{M}(\mathcal{A}'(T))$ $\mathcal{A}'(T)/\mathcal{J}\cong M_l(C)$, where $l=1$ or $l=2$.

Without loss of generality, we can assume that there is a $\hat{\mathcal{J}}\in\mathcal{M}(\mathcal{A}'(T))$ satisfying $\mathcal{A}'(T)/\hat{\mathcal{J}}\cong M_2(\mathbf{C})$. Then

$$\mathcal{A}'(T^{(n)})/M_n(\hat{\mathcal{J}})\cong M_n(\mathcal{A}'(T))/\hat{\mathcal{J}})\cong M_{2n}(C).$$

Since $m_i=n-m$,

$$T^{(n)} \sim A_1^{(m_1)}\oplus A_2^{(m_2)}\oplus B_1\oplus\cdots\oplus B_m$$

$$= A_1^{(n-m)}\oplus A_2^{(n-m)}\oplus B_1\oplus\cdots\oplus B_m = T^{(n-m)}\oplus B_1\oplus\cdots\oplus B_m,$$

Thus $\mathcal{A}'(T^{(n)}) = \mathcal{A}'(T^{(n)}) = diag(\mathcal{A}'(T^{(n-m)}),\mathcal{A}'(B_1),\cdots,\mathcal{A}'(B_m))$. Therefore, for every $\hat{\mathcal{J}}\in\mathcal{M}(\mathcal{A}'(T^{(n)}))$, $\Phi_{11}(\hat{\mathcal{J}}) = \mathcal{A}'(T^{(n-m)})$ or $\Phi_{11}(\hat{\mathcal{J}})$ is a maximal ideal of $\mathcal{A}'(T^{(n-m)})$. This implies that $\mathcal{A}'(T^{(n)})/\mathcal{J}\cong M_s(C)$ for each $\mathcal{J}\in\mathcal{M}(\mathcal{A}'(T^{(n)}))$, where $s\le 2(n-m)+m$. So $2n\le 2(n-m)+m$, i.e. $2m\le m$ and $m=0$.

Repeating the arguments in the proofs of Lemma 5.5.8, Lemma 5.5.9 and Lemma 5.5.10, we have the following theorem.

Theorem 5.5.11 *Let* $A_1, A_2,\cdots, A_k$ *be strongly irreducible Cowen-Douglas operators. Assume that* $A_i\not\sim A_j$ *for* $i\neq j$, *and* $T = A_1^{(n_1)}\oplus A_2^{(n_2)}\oplus\cdots\oplus A_k^{(n_k)}$, *where* $(n_1,\cdots,n_k)$ *is a tuple of natural numbers, then*

$$\bigvee(\mathcal{A}'(T))\cong N^{(k)}, \;\; K_0(\mathcal{A}'(T))\cong Z^{(k)}.$$

From Theorem 5.5.11, we can get the following theorem.

Theorem 5.5.12 *Let* T *be a Cowen-Douglas operator, then for each natural number* n, $T^{(n)}$ *has a unique (SI) decomposition up to similarity.*

Theorem 5.5.13 Let $A, B\in\mathcal{B}_n(\Omega)$ and assume that

$$A\sim A_1^{(n_1)}\oplus A_2^{(n_2)}\oplus\cdots\oplus A_k^{(n_k)},$$

where $0\neq n_i\in\mathbf{N}$, $A_i\in(SI)$ for $i=1,2,\cdots,k$ and $A_i\not\sim A_j$ for $i\neq j$, then $A\sim B$ if and only if the following two conditions are satisfied:

(i) $(K_0(\mathcal{A}'(A\oplus B)), \bigvee(\mathcal{A}'(A\oplus B)), I)\cong(\mathbf{Z}^{(k)}, \mathbf{N}^{(k)}, 1)$;

(ii) The isomorphism h from $\bigvee(\mathcal{A}'(A\oplus B))$ to $\mathbf{N}^{(k)}$ satisfies

$$h([I]) = 2n_1e_1 + 2n_2e_2 + \cdots + 2n_ke_k,$$

where I is the identity of $\mathcal{A}'(A\oplus B)$ and $\{e_i\}_{i=1}^k$ are the generators of $\mathbf{N}^{(k)}$.
Proof "⟸": Since B is a Cowen-Douglas operator, we know

$$B = B_1^{(s_1)}\oplus B_2^{(s_2)}\oplus\cdots\oplus B_m^{(s_m)},$$

where each B_i is strongly irreducible Cowen-Douglas operator for $i = 1, 2, \cdots, m$ and $B_i\not\sim B_j$ for $i\neq j$.
   Without loss of generality, we can assume that

$$A = A_1^{(n_1)}\oplus A_2^{(n_2)}\oplus\cdots\oplus A_k^{(n_k)},$$

$$B = B_1^{(s_1)}\oplus B_2^{(s_2)}\oplus\cdots\oplus B_m^{(s_m)}.$$

{Claim 1} $\forall B_i$, $i = 1, 2, \cdots, m$, there exist $A_j, j = 1, 2, \cdots, k$ such that $B_i\sim A_j$.
   Otherwise, without loss of generality, we may assume that $B_1, \cdots, B_l$, for $1\leq l\leq m$, are not similar to each $A_j, (1\leq j\leq k)$, and each $B_i(l < i\leq m)$ is similar to some A_j, $1\leq j\leq k$. Thus

$$(A\oplus B)\sim(A_1^{(t_1)}\oplus A_2^{(t_2)}\oplus\cdots\oplus A_k^{(t_k)}\oplus B_1^{(s_1)}\oplus\cdots\oplus B_l^{(s_l)}).$$

By Theorem 5.5.11,

$$\bigvee(\mathcal{A}'(A\oplus B)) = \bigvee(\mathcal{A}'(A_1^{(t_1)}\oplus A_2^{(t_2)}\oplus\cdots\oplus A_k^{(t_k)}\oplus B_1^{(s_1)}\oplus\cdots\oplus B_l^{(s_l)}))\cong N^{(k+l)}.$$

This contradicts (i).
{Claim 2} $m = k$.
   It follows from Claim 1 that $m\leq k$. If $m < k$, without loss of generality, we assume that $B_1\sim A_1, B_2\sim A_2, \cdots, B_m\sim A_m$. Then

$$(A\oplus B)\sim(A_1^{(n_1+s_1)}\oplus A_2^{(n_2+s_2)}\oplus\cdots\oplus A_m^{(n_m+s_m)}\oplus A_{m+1}^{(n_{m+1})}\oplus\cdots\oplus A_k^{(n_k)}).$$

By Theorem 4.2.1, the isomorphism h satisfies

$$h([I]) = (n_1 + s_1)e_1 + \cdots + (n_m + s_m)e_m + n_{m+1}e_{m+1} + \cdots + n_ke_k.$$

By (ii), $h([I]) = 2n_1e_1 + 2n_2e_2 + \cdots + 2n_ke_k$. A contradiction. Thus $m = k$.
   Now we may assume that $B_1\sim A_1, B_2\sim A_2, \cdots, B_k\sim A_k$, i.e.,

$$(A\oplus B)\sim(A_1^{(n_1+s_1)}\oplus A_2^{(n_2+s_2)}\oplus\cdots\oplus A_k^{(n_k+s_k)}).$$

Repeating the proof of Claim 2, we have $s_i = n_i, i = 1, 2, \cdots, k$.

"$\Rightarrow$" Since $A \sim B$, $B = B_1^{(n_1)} \oplus B_2^{(n_2)} \oplus \cdots \oplus B_k^{(n_k)}$, where $B_i \in (SI)$, $i = 1, 2, \cdots, k$ and $A_i \sim B_i$ for $i = 1, 2, \cdots, k$. Thus $(A \oplus B) \sim (A_1^{(2n_1)} \oplus \cdots \oplus A_k^{(2n_k)})$.

The remainder of the proof of the theorem is a consequence of Theorem 4.2.1.

Theorem 5.5.13 is the Jordan canonical theorem for Cowen-Douglas operators. By the proofs of Theorem 5.5.11, Theorem 5.5.12 and Theorem 5.5.13, we have the following result.

Theorem 5.5.14 *Let* $T_1, T_2 \in \mathcal{L}(\mathcal{H})$ *and* $T_1 \sim \bigoplus_{i=1}^{n} A_i^{(k_i)}$, *where* A_i *is a strongly irreducible Cowen-Douglas operator for every* i, *and* T_i *is not similar to* T_j *if* $i \neq j$. *Then* $T_1 \sim T_2$ *if and only if:*

(i) $(K_0(\mathcal{A}'(T_1 \oplus T_2)), \bigvee(\mathcal{A}'(T_1 \oplus T_2)), I) \cong (\mathbf{Z}^{(n)}, \mathbf{N}^{(n)}, 1)$;

(ii) *The isomorphism* h *from* $\bigvee(\mathcal{A}'(T_1 \oplus T_2))$ *to* $\mathbf{N}^{(k)}$ *satisfies*

$$h([I]) = 2k_1 e_1 + 2k_2 e_2 + \cdots + 2k_n e_n,$$

where I *is the identity of* $\mathcal{A}'(T_1 \oplus T_2)$ *and* $\{e_i\}_{i=1}^{n}$ *are the generators of* $\mathbf{N}^{(n)}$.

In the following we will consider a more general form. Let $T = \bigoplus_{i=1}^{n} T_i^{(k_i)}$, where $T_i \in (SI)$, $\mathcal{A}'(T_i)/rad\mathcal{A}'(T_i)$ is commutative, $\bigvee(\mathcal{A}'(T_i)) \cong \mathbf{N}, i = 1, 2, \cdots, n$, and T_i is not similar to T_j if $i \neq j$. By the arguments used in the proofs Theorem 5.5.11, Theorem 5.5.12 and Theorem 5.5.13, we have the following theorem.

Theorem 5.5.15 $\bigvee(\mathcal{A}'(T)) \cong \mathbf{N}^{(n)}$ *and* $K_0(\mathcal{A}'(T)) \cong \mathbf{Z}^{(n)}$. *Furthermore,* T *has a unique* (SI) *decomposition up to similarity and* $T \sim G \in \mathcal{L}(\mathcal{H})$ *if and only if the following are satisfied:*

(i) $(K_0(\mathcal{A}'(T \oplus G)), \bigvee(\mathcal{A}'(T \oplus G)), I) \cong (\mathbf{Z}^{(n)}, \mathbf{N}^{(n)}, 1)$;

(ii) *The isomorphism* h *from* $\bigvee(\mathcal{A}'(T \oplus G))$ *to* $\mathbf{N}^{(n)}$ *satisfies*

$$h([I]) = 2k_1 e_1 + 2k_2 e_2 + \cdots + 2k_n e_n.$$

Corollary 5.5.16 *Let* $T = \bigoplus_{k=1}^{m_1} A_k^{(n_k)} \oplus \bigoplus_{j=1}^{m_2} B_j^{(l_j)}$, *where* A_k's, B_j^*'s *are Cowen-Douglas operators,* $A_k, B_j \in (SI)$ *and* $A_k \not\sim A_{k'}$, $B_j \not\sim B_{j'}$ *if* $k \neq$

k', $j \neq j'$. Then

$$\bigvee(\mathcal{A}'(T)) \cong \mathbf{N}^{(m_1+m_2)} \quad and \quad K_0(\mathcal{A}'(T)) \cong \mathbf{Z}^{(m_1+m_2)}.$$

This corollary implies such T has a unique (SI) decomposition up to similarity.

5.6 Maximal Ideals of The Commutant of Cowen-Douglas Operators

In this section we will use the K_0-group to characterize the commutant of Cowen-Douglas operators. The main theorem of this section is as follows.

Theorem 5.6.1 *Let A_1, A_2 be strongly irreducible Cowen-Douglas operators. Assume that $A_1 \nsim A_2$ and $T = A_1^{(n_1)} \oplus A_2^{(n_2)}$, where n_1, n_2 are natural numbers. Then for each $\mathcal{J} \in \mathcal{M}(\mathcal{A}'(T))$, $\mathcal{J}$ must be one of the following two forms.*

(i)

$$\mathcal{J} = \begin{bmatrix} \mathcal{J}_{11} & ker\tau_{A_1^{(n_1)},A_2^{(n_2)}} \\ ker\tau_{A_2^{(n_2)},A_1^{(n_1)}} & \mathcal{A}'(A_2^{(n_2)}) \end{bmatrix} ;$$

(ii)

$$\mathcal{J} = \begin{bmatrix} \mathcal{A}'(A_1^{(n_1)}) & ker\tau_{A_1^{(n_1)},A_2^{(n_2)}} \\ ker\tau_{A_2^{(n_2)},A_1^{(n_1)}} & \mathcal{J}_{22} \end{bmatrix} ,$$

where $\mathcal{J}_{ii}$ is a maximal ideal of $\mathcal{A}'(A_i^{(n_i)})$, $i = 1, 2$.

Proof First, we assume that $T = A_1 \oplus A_2$.

Since A_1, A_2 are strongly irreducible Cowen-Douglas operators, by Theorem 4.4.3, $\mathcal{A}'(A_1)/rad\mathcal{A}'(A_1)$ and $\mathcal{A}'(A_2)/rad\mathcal{A}'(A_2)$ are commutative.

Assume that $\mathcal{J}$ has neither form (i) nor form (ii). Then

$$\mathcal{J} = \begin{bmatrix} \mathcal{J}_{11} & \mathcal{J}_{12} \\ \mathcal{J}_{21} & \mathcal{J}_{22} \end{bmatrix} .$$

According to the discussion in Section 5.4,

$$\mathcal{J}_{ii} \in \mathcal{M}(\mathcal{A}'(A_i)), (i = 1, 2), \mathcal{J}_{12} \subsetneqq ker\tau_{A_1, A_2}, \mathcal{J}_{21} \subsetneqq ker\tau_{A_2, A_1}.$$

Then $\mathcal{A}'(T)/\mathcal{J} \cong M_2(\mathbf{C})$ and we have the following exact sequence:

$$0 \longrightarrow \mathcal{J} \xrightarrow{i} \mathcal{A}'(T) \xrightarrow{\pi} \mathcal{A}'(T)/\mathcal{J} \longrightarrow 0,$$

which induces the following six-term cyclic exact sequence:

$$\begin{array}{ccccc}
K_0(\mathcal{J}) & \xrightarrow{i_*} & K_0(\mathcal{A}'(T)) & \xrightarrow{\pi_*} & K_0(\mathcal{A}'(T)/\mathcal{J}) \\
\partial\uparrow & & & & \partial\downarrow \\
K_1(\mathcal{A}'(T)/\mathcal{J}) & \longleftarrow & K_1(\mathcal{A}'(T)) & \longleftarrow & K_1(\mathcal{J})
\end{array}$$

Since $\mathcal{A}'(T)/\mathcal{J} \cong M_2(\mathbf{C})$, $K_0(\mathcal{A}'(T)/\mathcal{J}) \cong \mathbf{Z}$, $K_1(\mathcal{A}'(T)/\mathcal{J}) \cong 0$. By Theorem 5.5.11, $K_0(\mathcal{A}'(T)) \cong \mathbf{Z}^{(2)}$. Note that

$$\pi_*([diag(I_{\mathcal{A}'(A_1)}, 0)]) = 1, \ \pi_*([diag(0, I_{\mathcal{A}'(A_2)})]) = 1.$$

Thus π_* is a surjective map from $\mathbf{Z} \oplus \mathbf{Z}$ to $\mathbf{Z}$, and we get a split exact sequence:

$$0 \longrightarrow K_0(\mathcal{J}) \xrightarrow{i_*} \mathbf{Z} \oplus \mathbf{Z} \overset{\pi_*}{\rightleftharpoons} \mathbf{Z} \longrightarrow 0.$$

Since $\pi_*(K_0(\mathcal{A}'(T))) \cong \mathbf{Z}$, $K_0(\mathcal{J}) \cong \mathbf{Z}$.

Consider the following split exact sequence:

$$0 \longrightarrow \mathcal{J} \longrightarrow \mathcal{J} \dot{+} 1 \rightleftharpoons (\mathcal{J} \dot{+} 1)/\mathcal{J} \longrightarrow 0,$$

Using the six-term cyclic exact sequence again and by the fact that $\partial : K_0((\mathcal{J} \dot{+} 1)/\mathcal{J}) \longrightarrow K_1(\mathcal{J})$ is a zero map, we have the following split exact sequence:

$$0 \longrightarrow K_0(\mathcal{J}) \longrightarrow K_0(\mathcal{J} \dot{+} 1) \rightleftharpoons K_0((\mathcal{J} \dot{+} 1)/\mathcal{J}) \longrightarrow 0.$$

Since $(\mathcal{J} \dot{+} 1)/\mathcal{J} \cong \mathbf{C} \oplus \mathbf{C}$, $K_0((\mathcal{J} \dot{+} 1)/\mathcal{J}) \cong \mathbf{Z}^{(2)}$. Combining it with $K_0(\mathcal{J}) \cong \mathbf{Z}$, we have $K_0(\mathcal{J} \dot{+} 1) \cong \mathbf{Z} \oplus \mathbf{Z} \oplus \mathbf{Z}$.

Note that

$$P_1 = \begin{bmatrix} I_{\mathcal{A}'(A_1)} & 0 \\ 0 & 0 \end{bmatrix}$$

and

$$P_2 = \begin{bmatrix} 0 & 0 \\ 0 & I_{\mathcal{A}'(A_2)} \end{bmatrix}$$

are two minimal idempotents of $M_\infty(\mathcal{A}'(T))$ and $P_1 \not\sim_a P_2$ in $M_\infty(\mathcal{A}'(T))$. It is obvious that P_1, P_2 are also two minimal idempotents of $M_\infty(\mathcal{J}\dot{+}1)$ and $P_1 \not\sim_a P_2$ in $M_\infty(\mathcal{J}\dot{+}1)$. Since $K_0(\mathcal{J}\dot{+}1) \cong \mathbf{Z} \oplus \mathbf{Z} \oplus \mathbf{Z}$, there exists a minimal idempotent P in $M_\infty(\mathcal{J}\dot{+}1)$ such that $P \not\sim_a P_1$, $P \not\sim_a P_2$ in $M_\infty(\mathcal{J}\dot{+}1)$. Without loss of generality, we assume that $P \in \mathcal{A}'(T)$.

{**Claim**} $I - P \sim_a P$ in $M_\infty(\mathcal{J}\dot{+}1)$.

Otherwise, $I - P \not\sim_a P$ in $M_\infty(\mathcal{J}\dot{+}1)$. Then $I - P \sim_a P_1$ or $I - P \sim_a P_2$ in $M_\infty(\mathcal{J}\dot{+}1)$. Thus $P \sim_a P_2$ or $P \sim_a P_1$ in $M_\infty(\mathcal{J}\dot{+}1)$. This contradicts the choice of P.

Since $M_\infty(\mathcal{J}\dot{+}1) \subset M_\infty(\mathcal{A}'(T))$, we have $P \sim_a P_1$ and $(I - P) \sim_a P_1$ in $M_\infty(\mathcal{A}'(T))$ or $P \sim_a P_2$ and $(I - P) \sim_a P_2$ in $M_\infty(\mathcal{A}'(T))$. By Lemma 4.2.4, there exists a natural number n such that

$$T^{(n)}|_{ran P \oplus 0} = T|_{ran P} \sim T|_{ran P_1} \ (\text{or} \ T|_{ran P_2}) = A_1 \ (\text{or} \ A_2)$$

and

$$T^{(n)}|_{ran(I-P) \oplus 0} = T|_{ran(I-P)} \sim T|_{ran P_1} \ (\text{or} \ T|_{ran P_2}) = A_1 \ (\text{or} \ A_2).$$

This implies that $T = A_1 \oplus A_2$ or $T = A_2 \oplus A_1$. By Theorem 5.5.13 T has a unique (SI) decomposition up to similarity. Thus $A_1 \sim A_2$. The contradiction indicates that $\mathcal{J}$ must have the form (i) or (ii).

Now we consider the general case, i.e., $T = A_1^{(n_1)} \oplus A_2^{(n_2)}$.

It follows from the proof of Theorem 5.3.7 that $\mathcal{A}'(T)/\mathcal{J} \cong M_k(\mathbf{C})$ for every $\mathcal{J} \in \mathcal{M}(\mathcal{A}'(T))$, where $k = n_1$ or $k = n_2$ or $k = n_1 + n_2$. Thus we need only to show that $k \neq n_1 + n_2$. Otherwise, there exists $\mathcal{J} \in \mathcal{M}(\mathcal{A}'(T))$ such that

$$\mathcal{A}'(T)/\mathcal{J} \cong M_{n_1+n_2}(\mathbf{C}).$$

Observing

$$\mathcal{A}'(T) = \mathcal{A}'(T) \sim (diag(\mathcal{A}'(A_1 \oplus A_2)), diag(\mathcal{A}'(A_1^{(n_1-1)} \oplus A_2^{(n_2-1)}))),$$

$$\mathcal{J} = \mathcal{J} \sim (diag(\pi_{11}(\mathcal{J}), \pi_{22}(\mathcal{J}))).$$

Then $\pi_{11}(\mathcal{J}) \in \mathcal{M}(\mathcal{A}'(A_1 \oplus A_2))$ and

$$\pi_{11}(\mathcal{J}) = \begin{bmatrix} \mathcal{J}_{11} & \mathcal{J}_{12} \\ \mathcal{J}_{21} & \mathcal{J}_{22} \end{bmatrix},$$

where $\mathcal{J}_{11}\in\mathcal{M}(\mathcal{A}'(A_1))$, $\mathcal{J}_{22}\in\mathcal{M}(\mathcal{A}'(A_2))$, $\mathcal{J}_{12}\subsetneqq ker\tau_{A_1,A_2}$, $\mathcal{J}_{21}\subsetneqq ker\tau_{A_2,A_1}$. This contradicts the conclusion got in the first part. Thus, for each $\mathcal{J}\in\mathcal{M}(\mathcal{A}'(T))$, $\mathcal{A}'(T)/\mathcal{J}$ is not isomorphic to $M_{n_1+n_2}(\mathbf{C})$.

By the arguments similar to that in the proof of Theorem 5.6.1, we have the following theorem.

Theorem 5.6.2 *Let* $A_1, A_2, \cdots, A_k$ *be strongly irreducible Cowen-Douglas operators. If* A_i *is not similar to* A_j *if* $i \neq j$ *and* $T = A_1^{(n_1)}\oplus A_2^{(n_2)}\oplus\cdots\oplus A_k^{(n_k)}$, *where* $n_1, \cdots, n_k$ *are positive integers, then for each* $\mathcal{J}\in\mathcal{M}(\mathcal{A}'(T))$,

$$
\mathcal{J} = \begin{bmatrix} \mathcal{J}_{11} & \mathcal{J}_{12} & \cdots & \mathcal{J}_{1k} \\ \mathcal{J}_{21} & \mathcal{J}_{22} & \cdots & \mathcal{J}_{2k} \\ \cdots & \cdots & \cdots & \cdots \\ \mathcal{J}_{k1} & \mathcal{J}_{k2} & \cdots & \mathcal{J}_{kk} \end{bmatrix},
$$

where $\mathcal{J}_{ij}$ *satisfies the following properties.*

(i) $\mathcal{J}_{ij} = ker\tau_{A_i^{(n_i)}, A_j^{(n_j)}}, i \neq j;$

(ii) There exists a unique $i, 1\leq i\leq k$ *such that* $\mathcal{J}_{ii}\in\mathcal{M}(\mathcal{A}'(A_i^{(n_i)}))$ *and* $\mathcal{J}_{jj} = \mathcal{A}'(A_j^{(n_j)})$ *for* $j, j \neq i$.

Corollary 5.6.3 *Let* $T = A_1^{(n_1)}\oplus A_2^{(n_2)}\oplus\cdots\oplus A_k^{(n_k)}$, *where* $A_1, A_2, \cdots, A_k$ *are strongly irreducible Cowen-Douglas operators and* A_i *is not similar to* A_j *if* $i \neq j$. *Then for each* $\mathcal{J}\in\mathcal{M}(\mathcal{A}'(T))$, $\mathcal{A}'(T)/\mathcal{J}\cong M_l(\mathbf{C})$, *where* $l\in\{n_1, n_2, \cdots, n_k\}$.

5.7 Some Approximation Theorem

Let Ω be an analytic Cauchy domain, $\Gamma = \partial\Omega$. Denote $L^2(\Gamma)$ the Hilbert space of all square integrable complex functions with respect to the arc length measure on Γ. Denote $M_\lambda(\Gamma)$ the multiplication by λ operator acting on $L^2(\Gamma)$. Denote $H^2(\Gamma)$ the subspace of $L^2(\Gamma)$ generated by the rational functions with poles outside $\overline{\Omega}$. Then $H^2(\Gamma)\in LatM_\lambda(\Gamma)$ and

$$
M_\lambda(\Gamma) = \begin{bmatrix} M_+(\Gamma) & Z \\ 0 & M_-(\Gamma) \end{bmatrix} \begin{matrix} H^2(\Gamma) \\ L^2(\Gamma)\ominus H^2(\Gamma). \end{matrix}
$$

It is easy to see that $M_+^*(\Gamma)\in\mathcal{B}_1(\Omega^*)$ and $\sigma(M_+(\Gamma)) = \overline{\Omega}$. If Ω is the unit disk D, $M_+(\Gamma) = T_z$. Furthermore, $\mathcal{A}'(M_+(\Gamma)) = H^\infty(\Omega)$ [Conway, J.B. (1990)].

Proposition 5.7.1 *Let Ω be a connected analytic Cauchy domain, $\mathcal{H}_1, \mathcal{H}_2$ be two nontrivial invariant subspace of $M_+(\Gamma)$. Then $\mathcal{H}_1 \cap \mathcal{H}_2 \neq \{0\}$.*
Proof   Given $\lambda \in \Omega$.   Since $(M_+(\Gamma) - \lambda)H^2(\Gamma)$ is closed subspace with codimension 1, $(M_+(\Gamma) - \lambda)\mathcal{H}_1$ is closed.   Clearly, $(M_+(\Gamma) - \lambda)\mathcal{H}_1 \in Lat M_+(\Gamma) \in \mathcal{H}_1$. Thus

$$(M_+(\Gamma) - \lambda)\mathcal{H}_1 \neq \mathcal{H}_1.$$

In fact, if $(M_+(\Gamma) - \lambda)\mathcal{H}_1 = \mathcal{H}_1$, then for arbitrary $f \in \mathcal{H}_1, f(z) = (z - \lambda)^n g_n(z), g_n \in \mathcal{H}_1$. This implies that $f^{(n)}(\lambda) = 0, (n = 1, 2, \cdots)$. So $f = 0$, a contradiction.

Given a nonzero function $\phi \in \mathcal{H}_1 \ominus [(M_+(\Gamma) - \lambda)\mathcal{H}_1]$. For each $h \in R(\Gamma)$,

$$0 = <(M_+(\Gamma) - \lambda)h(M_+(\Gamma))\phi, \phi> = \int_\Gamma (z - \lambda)h(z)|\phi|^2 d\mu,$$

where $R(\Gamma)$ is the set of rational functions with poles outside $\overline{\Omega}$. This implies that $|\phi|^2 \perp \{g \in ReR(\Gamma) : g(\lambda) = 0\}(\subset ReL^2(\Gamma))$. By [Fisher, S.D. (1983)], the orthogonal complement of $ReR(\Gamma)$ in $ReL^2(\Gamma))$ is the linear combinations of finite functions, denoted by $Q_1, \cdots, Q_m$ in C^∞. Clearly, constant functions is orthogonal to $\{g \in ReR(\Gamma) : g(\lambda) = 0\}$. Thus $|\phi|^2$ is a linear combination of $Q_1, \cdots, Q_m$ and 1 and $\sup_{z \in \Gamma} |\phi(z)| < \infty$. Therefore $\phi h = h(M_+(\Gamma))\phi \in \mathcal{H}_1$ for each $h \in R(\Gamma)$. Since $R(\Gamma)$ is dense in $H^2(\Gamma)$ and since ϕ is bounded, $\phi H^2(\Gamma) \subset \mathcal{H}_1$.

Similarly, we can find a nonzero bounded function $\psi \in \mathcal{H}_2$ such that

$$\psi H^2(\Gamma) \subset \mathcal{H}_2.$$

This means that

$$\mathcal{H}_1 \cap \mathcal{H}_2 \supset \phi\psi H^2(\Gamma) \neq \{0\}.$$

Lemma 5.7.2 *Let Ω be a connected analytic Cauchy domain, then*

$$[ran \tau_{M_+(\Gamma), M_+(\Gamma)}] \cap [ker \tau_{M_+(\Gamma), M_+(\Gamma)}] = \{0\},$$

i.e., if there exists an $X \in \mathcal{L}(H^2(\Gamma))$ such that

$$X M_+(\Gamma) = M_+(\Gamma)X$$

and

$$M_+(\Gamma)Y - Y M_+(\Gamma) = X$$

for some $Y \in H_+^2(\Gamma)$, *then* $X = 0$.

Proof Since $\mathcal{A}'(M_+(\Gamma)) = H^\infty(\Omega)$, there exists $g \in H^\infty(\Omega)$ such that

$$X = M_g(\Gamma),$$

i.e., $M_g(\Gamma)f = gf$ for all $f \in H^2(\Gamma)$. Thus

$$M_+(\Gamma)Yf = YM_+(\Gamma)f = gf, \quad f \in H^2(\Gamma).$$

Set $f \equiv 1$, $\lambda Y(1) - Y(\lambda) = g$ or $Y(\lambda) = \lambda h - g$, where $h = Y(1) \in H^2(\Gamma)$.
Set $f = \lambda$, $\lambda Y(\lambda) - Y(\lambda^2) = \lambda g$ or $Y(\lambda^2) = \lambda(\lambda h - g) - \lambda g = \lambda^2 h - 2\lambda g$.
Generally, set $f = \lambda^{n-1}$, we have

$$\lambda Y(\lambda^{n-1}) - Y(\lambda^n) = \lambda^{n-1}g$$

or

$$Y(\lambda^n) = \lambda^n h - n\lambda^{n-1}g \quad (n = 1, 2, \cdots).$$

Without loss of generality, we can assume that $\Omega \subset D$. Thus it follows from $|\lambda|^{2n} \leq |\lambda|^{2(n-1)}$ $(\lambda \in \Omega, n = 1, 2, \cdots)$ that

$$\frac{n^2 \int_\Gamma |\lambda|^{2(n-1)} dm}{\int_\Gamma |\lambda|^{2n} dm} \longrightarrow \infty, \quad (n \longrightarrow \infty).$$

This implies $\dfrac{\|n\lambda^{n-1}\|^2_{H^2(\Gamma)}}{\|\lambda^n\|_{H^2(\Gamma)}} \longrightarrow \infty$, $(n \longrightarrow \infty)$. Therefore, if $g \neq 0$, then Y is unbounded, a contradiction. This means that $g = 0$ and $X = 0$.

For each natural number n, define

$$M_n(\Gamma) = \begin{bmatrix} M_+(\Gamma) & & & 0 \\ I & M_+(\Gamma) & & \\ & \ddots & \ddots & \\ 0 & & I & M_+(\Gamma) \end{bmatrix} \in \mathcal{L}((H_+^2(\Gamma))^{(n)}).$$

Theorem 5.7.3 *Let* $M_n(\Gamma)$ *be defined as above. Then for each* $A \in \mathcal{A}'(M_n(\Gamma))$,

(i)

$$A = \begin{bmatrix} M_{f_1} & & & \\ M_{f_2} & M_{f_1} & & \\ \vdots & \ddots & \ddots & \ddots \\ M_{f_n} & \cdots & M_{f_2} & M_{f_1} \end{bmatrix},$$

where $f_i \in H^\infty(\Gamma)$ $(i = 1, 2, \cdots, n)$. *Moreover,* $\mathcal{A}'(M_n(\Gamma))$ *is commutative;*

(ii) $M_n(\Gamma) \in (SI)$;

(iii) $\sigma(M_n(\Gamma)) = \overline{\Omega}$, $\sigma_e(M_n(\Gamma)) = \Gamma$, $\dim\ker(\lambda - M_n(\Gamma)) = 0$ *and* $\text{ind}(\lambda - M_n(\Gamma)) = -n$ *for all* $\lambda \in \Omega$;

(iv) $M_n^*(\Gamma) \in \mathcal{B}_n(\Omega^*)$;

(v) $\mathcal{A}'(A)$ *is commutative.*

Proof (i) Assume that

$$A = \begin{bmatrix} A_{11} & \cdots & A_{1n} \\ \cdots & \cdots & \cdots \\ A_{n1} & \cdots & A_{nn} \end{bmatrix} \in \mathcal{A}'(M_n(\Gamma)),$$

i.e.,

$$\begin{bmatrix} A_{11} & \cdots & A_{1n} \\ \cdots & \cdots & \cdots \\ A_{n1} & \cdots & A_{nn} \end{bmatrix} \begin{bmatrix} M_+(\Gamma) & & & 0 \\ I & M_+(\Gamma) & & \\ & \ddots & \ddots & \\ 0 & & I & M_+(\Gamma) \end{bmatrix}$$

$$= \begin{bmatrix} M_+(\Gamma) & & & 0 \\ I & M_+(\Gamma) & & \\ & \ddots & \ddots & \\ 0 & & I & M_+(\Gamma) \end{bmatrix} \begin{bmatrix} A_{11} & \cdots & A_{1n} \\ \cdots & \cdots & \cdots \\ A_{n1} & \cdots & A_{nn} \end{bmatrix}.$$

At the $(1, n), (1, n - 1)$ entries, we have

$$A_{1n}M_+(\Gamma) = M_+(\Gamma)A_{1n}$$

and

$$A_{1,n-1}M_+(\Gamma) + A_{1n} = M_+(\Gamma)A_{1,n-1}.$$

By Lemma 5.7.2, $A_{1n} = 0$ and $A_{1,n-1} \in \mathcal{A}'(M_+(\Gamma))$. Similarly, we can conclude that $A_{ij} = 0$, $(1 \leq i < j \leq n)$. Consider the (i, i) entries of both sides, we have

$$A_{i,i} \in \mathcal{A}'(M_+(\Gamma)), \quad (1 \leq i \leq n).$$

Compare the $(i + 1, i)$ entries $(1 \leq i \leq n - 1)$, we get

$$A_{i+1,i}M_+(\Gamma) + A_{i+1,i+1} = A_{i,i} + M_+(\Gamma)A_{i+1,i}.$$

By Lemma 5.7.2 again, $A_{i+1,i+1} = A_{i,i}$ and $A_{i+1,i} \in \mathcal{A}'(M_+(\Gamma))$.

At the $(i+2,i)$ entries $(1 \leq i \leq n-2)$,

$$A_{i+2,i}M_+(\Gamma) + A_{i+2,i+1} = A_{i+1,i} + M_+(\Gamma)A_{i+2,i}.$$

Thus $A_{i+2,i+1} = A_{i+1,i}$ and $A_{i+2,i} \in \mathcal{A}'(M_+(\Gamma))$. Using this argument and Lemma 5.7.2 respectively, we get the general form of A.

(ii) Let P be an idempotent commuting with $M_n(\Gamma)$. By (i),

$$P = \begin{bmatrix} M_{f_1} & & & \\ M_{f_2} & M_{f_1} & & \\ \vdots & \ddots & \ddots & \ddots \\ M_{f_n} & \cdots & M_{f_2} & M_{f_1} \end{bmatrix}.$$

Since $P^2 = P$, $f_1^2 = f_1$. Since Ω is connected, $f_1 \equiv 1$ or $f_1 \equiv 0$. In either case, $f_2 = f_3 = \cdots = f_n = 0$, i.e., $P = I$ or $P = 0$ and $M_n(\Gamma) \in (SI)$.

(iii), (iv) and (v) are obvious.

Lemma 5.7.4 *Suppose that*

$$R = \begin{bmatrix} A_1 & & & & 0 & Q_{11} & Q_{12} & \cdots & Q_{1m} \\ & A_2 & & & & Q_{21} & Q_{22} & \cdots & Q_{2m} \\ & & \ddots & & & \vdots & \vdots & \vdots & \vdots \\ & & & A_n & Q_{n1} & Q_{n2} & \cdots & Q_{nm} \\ & & & & c_1 & & & & 0 \\ & & & & & c_2 & & & \\ & 0 & & & & & \ddots & \\ & & & & & & & c_m \end{bmatrix}$$

with respect to the orthogonal direct sum

$$\mathcal{H} = \left(\bigoplus_{j=1}^{n} \mathcal{H}_j' \right) \oplus \left(\bigoplus_{i=1}^{m} \mathcal{H}_j^2 \right) \quad (n, m < \infty),$$

where

(i) $\{\sigma(A_j)\}_{j=1}^{n}$ *and* $\{\sigma(c_i)\}_{i=1}^{m}$ *are two families of pairwise disjoint compact sets such that*

$$\sigma(R) = \left[\bigcup_{j=1}^{n} \sigma(A_j) \right] \cup \left[\bigcup_{i=1}^{m} \sigma(c_i) \right]$$

and $\sigma(R)$ *is connected;*

(ii) $A_j, c_i \in (SI)$ *and* $\ker \tau_{c_i, A_j} = \{0\}$ $(1 \leq j \leq n, 1 \leq i \leq m)$;

(iii) Either $\sigma(A_j)\cap\sigma(c_i) = \emptyset$ and $Q_{ji} = 0$ or $\sigma(A_j)\cap\sigma(c_i) \neq \emptyset$ and $Q_{ji}\notin ran\tau_{A_j,c_i}$.

Then $R\in(SI)$.

Proof Let P be an idempotent commuting with R and assume that P admits the block matrix representation $P = (P_{ij})_{i,j=1}^{n+m}$ with respect to the decomposition of the space $\mathcal{H}$.

Since $ker\tau_{c_i,A_j} = \{0\}$ $(1\leq j\leq n, 1\leq i\leq m)$, computations indicate that

$$P_{ji} = 0 \ (n+1\leq j\leq n+m \ \text{ and } \ 1\leq i\leq n).$$

Since $\{\sigma(A_j)\}_{j=1}^{n}$ and $\{\sigma(c_i)\}_{i=1}^{m}$ are two families of pairwise disjoint compact sets, $ker\tau_{A_i,A_j} = \{0\}$ and $ker\tau_{c_i,c_j} = \{0\}$ $(i \neq j)$. Thus computation shows that

$$P_{ji} = 0 \ \ (n\geq j > i\geq 1 \text{ or } n+m\geq j > i\geq n+1),$$

i.e., P has an upper triangular matrix representation with respect to the given decomposition of the space. Moreover, the same arguments indicate that

$$P_{ji} = 0 \ \ (1\leq j < i\leq n \ \text{ or } n+1\leq j < i\leq n+m).$$

Since P_{jj} is an idempotent and $A_j, c_i\in(SI), P_{jj} = 0$ or 1, $(j = 1, 2, \cdots, n+m)$. Without loss of generality, we can suppose that $P_{11} = 0$. Since $\sigma(R)$ is connected, there must exist an integer i $(1\leq i\leq m)$ such that $\sigma(A_1)\cap\sigma(c_i) \neq \emptyset$. It follows from $PR = RP$ that

$$P_{1,n+i}c_i = A_1 P_{1,n+i} + Q_{1i}P_{n+i,n+i}$$

or

$$Q_{1i}P_{n+i,n+i} = \tau_{A_1,c_i}P_{1,n+i}.$$

Since $Q_{1i}\notin ran\tau_{A_1,c_i}$, $P_{n+i,n+i} = 0$. If $\sigma(A_j)\cap\sigma(c_i) \neq \emptyset$, by the same argument, $P_{jj} = 0$.

Since $\sigma(R)$ is connected, after a finite number of steps $P_{jj} = 0$ $(1\leq j\leq n+m)$. It follows from $P^2 = P$ that $P = 0$, i.e., $R\in(SI)$.

Lemma 5.7.5 *Let Ω be a bounded and connected Cauchy domain, then there exists an (SI) operator B satisfying*
   (i) $\sigma(B) = \overline{\Omega}$;
   (ii) $\rho_F(B)\cap\sigma(B) = \overline{\Omega}$ and $ind(\lambda - B) = 0$ $(\lambda \in \Omega)$;
   (iii) $min\text{-}ind(\lambda - B) = 1$ $(\lambda \in \Omega)$;
   (iv) $\mathcal{A}'(B)/rad\mathcal{A}'(B)$ is commutative.

Proof Denote

$$B = \begin{bmatrix} B_1 & I \\ 0 & B_2 \end{bmatrix} \begin{matrix} \mathcal{H} \\ \mathcal{H} \end{matrix},$$

where $B_1 = M_+^*(\Gamma^*), B_2 = M_+(\Gamma), \Gamma^* = \partial\Omega^*$ and $\Gamma = \partial\Omega$. Then B satisfies (i), (ii) and (iii). Suppose that P is an idempotent commuting with B. By Lemma 3.2.3,

$$P = \begin{bmatrix} P_1 & P_{12} \\ 0 & P_2 \end{bmatrix} \begin{matrix} \mathcal{H} \\ \mathcal{H} \end{matrix}.$$

Since $B_1, B_2 \in (SI), P_1 = 0$ or I, $P_2 = 0$ or I. If $P_1 = I, P_2 = 0$, it follows from $PB = BP$ that $I + P_{12}B_2 = B_1P_{12}$, or $I \in ran\tau_{B_1,B_2}$. But this is impossible. Thus $B \in (SI)$. Note that $\mathcal{A}'(B_1)$ commutes with $\mathcal{A}'(B_2)$ and for each $A \in \mathcal{A}'(B)$,

$$A = \begin{bmatrix} A_1 & A_{12} \\ 0 & A_2 \end{bmatrix},$$

where $A_1 \in \mathcal{A}'(B_1), A_2 \in \mathcal{A}'(B_2)$. Thus $\mathcal{A}'(B)/rad\mathcal{A}'(B)$ is commutative.

Lemma 5.7.6 *Let T be in $\mathcal{L}(\mathcal{H})$ with connected spectrum $\sigma(T)$. If $\sigma_{lre}(T)$ is the closure of an analytic Cauchy domain Ω. Then there exists an (SI) operator A such that the spectral picture $\Lambda(A)$ of A is the same as the spectral picture $\Lambda(T)$ of T and*

$$min\cdot ind(A - \lambda)^k \leq min\cdot ind(T - \lambda)^k, \quad (k \geq 1, \lambda \in \rho_{s-F}(A)).$$

Proof Let $(\Omega_1, k_1), (\Omega_2, k_2), \cdots, (\Omega_n, k_n)$ be the finitely many components of $\sigma(T)\backslash\overline{\Omega}$, where $k_i = ind(T - \lambda), \lambda \in \Omega_i (i = 1, 2, \cdots, n)$. Thus $\overline{\Omega}_i$ are pairwise disjoint and each Ω_i is a connected Cauchy domain.

If $|k_i| < \infty$, by Theorem 5.7.3 and Lemma 5.7.5, there is an (SI) operator $A_i = A(\Omega_i, k_i) \in \mathcal{L}(\mathcal{H})$ such that $\sigma(A_i) = \overline{\Omega}_i, ind(A_i - \lambda) = k_i \ (\lambda \in \Omega_i)$ and either $min\cdot ind(A_i - \lambda) = 0$ (if $k_i \neq 0$) or $min\cdot ind(A_i - \lambda) = i$ (if $k_i = 0$) and $\mathcal{A}'(A_i)/rad\mathcal{A}'(A_i)$ is commutative.

If $|k_i| = \infty$, by Proposition 3.13 of [Jiang, C.L. and Wang, Z.Y. (1998)], we can find an operator $A_i = A(\Omega_i, k_i) \in \mathcal{L}(\mathcal{H})$ such that

$$\sigma(A_i) = \overline{\Omega}_i, ind(A_i - \lambda) = k_i, min\cdot ind(A_i - \lambda) = 0 \ (\lambda \in \Omega_i),$$

$\mathcal{A}'(A_i)/rad\mathcal{A}'(A_i)$ is commutative and $\bigvee_{\lambda \in \Omega_i} ker(A_i - \lambda) = \mathcal{H}$.

Let $\Phi_1, \Phi_2, \cdots, \Phi_m$ be all the components of $\overline{\Omega}$. By Theorem 3.7 of [Jiang, C.L. and Wang, Z.Y. (1998)], we can find (SI) operators $c_1, c_2, \cdots, c_m$ such that $\sigma(c_i) = \sigma_{lre}(c_i) = \Phi_i$ and $\mathcal{A}'(c_i)$ is commutative. Set

$$
A = \begin{bmatrix}
A_1 & & & & Q_{11} & Q_{12} & \cdots & Q_{1m} \\
& A_2 & & & Q_{21} & Q_{22} & \cdots & Q_{2m} \\
& & \ddots & & \vdots & \vdots & \vdots & \vdots \\
& & & A_n & Q_{n1} & Q_{n2} & \cdots & Q_{nm} \\
& & & & c_1 & & & 0 \\
& & & & & c_2 & & \\
& 0 & & & & & \ddots & \\
& & & & & & & c_m
\end{bmatrix} \in \mathcal{L}(\mathcal{H}^{(n+m)}),
$$

where $\{Q_{ij} : 1 \le i \le n, 1 \le j \le m\}$ are defined as in Lemma 5.7.4. By Lemma 5.7.4, $A \in (SI)$ and satisfies all the requirements of the lemma.

Lemma 5.7.7 *Let A be given in Lemma 5.7.6, then $\mathcal{A}'(A)/rad\mathcal{A}'(A)$ is commutative.*

Proof Since $\mathcal{A}'(\bigoplus_{i=1}^{n} A_i)/rad\mathcal{A}'(\bigoplus_{i=1}^{n} A_i)$ and $\mathcal{A}'(\bigoplus_{j=1}^{m} c_j)/rad\mathcal{A}'(\bigoplus_{j=1}^{m} c_j)$ are commutative and since $ker\tau_{\bigoplus_{i=1}^{n} A_i, \bigoplus_{j=1}^{m} c_j} = \{0\}, \mathcal{A}'(A)/rad\mathcal{A}'(A)$ is commutative.

Theorem 5.7.8 *Let $T \in \mathcal{L}(\mathcal{H})$ with connected spectrum $\sigma(T)$, then there exists a sequence of (SI) operators $\{T_n\}_{n=1}^{\infty}$ satisfying*
 (i) *$\mathcal{A}'(T_n)/rad\mathcal{A}'(T_n)$ is commutative for all n;*
 (ii) *$\lim_{n \to \infty} \|T_n - T\| = 0$.*

Proof By Theorem 1.27 of [Jiang, C.L. and Wang, Z.Y. (1998)], we can find a sequence $\{A_n\}_{n=1}^{\infty}$ of operators such that for each n
 (i) $\sigma_{lre}(A_n)$ is the closure of an analytic Cauchy domain;
 (ii) $\sigma(A_n)$ is connected; and
 (iii) $\lim_{n \to \infty} \|A_n - T\| = 0$.
By Lemma 5.7.6 and Lemma 5.7.7, for each A_n, there exists a $B_n \in (SI)$ such that $\Lambda(B_n) = \Lambda(A_n)$ and

$$min{\cdot}ind(B_n - \lambda)^k \le min{\cdot}ind(A_n - \lambda)^k, \quad (k \ge 1, \lambda \in \rho_{s-F}(A)).$$

Moreover, $\mathcal{A}'(B_n)/rad\mathcal{A}'(B_n)$ is commutative.

It follows from Similarity Orbit Theorem that $A_n \in \mathcal{S}(B_n)^-$. Thus we can find a sequence $\{T_n\}_{n=1}^{\infty}$ of (SI) operators satisfying the requirements of the theorem.

Applying Theorem 5.7.3 and repeating the arguments above, we have the following theorem.

Theorem 5.7.9 *Given $A \in \mathcal{B}_n(\Omega)$, there exists a sequence of operators*

$$\{A_k\} \subset \mathcal{B}_n(\Omega) \cap (SI)$$

such that
  (i) $\mathcal{A}'(A_k)$ is commutative for each k;
  (ii) $\lim_{k \to \infty} \|A_k - A\| = 0$.

Definition 5.7.10 Let Ω_1, Ω_2 be two bounded connected open subsets of $\mathbf{C}$ and let n and m be two natural numbers. An operator $T \in \mathcal{L}(\mathcal{H})$ is said in the operator class $\mathcal{B}_{n,m}(\Omega_1, \Omega_2)$ if
  (i) $\Omega_i \subset \rho_F(T) \cap \sigma(T)$, $(i = 1, 2)$;
  (ii) $dimker(T - \lambda) = n$ for $\lambda \in \Omega_1$ and $dimker(T - \mu)^* = m$ for $\mu \in \Omega_2$;
  (iii) $\bigvee\limits_{\substack{\lambda \in \Omega_1 \\ \mu \in \Omega_2}} \{ker(T-\lambda), ker(T-\mu)^*\} = \mathcal{H}$. [M.J. Cowen and P.R. Douglas [1]].

Proposition 5.7.11 *Given $T \in \mathcal{B}_{1,1}(\Omega_1, \Omega_2)$, $\mathcal{A}'(T)/rad\mathcal{A}'(T)$ is commutative.*

Proof Denote $\mathcal{H}_r(T) = \bigvee\limits_{\lambda \in \Omega_1} ker(T - \lambda), \mathcal{H}_l(T) = \bigvee\limits_{\mu \in \Omega_2} ker(T - \mu)^*$. By Apostol's triangular representation, $\mathcal{H}_r(T) \perp \mathcal{H}_l(T)$ and

$$T = \begin{bmatrix} T_r & T_{12} \\ 0 & T_l \end{bmatrix} \begin{matrix} \mathcal{H}_r(T) \\ \mathcal{H}_l(T) \end{matrix},$$

where $T_r = T|_{\mathcal{H}_r(T)}$ and $T_l = (T^*|_{\mathcal{H}_l(T)})^*$. A simple computation shows that $T_r \in \mathcal{B}_1(\Omega_1)$ and $T_l^* \in \mathcal{B}_1(\Omega_2^*)$.
    Thus $\mathcal{A}'(T_r)$ and $\mathcal{A}'(T_l)$ are commutative. This implies $\mathcal{A}'(T)/rad\mathcal{A}'(T)$ is commutative.

Proposition 5.7.12 *Given $T \in \mathcal{B}_{m,m}(\Omega_1, \Omega_2)$, for each $\varepsilon > 0$, there exists a compact operator K with $\|K\| < \varepsilon$ such that $\mathcal{A}'(T + K)/rad\mathcal{A}'(T + K)$ is commutative.*

Proof By the argument of Proposition 5.7.11,

$$T = \begin{bmatrix} T_r & T_{12} \\ 0 & T_l \end{bmatrix} \begin{array}{l} \mathcal{H}_r(T) \\ \mathcal{H}_l(T) \end{array},$$

where $T_r \in \mathcal{B}_m(\Omega_1)$ and $T_l^* \in \mathcal{B}_n(\Omega_2^*)$. By Theorem 3.2.1, we can find compact operators K_1 and K_2 such that $max\{\|K_1\|, \|K_2\|\} < \frac{\varepsilon}{2}$ and

$$T_r + K_1 \in \mathcal{B}_m(\Omega_1) \cap (SI), \ (T_l + K_2)^* \in \mathcal{B}_n(\Omega_2^*) \cap (SI).$$

Set

$$K = \begin{bmatrix} K_1 & 0 \\ 0 & K_2 \end{bmatrix}.$$

By Theorem 4.4.3, $T + K$ satisfies the requirement of the proposition.

In this section we proved that for "almost every" strongly irreducible operator T, $\mathcal{A}'(T)/rad\mathcal{A}'(T)$ is commutative. For the matter of that, we conjecture that every (SI) operator T has the property, i.e., $\mathcal{A}'(T)/rad\mathcal{A}'(T)$ is commutative.

5.8 Remark

The results in Sections 5.1 and 5.2 are contributed by [Jiang, C.L. (1994)]. The work in Section 5.3 are due to [Fang, J.S.(2003)] and [Jiang, C.L. (1994)]. The work in Section 5.4 belongs to [Fang, J.S.(2003)]. The results in Section 5.5 were proved by [Fang, J.S.(2003)], [Jiang, C.L., Guo, X.Z. and Ji, K.], [He, H. and Ji, K.]. The work in Section 5.6 are due to [Jiang, C.L., Guo, X.Z. and Ji, K.]. The work in Section 5.7 are given by [Jin, Y.F. and Wang, Z.Y.(1)], [Jiang, C.L. (1994)] except that Proposition 5.7.1 is due to [Fong, C.K. and Jiang, C.L. (1993)]. Here we must point out that the work of classification of Cowen-Douglas operators was inspired by the work of [Elliott, G. and Gong, G. (1996)], [Elliott, G., Gong, G. and Li, L.], [Dadarlat, M. and Gong, G. (1997)].

5.9 Open Problem

1. Let $T \in \mathcal{B}_{n,m}(\Omega_1, \Omega_2)$, given a necessary and sufficient condition for $T \in (SI)$.
2. Let $T \in \mathcal{B}_{n,m}(\Omega_1, \Omega_2) \cap (SI)$. Is $\mathcal{A}'(T)/rad\mathcal{A}'(T)$ commutative?

3. If $T \in \mathcal{B}_{n,m}(\Omega_1, \Omega_2) \cap (SI)$, is $K_0(\mathcal{A}'(T))$ isomorphic to $\mathbf{Z}$?

4. Let $T \sim \{w_k\}$ be an injective unilateral weighted shift. If for each $\mathcal{M} \in LatT, T|_{\mathcal{M}} \sim T$, then is T similar to αT_z for some positive α?

5. Let T_f be an analytic Toeplitz operator. Is the following statement true? $T_f \in (SI)$ if and only if $\mathcal{A}'(T_f)/rad\mathcal{A}'(T_f)$ is commutative.

6. Let Ω be an analytic Cauchy domain. Does there exist an (SI) operator A satisfying the following conditions?

    (i) $A \in \mathcal{B}_{1,1}(\Omega, \Omega)$;

    (ii) $\mathcal{A}'(A)$ is commutative.

7. Let Ω be an analytic Cauchy domain. What is $K_1(H^\infty(\Omega))$?

8. Given $T \in \mathcal{L}(\mathcal{H})$ with connected $\sigma(T)$, does there exist a sequence $\{T_n\}$ of (SI) operators satisfying

    (i) $\lim_{n \to \infty} \|T_n - T\| = 0$;

    (ii) $T_n - T$ is compact;

    (iii) $\mathcal{A}'(T_n)/rad\mathcal{A}'(T_n)$ is commutative.

9. Let T_f be an analytic Toeplitz operator. Is the following statement true? $T_f \in (SI)$ is equivalent to $T_f \in (RI)$.

10. If $\mathcal{A}$ is a unital subalgebra of H^∞, is $K_0(\mathcal{A})$ isomorphic to the integer group $\mathbf{Z}$?

Chapter 6

Some Other Results About Operator Structure

6.1 K_0-Group of Some Banach Algebra

Theorem 6.1.1 *Let Ω be a bounded connected open subset of $\mathbf{C}$ with $(\overline{\Omega})^0 = \Omega$. Let $H^\infty(\Omega)$ be the unital Banach algebra consisting of all bounded analytic functions on Ω. Then $K_0(H^\infty(\Omega)) \cong \mathbf{Z}$ and $\bigvee(H^\infty(\Omega)) \cong \mathbf{N}$, where $(\overline{\Omega})^0$ denotes the interior of closure $\overline{\Omega}$ of Ω.*

To prove Theorem 6.1.1, we need some lemmas. For a bounded connected open set Ω, if $(\overline{\Omega})^0 = \Omega$, then there exists a probability measure μ such that support $\mu = \Gamma := \partial\Omega$ satisfying

$$f(z) = \int_\Gamma f d\mu \ \text{ for all } f \text{ analytic on } \overline{\Omega} \quad \text{[Herrero, D.A. (1974)]}.$$

Denote $N(\Gamma) :=$ the "multiplication by λ" on $L^2(\Gamma, \mu)$ and $H^2(\Gamma) :=$ "the subspace generated by all analytic functions on $\overline{\Omega}$". Then $H^2(\Gamma) \in Lat N(\Gamma)$ and

$$N(\Gamma) = \begin{bmatrix} N_+(\Gamma) & Z \\ 0 & N_-(\Gamma) \end{bmatrix} \begin{matrix} H^2(\Gamma) \\ L^2(\Gamma, \mu) \ominus H^2(\Gamma) \end{matrix}.$$

Lemma 6.1.2 *(i) $N(\Gamma)$ is normal, and $N_+(\Gamma), N_-(\Gamma)$ are essentially normal;*

(ii) $N_+^*(\Gamma) \in \mathcal{B}_1(\Omega)$;

(iii) $\mathcal{A}'(N_+(\Gamma)) \cong H^\infty(\Omega)$. [Conway, J.B. (1978)]

Lemma 6.1.3 *Let $G = \mathbf{Z}$ and (G, G_+) be an ordered group. Then there exists an isomorphism $\phi : G \longrightarrow \mathbf{Z}$ such that $\phi(G_+) \subset \mathbf{N}$.*
Proof By the definition of ordered group, we can assume that there is an $n, 0 < n \in G_+$. If $0 > m \in G_+$, then $(-m)G \in G_+$ and $mn \in G_+$.

Note that $(-m)n + mn = 0$, by the definition of ordered group again, we have $(-m)n = 0$ and $mn = 0$. Thus $G_+ \subset \mathbf{N}$. Suppose that ϕ is the identity isomorphism from G onto $\mathbf{Z}$, then $\phi(G_+) \subset \mathbf{N}$.

Now we are in a position to prove Theorem 6.1.1.

Proof of Theorem 6.1.1 For the given Ω, by Theorem 5.1.6 $K_0(\mathcal{A}'(N_+(\Gamma))) \cong \mathbf{Z}$. It follows from Lemma 6.1.2 that $K_0(H^\infty(\Omega)) \cong \mathbf{Z}$. By Lemma 6.1.3,

$$\bigvee(\mathcal{A}'(N_+(\Gamma))) = \bigvee(H^\infty(\Omega)) \subset \mathbf{N}.$$

We need only to prove that $\bigvee(\mathcal{A}'(N_+(\Gamma))) \cong \mathbf{N}$.

Set $P = diag(I, 0, 0, \cdots) \in M_\infty(\mathcal{A}'(N_+(\Gamma)))$ and $r = [P] \in \bigvee(\mathcal{A}'(N_+(\Gamma)))$. Then r is a positive integer, since $\mathcal{A}'(N_+(\Gamma))$ is stably finite.

Suppose that $q \in M_n(\mathcal{A}'(N_+(\Gamma)))$ is a nonzero idempotent, then

$$0 \neq [q] = s \in \bigvee(\mathcal{A}'(N_+(\Gamma))).$$

Suppose that $B = (N_+(\Gamma))^{(n)}|_{P\mathcal{H}_n}$, then $\mathcal{A}'(B)$ is k-homogeneous and $k \geq 1$. Note that $rs = r[q] = s[p]$, there exists $n' \geq n$ such that

$$Q = diag(q, q_1, \cdots, q_r, 0, \cdots, 0) \sim_a diag(P, P_1, \cdots, P_s, 0, \cdots, 0) = P$$

in $M_n(\mathcal{A}'(N_+(\Gamma)))$. By Lemma 4.2.4,

$$B^{(r)} = A^{(n')}|_{Q\mathcal{H}^{(n')}} \sim A^{(n')}|_{Q\mathcal{H}^{(n')}} = A^{(s)}.$$

Thus $\mathcal{A}'(B^{(r)}) \cong \mathcal{A}'(A^{(s)})$, i.e., $M_r(\mathcal{A}'(B)) \cong M_s(\mathcal{A}'(A))$.

Since $M_r(\mathcal{A}'(B))$ is rk-homogeneous and $M_s(\mathcal{A}'(A))$ is s-homogeneous, $s = rk$. Therefore,

$$\bigvee(\mathcal{A}'(A)) = \{kr : k = 0, 1, 2, \cdots\}.$$

Since $(K_0(\mathcal{A}'(A)), \bigvee(\mathcal{A}'(A)))$ is an ordered group, $r = 1$. Thus $\bigvee(\mathcal{A}'(N_+(\Gamma))) \cong \mathbf{N}$.

In Theorem 5.1.6, we have directly proved that $\bigvee(\mathcal{A}'(N_+(\Gamma))) \cong \mathbf{N}$. In the proof of Theorem we give a new proof by using a different method.

Let Ω be an analytic connected Cauchy domain and $W^{22}(\Omega)$ denote the Sobolev space $W^{22}(\Omega) = \{f \in L^2(\Omega, dm) :$ the distributional partial derivatives of first and second order of f in $L^2(\Omega, dm)\}$, where dm us the planar Lebesgue measure.

Set

$$W(\Omega) = \{M_f : f \in W^{22}(\Omega)\},$$

where M_f is "multiplication by f" on $W^{22}(\Omega)$. Denote $R(\Omega)$ the subspace of $W^{22}(\Omega)$ generated by the set of all rational functions with poles outside $\overline{\Omega}$. Note that $R(\Omega) \in LatM_\lambda$. Denote $M_\lambda(\Omega) = M_\lambda|_{R(\Omega)}$. By Proposition 4.5.1,

$$\mathcal{A}'(M_\lambda(\Omega)) \cong R(\Omega).$$

Theorem 6.1.4 $K_0(R(\Omega)) \cong \mathbf{Z}$ *and* $\bigvee(R(\Omega)) \cong \mathbf{N}$.
Proof Since $M_\lambda^* \in \mathcal{B}_1(\Omega^*)$, by a similar argument of the proof Lemma 6.1.2 we can prove Theorem 6.1.4.

For an analytic connected Cauchy domain Ω, denote

$$A(\Omega) = \{f : f \text{ is analytic in } \Omega \text{ and } f \in C(\overline{\Omega}).$$

Then $A(\Omega)$ is a unital Banach algebra with the norm $\|f\| = \max\limits_{z \in \overline{\Omega}} |f(z)|$. $A(D)$, when $\Omega = D$, is called disk algebra.

Given an $f \in A(\Omega)$, there exists a sequence r_n of rational functions with poles outside $\overline{\Omega}$ such that

$$\|r_n - f\|_{A(\Omega)} \longrightarrow 0 \quad (n \longrightarrow \infty).$$

Theorem 6.1.5 $K_0(A(D)) \cong \mathbf{Z}$ *and* $\bigvee(A(D)) \cong \mathbf{N}$.
Proof We need only to show that for each idempotent $P \in M_n(A(D))$,

$$P \sim_a diag(1, 1, \cdots, 1_k, 0, \cdots)$$

in $M_\infty(A(D))$, where $k \geq n$. Since $P \in M_n(A(D))$,

$$P = \begin{bmatrix} f_{11}(z) & f_{12}(z) & \cdots & f_{1n}(z) \\ \cdots & \cdots & \cdots & \cdots \\ f_{n1}(z) & f_{n2}(z) & \cdots & f_{nn}(z) \end{bmatrix}.$$

Let $y = \frac{1}{m} + z$ and define

$$P' = \begin{bmatrix} f_{11}(\frac{1}{m} + z) & f_{12}(\frac{1}{m} + z) & \cdots & f_{1n}(m + z) \\ \cdots & \cdots & \cdots & \cdots \\ f_{n1}(\frac{1}{m} + z) & f_{n2}(\frac{1}{m} + z) & \cdots & f_{nn}(m + z) \end{bmatrix}$$

$$= \begin{bmatrix} f_{11}(y) & f_{12}(y) & \cdots & f_{1n}(y) \\ \cdots & \cdots & \cdots & \cdots \\ f_{n1}(y) & f_{n2}(y) & \cdots & f_{nn}(y) \end{bmatrix}.$$

Since $\|P' - P\|$ can be arbitrary small when m is big enough, P' is homotopic to P, denoted by $P' \sim_h P$ in $M_n(A(D))$.

Note that $P' \in M_n(A(D))$. It follows from $\bigvee(R(D)) \cong \mathbf{N}$ and the K-theory that

$$P' \sim_h diag(1, 1, \cdots, 1_k, 0, \cdots, 0)$$

in $M_{4n}(A(D))$.

Let $h_1(t) : [0, 1] \longrightarrow M_n(A(D))$ be a continuous map satisfying

$$h_1(0) = P, \ h_1(1) = P'$$

and $h_2(t) : [0, 1] \longrightarrow M_{4n}(A(D))$ be a continuous map satisfying

$$h_2(0) = P', \ h_2(1) = diag(1, 1, \cdots, 1_k, 0, \cdots, 0).$$

Denote $h_1(t) = (h(t) \oplus 0^{(3n)})$, then

$$h_1(t) : [0, 1] \longrightarrow M_{4n}(A(D))$$

is a continuous map and

$$h_1(0) = P \oplus 0^{(3n)}, \ h_1(1) = P' \oplus 0^{(3n)}.$$

Thus $h_3(t) = (h_2 \circ h_1)(t) : [0, 1] \longrightarrow M_{4n}(A(D))$ satisfies

$$h_3(0) = P, \ h_3(1) = diag(1, 1, \cdots, 1_k, 0, \cdots, 0).$$

Therefore, $\bigvee(A(D)) \cong \mathbf{N}$ and $K_0(A(D)) \cong \mathbf{Z}$.

6.2 Similarity and Quasisimilarity

We have the following question: Is $A \sim B$ if $A^{(n)} \sim B^{(n)}$, where n is a natural number.

Theorem 6.2.1 *Let $A, B \in M_k(\mathbf{C})$. Then $A \sim B$ if and only if $A^{(n)} \sim B^{(n)}$.*
Proof By matrix theory, we can easily obtain the theorem.

Theorem 6.2.2 *Let $A, B \in \mathcal{L}(\mathcal{H}) \cap (SI)$. If*

$$\bigvee(A'(A)) \cong \bigvee(A'(B)) \cong \mathbf{N},$$

$A'(A)/radA'(A)$ and $A'(B)/radA'(B)$ are commutative, then $A \sim B$ if and only if $A^{(n)} \sim B^{(n)}$ for each natural number n.
Proof By Theorem 5.5.15, $A \sim B$ if and only if $K_0(\mathcal{A}'(A \oplus B)) \cong \mathbf{Z}$ and by Theorem 5.5.15 again,

$$\bigvee(\mathcal{A}'(A \oplus B)) \cong \bigvee(\mathcal{A}'(A \oplus B)^{(n)}) \cong \mathbf{N}^{(l)},$$

where $l = \begin{cases} 1 \text{ if } A \nsim B \\ 0 \text{ if } A \sim B \end{cases}$. Therefore, $A \sim B$ if and only if $A^{(n)} \sim B^{(n)}$.

Theorem 6.2.3 *Suppose that $A = \bigoplus_{i=1}^{k} A_i^{(n_i)}, B = \bigoplus_{j=1}^{l} B_j^{(m_j)}$, where $\{A_i\}$ and $\{B_j\}$ are two families of (SI) and pairwise not similar operators. If for each i and j, $(1 \leq i \leq k, 1 \leq j \leq l)$, $A'(A_i)/radA'(A_i)$ and $A'(B_j)/radA'(B_j)$ are commutative, and $\bigvee(\mathcal{A}'(A_i)) \cong \bigvee(\mathcal{A}'(B_j)) \cong \mathbf{N}$, then $A \sim B$ if and only if $A^{(n)} \sim B^{(n)}$ for each natural number n.*
Proof Applying Theorem 5.5.15 and repeating the arguments in the proof of Theorem 6.2.2, we can prove the theorem.

Corollary 6.2.4 *Given $A, B \in \mathcal{B}_n(\Omega)$, $A \sim B$ if and only if $A^{(n)} \sim B^{(n)}$ for each natural number n.*
Proof By Theorem 4.4.3, Theorem 5.5.15 and Theorem 6.2.3, we get the corollary.

K. Davidson and D.A. Herrero raised the following question in [Davidson, K.R. and Herrero, D.A. (1990)].
Suppose that $A \in \mathcal{L}(\mathcal{H}) \cap (SI)$ and $A \sim_{q.s.} B$, is $B \in (SI)$?
In general, the answer to the question above is "no".

Proposition 6.2.5 *Suppose that $A \in \mathcal{B}_n(\Omega) \cap (SI)$ with $\sigma(A) = \overline{D}$, and*

$$W(I_\mathcal{H} + A) + (I_\mathcal{H} + A)X \neq I_\mathcal{H}$$

for each $X \in \mathcal{L}(\mathcal{H})$, then

$$T = \begin{bmatrix} I_\mathcal{H} + A & I_\mathcal{H} \\ 0 & -I_\mathcal{H} - A \end{bmatrix} \begin{matrix} \mathcal{H} \\ \mathcal{H} \end{matrix} \in (SI)$$

and if $ker(I_{\mathcal{H}} + A) = \{0\}$, *then* $T \sim_{q.s.} (I_{\mathcal{H}} + A) \oplus (-I_{\mathcal{H}} - A)$.

Proof $I_{\mathcal{H}} + A \in \mathcal{B}_n(D(1,1)), -(I_{\mathcal{H}} + A) \in \mathcal{B}_n(D(-1,1))$, where

$$D(1,1) = \{z \in \mathbf{C} : |z - 1| < 1\}, \quad D(-1,1) = \{z \in \mathbf{C} : |z + 1| < 1\}.$$

Thus

$$D(1,1) \subset \sigma_p(I_{\mathcal{H}} + A), \quad D(-1,1) \subset \sigma_p(-(I_{\mathcal{H}} + A)).$$

Since $D(1,1) \cap D(-1,1) = \emptyset$ and

$$\bigvee_{z \in D(1,1)} ker(I_{\mathcal{H}} + A - z) = \bigvee_{z \in D(-1,1)} ker(-I_{\mathcal{H}} - A - z) = \mathcal{H},$$

by Lemma 3.2.3,

$$ker \tau_{I+A,-(I+A)} = \{0\}.$$

Let P be a nontrivial idempotent in $\mathcal{A}'(T)$ and

$$P = \begin{bmatrix} P_{11} & P_{12} \\ P_{21} & P_{22} \end{bmatrix} \begin{matrix} \mathcal{H} \\ \mathcal{H} \end{matrix}.$$

Since $(I+A)P_{21} + P_{21}(I+A) = 0$ and $P_{21} = 0$, P_{11} and P_{22} are idempotents in $\mathcal{A}'(A)$. Since $A \in (SI)$, $P_{11} = 0$ and $P_{22} = I$ or $P_{11} = I$ and $P_{22} = 0$. Assume that $P_{11} = 0$ (otherwise consider $I - P$). If follows from $PT = TP$ that

$$P_{12}(I + A) + (I + A)P_{12} = -P_{22}.$$

By the assumption of the proposition $P_{22} = 0$. This shows that $T \in (SI)$.

Set

$$X = \begin{bmatrix} 2(I + A) & I \\ 0 & I \end{bmatrix} \begin{matrix} \mathcal{H} \\ \mathcal{H} \end{matrix}$$

and

$$Y = \begin{bmatrix} I & I \\ 0 & -2(I + A) \end{bmatrix} \begin{matrix} \mathcal{H} \\ \mathcal{H} \end{matrix}.$$

Then $XT = ((I + A) \oplus (-I - A))X, TY = Y((I + A) \oplus (-I - A))$. This implies that $T \sim_{q.s.} (I + A) \oplus (-(I + A))$.

Example 6.2.6 *Let S be the injective unilateral shift on $\mathcal{H} = l^2$. Then $S^* \in \mathcal{B}_1(D)$. Let $\{e_n\}_{n=1}^{\infty}$ be the ONB of $\mathcal{H}$ and $Se_n = e_{n+1}$ $(n = 1, 2, \cdots)$.*

If for some $X \in \mathcal{L}(\mathcal{H})$, $X(I + S^*) + (I + S^*)X = I$, then

$$XS^* + S^*X + 2X = I.$$

A simple computation indicates that $X e_n = \frac{1}{2} \sum_{k=1}^{n} (-1)^{k-1} e_{n+1-k}$ ($n =$

$1, 2, \cdots$) *and* $\|X e_n\| = \frac{\sqrt{n}}{2} \longrightarrow \infty$ ($n \longrightarrow \infty$). *Thus $X(I+S^*)+(I+S^*)X \neq I$ for all $X \in \mathcal{L}(\mathcal{H})$. By Proposition 6.2.5,*

$$T = \begin{bmatrix} I + S^* & I \\ 0 & -(I + S^*) \end{bmatrix} \begin{matrix} \mathcal{H} \\ \mathcal{H} \end{matrix}$$

and

$$Y = \begin{bmatrix} I & I \\ 0 & -2(I + A) \end{bmatrix} \begin{matrix} \mathcal{H} \\ \mathcal{H} \end{matrix} \in (SI).$$

Theorem 6.2.7 *Suppose that $A \in \mathcal{B}_n(\Omega) \cap (SI)$ and $B \sim_{q.s.} A$, then $B \in (SI)$.*
Proof Without loss of generality, we assume that the minimal index of A is n. Since $B \sim_{q.s.} A$, there exist operators X and Y, with trivial kernels and dense ranges, such that $AX = XB$ and $YA = BY$.

A simple computation indicates that $\Omega \in \sigma_p(B)$ and $dimker(B-z) = n$ for all $z \in \Omega$. Note that $AXTY = XBTY = XTYA$ for each $T \in \mathcal{A}'(B)$. This implies that $XTY \in \mathcal{A}'(A)$. Since n is the minimal index of A, by Theorem 4.4.3, $\sigma(XTY|_{ker(A-z)})$ is connected for each $z \in \Omega$. Since $YA = BY, Y(z)$ is a linear transformation from $ker(A - z)$ to $ker(B - z)$, where $Y(z) = Y|_{ker(A-z)}$.

Similarly, $X(z) = X|_{ker(B-z)}$ is a linear transformation from $ker(B-z)$ to $ker(A - z)$. Since

$$AXY = XBY = XYA,$$

$$XY \in \mathcal{A}'(A)$$

and

$$\Gamma(XY)(z) = X(z)Y(z)$$

for all $z \in \Omega$. Similarly, $YX \in \mathcal{A}'(B)$ and $\Gamma(YX)(z) = Y(z)X(z)$ for all $z \in \Omega$.

By Theorem 4.4.3, $\sigma(X(z)Y(z)) = \{\lambda(z)\}, \lambda(z) \neq 0$. Since X, Y are injective, $Y(z)X(z)$ is invertible. Thus $\sigma(Y(z)X(z)) = \{\lambda(z)\}$.

For arbitrary $T \in \mathcal{A}'(B)$, since $XTY \in \mathcal{A}'(A)$,

$$\Gamma(XTY)(z) = X(z)(T|_{ker(B-z)})Y(z)$$

and $\sigma(X(z)(T|_{ker(B-z)})Y(z))$ is connected. Assume that

$$\sigma(X(z)(T|_{ker(B-z)})Y(z)) = \{\mu(z)\},$$

then

$$\sigma(Y(z)X(z)(T|_{ker(B-z)})) = \{\mu(z)\} \cup \{0\}.$$

{Claim 1} If $\mu(z) = 0$ for some $z \in \Omega$, then $\sigma(T|_{ker(B-z)}) = 0$.

The claim can be proved as follows.

Since $Y(z)$ and $X(z)$ are invertible, $0 \in \sigma(T|_{ker(B-z)})$. If $0 \neq \lambda \in \sigma(T|_{ker(B-z)})$, then $\lambda I_{ker(B-z)} - T|_{ker(B-z)}$ is not invertible. Repeating the argument above, we have

$$\sigma(X(z)(\lambda I_{ker(B-z)} - T|_{ker(B-z)})Y(z)) = \{0\}.$$

Thus

$$tr(X(z)(\lambda I_{ker(B-z)} - T|_{ker(B-z)})Y(z)) = \{0\}.$$

But

$$tr(X(z)(\lambda I_{ker(B-z)} - T|_{ker(B-z)})Y(z))$$

$$= tr(\lambda X(z)Y(z) - X(z)(T|_{ker(B-z)})Y(z))$$

$$= \lambda tr(X(z)Y(z)) - tr(X(z)(T|_{ker(B-z)})Y(z))$$

$$= n\lambda\lambda(z) \neq 0$$

A contradiction.

{Claim 2} If there exists $z \in \Omega$ such that $\mu(z) \neq 0$, then $\sigma(T|_{ker(B-z)})$ is connected. Suppose that $\beta(z) \in \sigma(T|_{ker(B-z)})$, then

$$\beta(z)I_{ker(B-z)} - T|_{ker(B-z)}$$

is not invertible. Repeating the proof of Claim 1, we have

$$\sigma(X(z)(\beta(z)I_{ker(B-z)} - T|_{ker(B-z)})Y(z)) = \{0\}.$$

By Claim 1, $\sigma(T|_{ker(B-z)}) = \{\beta(z)\}$. It follows from $\bigvee\limits_{z\in D} ker(B-z) = \mathcal{H}$ that $\bigvee\limits_{z\in D} ker(T-\beta(z))^n = \mathcal{H}$. This implies that $\sigma(T)$ is connected. Therefore $\sigma(T)$ is connected for each $T\in\mathcal{A}'(B)$.

{Claim 3} $B\in(SI)$.

Otherwise, there exist $\mathcal{M},\mathcal{N}\in LatB$ such that $\mathcal{M}\cap\mathcal{N} = \{0\}$ and $\mathcal{M} + \mathcal{N} = \mathcal{H}$.

Denote $B_1 = T|_{\mathcal{M}}, B_2 = T|_{\mathcal{N}}$, then $B = B_1\dot{+}B_2$. Set $T = I_{\mathcal{M}}\dot{+}2I_{\mathcal{N}}$, then $T\in\mathcal{A}'(B)$, but $\sigma(T)$ is disconnected. This contradiction implies that $B\in(SI)$.

Proposition 6.2.8 *Suppose that* $A\in\mathcal{B}_n(\Omega)\cap(SI)$ *and* $B\sim_{q.s.}A$, *then*

$$\mathcal{A}'(B)/rad\mathcal{A}'(B)$$

is commutative.

Proof Without loss of generality, we still assume that the minimal index of A is n. Then $\Omega\subset\sigma_p(B), dimker(B-z) = n$ $(z\in\Omega)$ and $\bigvee\limits_{z\in\Omega} ker(B-z) = \mathcal{H}$.

For $X, Y\in\mathcal{A}'(B)$,

$$(XY-YX)|_{ker(B-z)} = X|_{ker(B-z)}Y|_{ker(B-z)} - Y|_{ker(B-z)}X|_{ker(B-z)}$$

for all $z\in\Omega$. It follows from Claim 2 in the proof of Theorem 6.2.7 that

$$\sigma((XY-YX)|_{ker(B-z)}) = \{0\}$$

for all $z\in\Omega$. Thus $((XY-YX)|_{ker(B-z)})^n = 0$ for all $z\in\Omega$ and $\mathcal{A}'(B)/rad\mathcal{A}'(B)$ is commutative.

Next we will discuss the similarity of irreducible operators. Since irreducibility is unitarily invariant, some important behavior of C^*-algebras and irreducible operators can be described in terms of irreducible C^*-algebras and irreducible operators. F. Gilfeather proved that if $N\in\mathcal{L}(\mathcal{H})$ is a normal operator with empty point spectrum, then N is similar to a irreducible operator. [Jiang, Z.J. and Sun, S.L. (1992)] proved that each self-adjoint operator with infinitely many point spectrum is similar to a irreducible operator. D.A. Herrero posed the following conjecture in [Herrero, D.A. (1979)]:

An operator $Q\in\mathcal{L}(\mathcal{H})$ is similar to an irreducible operator if and only if
(i) $\lambda - Q$ is not finite rank for each complex number λ;

(ii) Q does not satisfy any quadratic equation $ax^2 + bx + c = 0$, where $|a| + |b| + |c| \neq 0$;

(iii) $\lambda - Q$ is not a direct sum of a finite rank operator and an operator satisfying a quadratic equation for $\lambda \in \mathbf{C}$.

In the following we will answer Herrero's conjecture in the case of nilpotent operators, normal operators and Cowen-Douglas operators.

Denote $\mathcal{N}_k = \{T \in \mathcal{L}(\mathcal{H}) : T^k = 0 \text{ and } T^{k-1} \neq 0\}$. For $T \in \mathcal{N}_k(\mathcal{H})$, T admits a $k \times k$ operator matrix representation of the following form:

$$
T = \begin{bmatrix} 0 & T_{12} & T_{13} & \cdots & T_{1\,k-1} & T_{1k} \\ & 0 & T_{23} & \cdots & T_{2\,k-1} & T_{2k} \\ & & 0 & \ddots & T_{3\,k-1} & T_{3k} \\ & & & \ddots & \ddots & \vdots \\ 0 & & & & 0 & T_{k-1\,k} \\ & & & & & 0 \end{bmatrix} \begin{array}{l} kerT \ominus kerT^0 \\ kerT^2 \ominus kerT^1 \\ kerT^3 \ominus kerT^2 \\ \vdots \\ kerT^{k-1} \ominus kerT^{k-2} \\ kerT^k \ominus kerT^{k-1} \end{array} \ ,
$$

where $kerT^0 = \{0\}, kerT^k = \mathcal{H}, \ kerT_{j-1\,j} = \{0\}, j = 1, 2, \cdots, k$ [Davidson, K.R. and Herrero, D.A. (1990)]. Using Lemma 7.8 of [Herrero, D.A. (1990)], we can prove that $T \in \mathcal{N}_k(\mathcal{H})$ $(k \geq 3)$ is similar to an operator of the form

$$
T \sim \begin{bmatrix} 0 & T_1 & T_{12} & \cdots & T_{1\,k-2} & T_{1\,k-1} \\ & 0 & T_2 & \cdots & T_{2\,k-2} & T_{2\,k-1} \\ & & 0 & \ddots & T_{3\,k-2} & T_{3\,k-1} \\ & & & \ddots & \ddots & \vdots \\ 0 & & & & 0 & T_{k-1} \\ & & & & & 0 \end{bmatrix} \begin{array}{l} \mathcal{H}_1 \\ \mathcal{H}_2 \\ \mathcal{H}_3 \\ \vdots \\ \mathcal{H}_{k-1} \\ \mathcal{H}_k \end{array} . \tag{6.2.1}
$$

where $kerT_i = \{0\}$ $(i = 1, 2, \cdots, k-1)$ and $ranT_1, ranT_2$ are dense.

Or

$$
T \sim 0_{\mathcal{H}_1} \oplus \begin{bmatrix} 0 & T_1 & T_{12} & \cdots & T_{1\,k-2} & T_{1\,k-1} \\ & 0 & T_2 & \cdots & T_{2\,k-2} & T_{2\,k-1} \\ & & 0 & \ddots & T_{3\,k-2} & T_{3\,k-1} \\ & & & \ddots & \ddots & \vdots \\ 0 & & & & 0 & T_{k-1} \\ & & & & & 0 \end{bmatrix} . \tag{6.2.2}
$$

where $0_{\mathcal{H}_1}$ is the zero operator acting on $\mathcal{H}_1, dim\mathcal{H}_1 = n, (1\leq n\leq\infty)$, $kerT_i = \{0\}$ $(i = 1, 2, \cdots, k - 1)$ and $ranT_1, ranT_2$ are dense.

Or

$$T{\sim}T_1 + F, \qquad (6.2.3)$$

where $T_1\in\mathcal{N}_2(\mathcal{H})$ and F has finite rank.

Theorem 6.2.9   *Given* $T\in\mathcal{N}_k(\mathcal{H})$, T *is similar to an irreducible operator if and only if the following conditions are satisfied:*
  (i) T *is not finite rank;*
  (ii) $T^2 \neq 0$;
  (iii) $T \neq T_1{\oplus}F$, *where* $T_1\in\mathcal{N}_k(\mathcal{H})$ *and* F *is finite rank.*

Before we prove Theorem 6.2.9, we need several lemmas.

Lemma 6.2.10 [Kato, T. (1984)]   *Suppose that* $A, B\in\mathcal{L}(\mathcal{H})$. *If* A *is positive with* $kerA = \{0\}$, *then there exists a positive number* δ *such that* $ker(A + \lambda B) = \{0\}$ *for* $|\lambda| < \delta$.

Lemma 6.2.11   *Given* $A, B\in\mathcal{L}(\mathcal{H})$. *If* A *is positive with* $kerA = \{0\}$, *then there exists a positive number* λ *such that* $ran(\lambda A + B)$ *is dense in* $\mathcal{H}$.
Proof   Since $ran[\lambda A+B]^- = [ker(\lambda A+B^*)]^\perp$, we need only to show that $ker(\lambda A + B^*) = \{0\}$ for some positive number λ. Note that $\lambda A + B^* = \lambda(A + \frac{1}{\lambda}B^*)$. Thus Lemma 6.2.11 follows from Lemma 6.2.10.

Lemma 6.2.12   *Suppose that* $A, B\in\mathcal{L}(\mathcal{H})$, A *is a positive operator with* $kerA = \{0\}$ *and* $\sigma(A)$ *is uncountable. Then there exists an operator* $E\in\mathcal{L}(\mathcal{H})$ *such that the* C^*-*algebra generated by* $A, AE + B$ *and* I *is irreducible.*
Proof   Since $\sigma(A)$ is uncountable, there exist disjoint uncountable Borel uncountable sets σ_1, σ_2 such that $\sigma(A) = \sigma_1\cup\sigma_2$ and $0\notin\bar{\sigma}_2$. Suppose that E is the spectral measure of A satisfying $E(\sigma_1)E(\sigma_2) = 0$. Then

$$dim(E(\sigma_1)\mathcal{H}) = \infty, \quad \text{and} \quad dim(E(\sigma_2)\mathcal{H}) = \infty.$$

Denote

$$A = \begin{bmatrix} A_1 & 0 \\ 0 & A_2 \end{bmatrix} \begin{matrix} E(\sigma_1)\mathcal{H} \\ E(\sigma_2)\mathcal{H} \end{matrix},$$

where $A_1 = A|_{E(\sigma_1)\mathcal{H}}$, $A_2 = A|_{E(\sigma_2)\mathcal{H}}$. Then A_1, A_2 are positive operators and $kerA_1 = kerA_2 = \{0\}$. Since $0\notin\bar{\sigma}_2$, A_2 is invertible.

Assume that

$$B = \begin{bmatrix} B_{11} & B_{12} \\ B_{21} & B_{22} \end{bmatrix} \begin{matrix} E(\sigma_1)\mathcal{H} \\ E(\sigma_2)\mathcal{H} \end{matrix}.$$

By Lemma 6.2.11 $ran(\lambda A_1 + B_{12})$ is dense in $E(\sigma_1)\mathcal{H}$ for some $\lambda > 0$.
Set

$$E = \begin{bmatrix} 0 & \lambda \\ -A_2^{-1} B_{21} & -A_2^{-1}(B_{22} - (d + V)) \end{bmatrix},$$

where $d > \|B_{11}\| + 1$, V is the Volterra operator. Thus V is irreducible and $\sigma(d + V) = \{d\}$.

$$AE + B = \begin{bmatrix} 0 & \lambda A_1 \\ -B_{21} & d + V - B_{22} \end{bmatrix} + \begin{bmatrix} B_{11} & B_{12} \\ B_{21} & B_{22} \end{bmatrix}$$

$$= \begin{bmatrix} B_{11} & \lambda A_1 + B_{12} \\ 0 & d + V \end{bmatrix}.$$

We are now to prove that the C^*-algebra generated by I, A and $AE + B$ is irreducible.

Suppose that P is a projection commuting with A and $AE + B$, and

$$P = \begin{bmatrix} P_{11} & P_{12} \\ P_{21} & P_{22} \end{bmatrix}.$$

Since $PA = AP$, $P_{21}A_1 = A_2 P_{21}$ and $A_1 P_{12} = P_{12} A_2$. It follows from $E(\sigma_1)E(\sigma_2) = 0$ that $P_{21} = P_{12} = 0$ and

$$P = \begin{bmatrix} P_{11} & 0 \\ 0 & P_{22} \end{bmatrix}.$$

Since $P(AE + B) = (AE + B)P$, $P_2(d + V) = (d + V)P_2$. Since $d + V$ is irreducible, $P_2 = 0$ or I. Without loss of generality, we assume that $P_2 = 0$ (Otherwise consider $I - P$). Since $P(AE + B) = (AE + B)P$,

$$\begin{bmatrix} P_1 & 0 \\ 0 & 0 \end{bmatrix} \begin{bmatrix} B_{11} & \lambda A_1 + B_{12} \\ 0 & d + V \end{bmatrix} = \begin{bmatrix} B_{11} & \lambda A_1 + B_{12} \\ 0 & d + V \end{bmatrix} \begin{bmatrix} P_1 & 0 \\ 0 & 0 \end{bmatrix}.$$

Thus $P_1(\lambda A_1 + B_{12}) = 0$. Since the range of $\lambda A_1 + B_{12}$ is dense, $P_1 = 0$ and $P = 0$. Therefore the C^*-algebra generated by $A, AE + B$ and I is irreducible.

Lemma 6.2.13 *Suppose that $A, B \in \mathcal{L}(\mathcal{H})$, A is a positive operator with $ker A = \{0\}$ and $\sigma(A)$ is uncountable. Then there exists an operator*

$E \in \mathcal{L}(\mathcal{H})$ *such that the C^*-algebra generated by $EA + B, A$ and I is irreducible.*

Proof The lemma follows from Lemma 6.2.12.

Lemma 6.2.14 *Let $A_1, A_2, B \in \mathcal{L}(\mathcal{H})$. If A_2 is a positive diagonal self-adjoint operator, A_1 is a positive operator and $\ker A_1 = \ker A_2 = \{0\}$. Then there exist $X, E \in \mathcal{L}(\mathcal{H})$, X is invertible, such that:*

(i) $A_2 X$ is a positive diagonal self-adjoint operator with trivial kernel

$$\ker(A_2 X) = \{0\};$$

(ii) The C^*-algebra generated by $A_1 E + BX, A_2 X$ and A_1 is irreducible.
Proof Since A_2 is a diagonal operator, $A_2 e_n = \lambda_n e_n$ for some ONB $\{e_n\}_{n=1}^{\infty}$. Since $\ker A_2 = \{0\}$, $\lambda_n \neq 0$ for all $n \geq 1$. Assume that A_1 admits the following matrix representation with respect to the ONB $\{e_n\}_{n=1}^{\infty}$,

$$A_1 = \begin{bmatrix} a_{11} & a_{12} & \cdots & a_{1n} & \cdots \\ a_{21} & a_{22} & \cdots & a_{2n} & \cdots \\ \vdots & \vdots & \ddots & \vdots & \vdots \\ a_{n1} & a_{n2} & \cdots & a_{nn} & \cdots \\ \vdots & \vdots & \ddots & \vdots & \ddots \end{bmatrix}$$

and each row vector $(a_{j1}, a_{j2}, \cdots, a_{jn}, \cdots) \neq 0$.

Otherwise, if $(a_{j1}, a_{j2}, \cdots, a_{jn}, \cdots) = 0$. Since $A_1^* = A_1$,

$$(a_{1j}, a_{2j}, \cdots, a_{nj}, \cdots) = 0.$$

This implies that $0 \in \sigma_p(A_1)$ and contradicts $\ker A_1 = \{0\}$. Without loss of generality, assume that $a_{1j_1} \neq 0$.

Set

$$X = \begin{bmatrix} t_1 & & & \\ & t_2 & & \mathbf{0} \\ & & \ddots & \\ & \mathbf{0} & & t_n \\ & & & & \ddots \end{bmatrix} \begin{matrix} e_1 \\ e_2 \\ \vdots \\ e_n \\ \vdots \end{matrix},$$

where $t_i \in [1, 2]$ satisfying $\lambda_i t_i \neq \lambda_j t_j$ $(i \neq j)$.

Clearly, X is bounded.

Assume that

$$BX = \begin{bmatrix} b_{11} & b_{12} & \cdots & b_{1n} & \cdots \\ b_{21} & b_{22} & \cdots & b_{2n} & \cdots \\ \vdots & \vdots & \ddots & \vdots & \vdots \\ b_{n1} & b_{n2} & \cdots & b_{nn} & \cdots \\ \vdots & \vdots & \ddots & \vdots & \ddots \end{bmatrix}.$$

Set

$$E = \begin{bmatrix} e_{11} & e_{12} & \cdots & e_{1n} & \cdots \\ e_{21} & e_{22} & \cdots & e_{2n} & \cdots \\ \vdots & \vdots & \ddots & \vdots & \vdots \\ e_{n1} & e_{n2} & \cdots & e_{nn} & \cdots \\ \vdots & \vdots & \ddots & \vdots & \ddots \end{bmatrix},$$

where $e_{j_1 n} = \frac{\alpha_n}{(j_1+n)!}$, $\alpha_n \in (0,1]$ such that $\alpha_{j_1} e_{j_1 n} + b_{1n} \neq 0$. If $(j,k) \neq (j_1, n)$, then $e_{jk} = 0$. It is easy to see that E is bounded. Since

$$A_2 X = \begin{bmatrix} a_1 & & & \\ & a_2 & & \mathbf{0} \\ & & \ddots & \\ & \mathbf{0} & & a_n \\ & & & & \ddots \end{bmatrix},$$

where $a_n = \lambda_n t_n$ and $\{a_n\}$ are pairwise distinct. Thus $ker(A_2 X) = \{0\}$ and $A_2 X$ is a diagonal self-adjoint operator.

Assume that

$$A_1 E + BX = \begin{bmatrix} l_{11} & l_{12} & \cdots & l_{1n} & \cdots \\ l_{21} & l_{22} & \cdots & l_{2n} & \cdots \\ \vdots & \vdots & \ddots & \vdots & \vdots \\ l_{n1} & l_{n2} & \cdots & l_{nn} & \cdots \\ \vdots & \vdots & \ddots & \vdots & \ddots \end{bmatrix}.$$

where $l_{1n} = a_{1j_1} e_{j_1 n} + b_{1n} \neq 0$ $n = 1, 2, \cdots$.

We now prove that the C^*-algebra generated by $A_1, A_2 X$ and $A_1 E + BX$ is irreducible.

Suppose that P is a projection commuting with A_2X, A_1 and $A_1E + BX$. Since P commutates with A_2X,

$$P = \begin{bmatrix} p_1 & & & \\ & p_2 & & \mathbf{0} \\ & & \ddots & \\ & \mathbf{0} & & p_n \\ & & & & \ddots \end{bmatrix}.$$

Since $P^2 = P$, $p_i = 0$ or I, $i = 1, 2, \cdots$. Without loss of generality, we assume that $p_1 = 0$ (Otherwise, consider $I - P$). From P commutates with $A_1E + BX$, i.e.,

$$
\begin{bmatrix} p_1 & & & \\ & p_2 & & \mathbf{0} \\ & & \ddots & \\ & \mathbf{0} & & p_n \\ & & & & \ddots \end{bmatrix}
\begin{bmatrix} l_{11} & l_{12} & \cdots & l_{1n} & \cdots \\ l_{21} & l_{22} & \cdots & l_{2n} & \cdots \\ \vdots & \vdots & \ddots & \vdots & \vdots \\ l_{n1} & l_{n2} & \cdots & l_{nn} & \cdots \\ \vdots & \vdots & \ddots & \vdots & \ddots \end{bmatrix}
$$

$$
= \begin{bmatrix} l_{11} & l_{12} & \cdots & l_{1n} & \cdots \\ l_{21} & l_{22} & \cdots & l_{2n} & \cdots \\ \vdots & \vdots & \ddots & \vdots & \vdots \\ l_{n1} & l_{n2} & \cdots & l_{nn} & \cdots \\ \vdots & \vdots & \ddots & \vdots & \ddots \end{bmatrix}
\begin{bmatrix} p_1 & & & \\ & p_2 & & \mathbf{0} \\ & & \ddots & \\ & \mathbf{0} & & p_n \\ & & & & \ddots \end{bmatrix}.
$$

We have $p_1 l_{12} = l_{12} p_2$. Since $p_1 = 0$ and $l_{12} \neq 0$, we get $p_2 = 0$. Similarly, since $p_1 l_{1n} = l_{1n} p_n$ and $l_{1n} \neq 0$, $p_n = 0$ ($n = 1, 2, \cdots$). Thus $P = 0$. This implies that the C^*-algebra generated by A_1, A_2X and $A_1E + BX$ is irreducible.

Lemma 6.2.15 *Assume that* $T = \begin{bmatrix} 0 & A_1 & B \\ 0 & 0 & A_2 \\ 0 & 0 & 0 \end{bmatrix} \begin{matrix} \mathcal{H} \\ \mathcal{H} \\ \mathcal{H} \end{matrix}$, *where A_1 and A_2 are positive operators and $\ker A_1 = \ker A_2 = \{0\}$. Then T is similar to an irreducible operator.*

Proof (1) If $\sigma(A_1)$ is uncountable, by Lemma 6.2.12 there exists an operator $E \in \mathcal{L}(\mathcal{H})$ such that the C^*-algebra generated by $A_1, A_1E + B$ and I is irreducible.

Define

$$X = \begin{bmatrix} I & 0 & 0 \\ 0 & I & -E \\ 0 & 0 & I \end{bmatrix},$$

then

$$X^{-1} = \begin{bmatrix} I & 0 & 0 \\ 0 & I & E \\ 0 & 0 & I \end{bmatrix}.$$

Denote

$$T_1 = XTX^{-1} = \begin{bmatrix} 0 & A_1 & A_1E + B \\ 0 & 0 & A_2 \\ 0 & 0 & 0 \end{bmatrix}.$$

We are now to prove that T_1 is irreducible.

Suppose that P is a projection commuting with T_1 and

$$P = \begin{bmatrix} P_{11} & P_{12} & P_{13} \\ P_{21} & P_{22} & P_{23} \\ P_{31} & P_{32} & P_{33} \end{bmatrix}.$$

It is obvious that $kerT_1 \in LatP$, $kerT_1 = \mathcal{H}\oplus 0\oplus 0$, $kerT_1^2 = \mathcal{H}\oplus\mathcal{H}\oplus 0$. Thus

$$P_{21} = P_{31} = P_{32} = 0.$$

Since P is a projection, $P_{12} = P_{13} = P_{23} = 0$. Therefore we may assume

$$P = \begin{bmatrix} P_1 & 0 & 0 \\ 0 & P_2 & 0 \\ 0 & 0 & P_3 \end{bmatrix}.$$

Since $PT_1 = T_1P$, $P_1A_1 = A_1P_2$. So $A_1P_1 = P_2A_1$. Thus

$$P_1A_1^2 = P_1A_1A_1 = A_1P_2A_1 = A_1^2P_1.$$

By functional calculus, $P_1A_1 = A_1P_1$. Thus $A_1P_1 = A_1P_2$, i.e., $A_1(P_1 - P_2) = 0$. Since A_1 is positive and $kerA_1 = \{0\}$, $P_1 - P_2 = 0$. Thus $p_1 = p_2$. Similarly, we have $P_1 = P_2 = P_3$ and

$$P = \begin{bmatrix} P_1 & 0 & 0 \\ 0 & P_1 & 0 \\ 0 & 0 & P_1 \end{bmatrix}.$$

Since the C^*-algebra generated by $A_1, A_1E + B$ and I is irreducible, $P_1 = 0$ or I, and therefore $P = 0$ or I. Thus T is similar to the irreducible operator T_1.

(2)   If $\sigma(A_2)$ is countable, by Lemma 6.2.14, there exist an invertible operator $X \in \mathcal{L}(\mathcal{H})$ and an operator $E \in \mathcal{L}(\mathcal{H})$ such that the C^*-algebra generated by A_1, A_2X and $A_1E + BX$ is irreducible and A_2X is diagonal.

Set

$$G_1 = \begin{bmatrix} I & 0 & 0 \\ 0 & I & 0 \\ 0 & 0 & X^{-1} \end{bmatrix},$$

then

$$G_1^{-1} = \begin{bmatrix} I & 0 & 0 \\ 0 & I & 0 \\ 0 & 0 & X \end{bmatrix}.$$

Thus,

$$G_1 T G_1^{-1} = \begin{bmatrix} 0 & A_1 & BX \\ 0 & 0 & A_2X \\ 0 & 0 & 0 \end{bmatrix}.$$

Set

$$G_2 = \begin{bmatrix} I & 0 & 0 \\ 0 & I & -E \\ 0 & 0 & I \end{bmatrix},$$

then

$$G_2^{-1} = \begin{bmatrix} I & 0 & 0 \\ 0 & I & E \\ 0 & 0 & I \end{bmatrix}.$$

Therefore,

$$G_2 G_1 T G_1^{-1} G_2^{-1} = \begin{bmatrix} 0 & A_1 & A_1E + BX \\ 0 & 0 & A_2X \\ 0 & 0 & 0 \end{bmatrix}.$$

Clearly, $G_2 G_1 T G_1^{-1} G_2^{-1}$ is irreducible. Thus T is similar to an irreducible operator.

Lemma 6.2.16 *Let*

$$T = \begin{bmatrix} 0 & A_1 & A_{12} & A_{13} & \cdots & A_{1n} \\ & 0 & A_2 & A_{23} & \cdots & A_{2n} \\ & & 0 & A_3 & \cdots & A_{3n} \\ & & & \ddots & \ddots & \vdots \\ & \mathbf{0} & & & \ddots & A_n \\ & & & & & 0 \end{bmatrix}.$$

If A_1, A_2 are positive operators and $\ker A_k = \{0\}$ $(k = 1, 2, \cdots, n)$. Then T is similar to an irreducible operator.

Proof Set

$$T_1 = \begin{bmatrix} 0 & A_1 & A_{12} \\ 0 & 0 & A_2 \\ 0 & 0 & 0 \end{bmatrix}.$$

By Lemma 6.2.14 and Lemma 6.2.15, there exists an invertible operator X_1 such that

$$X_1 T_1 X_1^{-1} = \begin{bmatrix} 0 & \overline{A}_1 & \overline{A}_{12} \\ 0 & 0 & \overline{A}_2 \\ 0 & 0 & 0 \end{bmatrix} \in (SI),$$

where $\overline{A}_1$ and $\overline{A}_2$ are positive injective operators. Therefore, there exists an invertible operator X such that

$$XTX^{-1} = \begin{bmatrix} 0 & \overline{A}_1 & \overline{A}_{12} & \overline{A}_{13} & \cdots & \overline{A}_{1n} \\ & 0 & \overline{A}_2 & \overline{A}_{23} & \cdots & \overline{A}_{2n} \\ & & 0 & \overline{A}_3 & \ddots & \overline{A}_{3n} \\ & & & \ddots & \ddots & \vdots \\ & \mathbf{0} & & & \ddots & \overline{A}_n \\ & & & & & 0 \end{bmatrix},$$

where $\overline{A}_1$ and $\overline{A}_2$ are positive and $\ker \overline{A}_k = \{0\}$, $k = 1, 2, \cdots, n$.

Now we prove that $XTX^{-1} \in (RI)$. Suppose that P is a projection commuting with XTX^{-1}. By the argument similar to that in the proof of

Lemma 6.2.15, we can prove that

$$
P = \begin{bmatrix}
P_1 & & & & \\
& P_2 & & \mathbf{0} & \\
& & P_3 & & \\
& \mathbf{0} & & \ddots & \\
& & & & P_{n+1}
\end{bmatrix}
$$

Since $XT_1X_1^{-1}\in(RI)$, $P_1 = P_2 = P_3 = 0$ or I. Without loss of generality, assume that $P_1 = P_2 = P_3 = 0$. Since $PXTX^{-1} = XTX^{-1}P$, $P_3\overline{A}_3 = \overline{A}_3P_4$ and $\overline{A}_3P_4 = 0$. Thus $ker\overline{A}_3 = \{0\}$. This implies that $P_4 = 0$.

Similarly, $P_1 = P_2 = P_3 = P_4 = \cdots = P_{n+1} = 0$. Therefore T is similar to an irreducible operator.

Lemma 6.2.17 *Suppose that $T\in\mathcal{N}_k(\mathcal{H})$ is of the form (6.2.1), then T is similar to an irreducible operator.*

Proof Without loss of generality, assume that

$$
T = \begin{bmatrix}
0 & T_1 & & & & \\
& 0 & T_2 & & \star & \\
& & 0 & T_3 & & \\
& & & \ddots & \ddots & \\
& \mathbf{0} & & & \ddots & T_k \\
& & & & & 0
\end{bmatrix},
$$

where $kerT_j = \{0\}$ $(j = 1, 2, \cdots, k)$ and $ranT_1$ and $ranT_2$ are dense. By the Polar Decomposition Theorem, $T_1 = A_1U_1$ and $U_1T_2 = A_2U_2$, where A_i is positive, $kerA_i = \{0\}$ and U_i is a unitary operator $(i = 1, 2)$.

Set

$$
G = \begin{bmatrix}
I & & & & \\
& U_1 & & \mathbf{0} & \\
& & U_2 & & \\
& & & I & \\
& \mathbf{0} & & & \ddots \\
& & & & & I
\end{bmatrix},
$$

then

$$GTG^{-1} = \begin{bmatrix} 0 & A_1 & & & & & \\ 0 & 0 & A_2 & & \star & & \\ 0 & 0 & 0 & T_3 & & & \\ 0 & 0 & 0 & 0 & \ddots & & \\ \vdots & \vdots & \vdots & \ddots & \ddots & T_{k-1} \\ 0 & 0 & 0 & 0 & \cdots & 0 \end{bmatrix}.$$

By Lemma 6.2.16, T is similar to an irreducible operator.

Lemma 6.2.18   *Suppose that T is similar to $0_{\mathcal{H}_1} \oplus T_1$, where $\dim\mathcal{H}_1 = \infty$. If*

$$T_1 = \begin{bmatrix} 0 & A_1 & B \\ 0 & 0 & A_2 \\ 0 & 0 & 0 \end{bmatrix} \begin{matrix} \mathcal{H} \\ \mathcal{H}, \\ \mathcal{H} \end{matrix}$$

where A_1 and A_2 are positive operators, and $\ker A_1 = \ker A_2 = \{0\}$, then T is similar to an irreducible operator.

Proof   Without loss of generality, we can assume that

$$T = 0_{\mathcal{H}_1} \oplus T_1 = \begin{bmatrix} 0 & 0 & 0 & 0 \\ 0 & 0 & A_1 & B \\ 0 & 0 & 0 & A_2 \\ 0 & 0 & 0 & 0 \end{bmatrix}.$$

Set

$$E = \begin{bmatrix} I & -I & 0 & 0 \\ 0 & I & 0 & 0 \\ -I & 0 & I & 0 \\ 0 & 0 & 0 & I \end{bmatrix}.$$

Then

$$E^{-1} = \begin{bmatrix} I & I & 0 & 0 \\ 0 & I & 0 & 0 \\ I & I & I & 0 \\ 0 & 0 & 0 & I \end{bmatrix}.$$

and

$$ETE^{-1} = \begin{bmatrix} -A_1 & -A_1 & -A_1 & -B \\ A_1 & A_1 & A_1 & B \\ 0 & 0 & 0 & A_2 \\ 0 & 0 & 0 & 0 \end{bmatrix}.$$

Suppose P be a projection commuting with ETE^{-1} and

$$P = \begin{bmatrix} P_{11} & P_{12} & P_{13} & P_{14} \\ P_{21} & P_{22} & P_{23} & P_{24} \\ P_{31} & P_{32} & P_{33} & P_{34} \\ P_{41} & P_{42} & P_{43} & P_{44} \end{bmatrix}.$$

It follows from $PETE^{-1} = ETE^{-1}P$ that

$$\begin{bmatrix} -P_{11}A_1 + P_{12}A_1 & -P_{11}A_1 + P_{12}A_1 & -P_{11}A_1 + P_{12}A_1 & -P_{11}B + P_{12}B + P_{13}A_2 \\ -P_{21}A_1 + P_{22}A_1 & -P_{21}A_1 + P_{22}A_1 & -P_{21}A_1 + P_{22}A_1 & -P_{21}B + P_{22}B + P_{23}A_2 \\ -P_{31}A_1 + P_{32}A_1 & -P_{31}A_1 + P_{32}A_1 & -P_{31}A_1 + P_{32}A_1 & -P_{31}B + P_{32}B + P_{33}A_2 \\ -P_{41}A_1 + P_{42}A_1 & -P_{41}A_1 + P_{42}A_1 & -P_{41}A_1 + P_{42}A_1 & -P_{41}B + P_{42}B + P_{43}A_2 \end{bmatrix}$$

$$= \begin{bmatrix} S & T \\ W & V \end{bmatrix}. \tag{6.2.4}$$

Where

$$S = \begin{bmatrix} -A_1P_{11} - A_1P_{21} - A_1P_{31} - BP_{41} & -A_1P_{12} - A_1P_{22} - A_1P_{32} - BP_{42} \\ A_1P_{11} + A_1P_{21} + A_1P_{31} + BP_{41} & A_1P_{12} + A_1P_{22} + A_1P_{32} + BP_{42} \end{bmatrix},$$

$$T = \begin{bmatrix} -A_1P_{13} - A_1P_{23} - A_1P_{33} - BP_{43} & -A_1P_{14} - A_1P_{24} - A_1P_{34} - BP_{44} \\ A_1P_{13} + A_1P_{23} + A_1P_{33} + BP_{43} & A_1P_{14} + A_1P_{24} + A_1P_{34} + BP_{44} \end{bmatrix},$$

$$W = \begin{bmatrix} A_2P_{41} & A_2P_{42} \\ 0 & 0 \end{bmatrix},$$

$$V = \begin{bmatrix} A_2P_{43} & A_2P_{44} \\ 0 & 0 \end{bmatrix}.$$

Compare the $(4, 1)$ entry of $(6.2.4)$, we have

$$-P_{41}A_1 + P_{42}A_1 = 0.$$

It follows from $[ran A_1]^- = \mathcal{H}$ that

$$P_{41} = P_{42}. \tag{6.2.5}$$

Compare the (4, 4) entry of (6.2.4), we have $-P_{41}B + P_{42}B + P_{43}A_2 = 0$. By (6.2.5), $P_{43}A_2 = 0$. Since $[ran A_2]^- = \mathcal{H}$,

$$P_{43} = 0. \tag{6.2.6}$$

Compare the (3, 3) entry of (6.2.4), we have $-P_{31}A_1 + P_{32}A_1 = A_2P_{43} = 0$. Thus

$$P_{31} = P_{32}. \tag{6.2.7}$$

This indicates that

$$W = \begin{bmatrix} 0 & 0 \\ 0 & 0 \end{bmatrix},$$

$A_2P_{41} = A_2P_{42} = 0$ and

$$P_{41} = P_{42} = 0. \tag{6.2.8}$$

Note that the sum of the (1, 3) entry and (2, 3) entry of the right side of (6.2.4) is 0. Thus

$$-(P_{11} + P_{12} - P_{21} + P_{22})B + P_{13}A_2 + P_{23}A_2 = 0$$

and

$$P_{13}A_2 + P_{23}A_2 = 0,$$

i.e.,

$$(P_{13} + P_{23})A_2 = 0$$

or

$$P_{13} + P_{23} = 0.$$

By (6.2.7) and $A^* = A$, $P_{13} = P_{31}, P_{23} = P_{32} = P_{31}$. Thus $P_{13} = P_{23} = 0$.

Therefore the (6.2.4) can be written as follows:

(α_{ij})

$$:= \begin{bmatrix} -P_{11}A_1 + P_{12}A_1 & -P_{11}A_1 + P_{12}A_1 & -P_{11}A_1 + P_{12}A_1 & -P_{11}B + P_{12}B \\ -P_{21}A_1 + P_{22}A_1 & -P_{21}A_1 + P_{22}A_1 & -P_{21}A_1 + P_{22}A_1 & -P_{21}B + P_{22}B \\ 0 & 0 & 0 & P_{33}A_2 \\ 0 & 0 & 0 & 0 \end{bmatrix}$$

$$= \begin{bmatrix} -A_1P_{11} - A_1P_{21} & -A_1P_{12} - A_1P_{22} & -A_1P_{33} & -BP_{44} \\ A_1P_{11} + A_1P_{21} & A_1P_{12} + A_1P_{22} & A_1P_{33} & BP_{44} \\ 0 & 0 & 0 & A_2P_{44} \\ 0 & 0 & 0 & 0 \end{bmatrix} := (\beta_{ij})$$

Since $\alpha_{21} = \beta_{21}$, $-P_{21}A_1 + P_{22}A_1 = A_1P_{11} + A_1P_{21}$ or

$$A_1P_{21} + P_{21}A_1 = P_{22}A_1 - A_1P_{11}. \tag{6.2.9}$$

It follows from $\alpha_{12} = \beta_{12}$ that

$$-P_{11}A_1 + P_{12}A_1 = -A_1P_{12} - A_1P_{22}.$$

Taking the adjoint of both sides, we get $-A_1P_{11} + A_1P_{21} = -P_{21}A_1 - P_{22}A_1$ or

$$A_1P_{21} + P_{21}A_1 = -P_{22}A_1 + A_1P_{11}. \tag{6.2.10}$$

From (6.2.9) and (6.2.10) we get $P_{22}A_1 = A_1P_{11}$.
Repeating the proof of Lemma 6.2.15, we can prove that $P_{11} = P_{22}$. Thus

$$P_{22}A_1 = A_1P_{22}. \tag{6.2.11}$$

Since $\beta_{12} + \beta_{22} = 0$, $\alpha_{12} + \alpha_{22} = 0$ or $-P_{11}A_1 + P_{12}A_1 - P_{21}A_1 + P_{22}A_1 = 0$, i.e., $P_{12}A_1 - P_{21}A_1 = 0$. Thus

$$P_{21} = P_{12}. \tag{6.2.12}$$

Since $\alpha_{22} = \beta_{22}$, $-P_{21}A_1 + P_{22}A_1 = A_1P_{12} + A_1P_{22}$. By (6,2,11),

$$P_{21}A_1 + A_1P_{12} = 0.$$

By (6.2.12),

$$A_1P_{12} + P_{12}A_1 = 0. \tag{6.2.13}$$

Note that $P^* = P$. Thus $P_{12}^* = P_{21}$ and P_{12} is self-adjoint. Thus P_{12}^2 is a positive operator. From (6.2.13) we know $P_{12}^2 A_1 = -P_{12} A_1 P_{12} = A_1 P_{12}^2$. Thus $P_{12} A_1 = A_1 P_{12}$. From (6.2.13), $P_{12} A_1 = 0$. Thus $P_{12} = 0$ and $P_{21} = 0$. The (6.2.4) can be written as

$$\begin{bmatrix} -P_{11}A_1 & -P_{11}A_1 & -P_{11}A_1 & -P_{11}B \\ P_{22}A_1 & P_{22}A_1 & P_{22}A_1 & P_{22}B \\ 0 & 0 & 0 & P_{33}A_2 \\ 0 & 0 & 0 & 0 \end{bmatrix}$$

$$= \begin{bmatrix} -A_1P_{11} & -A_1P_{22} & -A_1P_{33} & -BP_{44} \\ A_1P_{11} & A_1P_{22} & A_1P_{33} & BP_{44} \\ 0 & 0 & 0 & A_2P_{44} \\ 0 & 0 & 0 & 0 \end{bmatrix}.$$

By a proof similar to the proof of Lemma 6.2.15, $P_{11} = P_{22} = P_{33} = P_{44}$ and $XTX^{-1} \in (RI)$ for some invertible X.

Lemma 6.2.19 *Suppose that*

$$T = 0_{\mathcal{H}_1} \oplus \begin{bmatrix} 0 & A_1 & A_{12} \\ 0 & 0 & A_2 \\ 0 & 0 & 0 \end{bmatrix} \begin{matrix} \mathcal{H} \\ \mathcal{H}, \\ \mathcal{H} \end{matrix}$$

where A_1 and A_2 are positive operators, and $\ker A_1 = \ker A_2 = \{0\}$. Then T is similar to an irreducible operator.

Proof If $\dim \mathcal{H}_1 = \infty$, then by Lemma 6.2.18 the conclusion is true.
   If $\dim \mathcal{H}_1 < \infty$, say $\dim \mathcal{H}_1 = n$. Denote

$$A = \begin{bmatrix} 0 & A_1 & A_{12} \\ 0 & 0 & A_2 \\ 0 & 0 & 0 \end{bmatrix}.$$

Then by Lemma 6.2.15 there is an invertible operator X_1 such that

$$\overline{A} = X_1 A X_1^{-1} = \begin{bmatrix} 0 & \overline{A}_1 & \overline{A}_{12} \\ 0 & 0 & \overline{A}_2 \\ 0 & 0 & 0 \end{bmatrix} \in (RI),$$

where $\overline{A}_1$ and $\overline{A}_2$ are positive operators, and $\ker \overline{A}_1 = \ker \overline{A}_2 = \{0\}$.

Thus

$$XTX^{-1} = \begin{bmatrix} 0 & 0 & 0 & 0 \\ 0 & 0 & \overline{A}_1 & \overline{A}_{12} \\ 0 & 0 & 0 & \overline{A}_2 \\ 0 & 0 & 0 & 0 \end{bmatrix} \begin{matrix} \mathcal{H}_1 \\ \mathcal{H} \\ \mathcal{H} \\ \mathcal{H} \end{matrix} = T_1 = \begin{bmatrix} 0 & 0 \\ 0 & \overline{A} \end{bmatrix} \begin{matrix} \mathcal{H}_1 \\ \mathcal{H}^{(3)} \end{matrix}$$

for some invertible X.

Let $e_1, e_2, \cdots, e_n$
be an ONB of $\mathcal{H}_1$, $f_1, f_2, \cdots, f_n \in ker\overline{A}^2 \ominus ker\overline{A}$, $E \in L(\mathcal{H}^{(3)}, \mathcal{H}_1)$ such that
$E\overline{A}f_1 = e_1, E\overline{A}f_2 = e_2, \cdots, E\overline{A}f_n = e_n$ and $E([\bigvee_{k=1}^{n} \overline{A}f_k]^{\perp}) = 0$.

Denote

$$Y = \begin{bmatrix} I & E \\ 0 & I \end{bmatrix} \begin{matrix} \mathcal{H}_1 \\ \mathcal{H}^{(3)} \end{matrix}$$

and

$$T_2 = YT_1Y^{-1} = \begin{bmatrix} 0 & E\overline{A} \\ 0 & \overline{A} \end{bmatrix} \begin{matrix} \mathcal{H}_1 \\ \mathcal{H}^{(3)} \end{matrix}.$$

Now we are to prove that $T_2 \in (RI)$.

Suppose that P is a projection commuting with T_2. Denote

$$\mathcal{M} = (0 \oplus ker\overline{A}) \cap Pker T_2, \quad \mathcal{N} = (0 \oplus ker\overline{A}) \cap (I - P)ker T_2,$$

$$\mathcal{M}' = Pker T_2 \ominus \mathcal{M}, \quad \mathcal{N}' = (I - P)ker T_2 \ominus \mathcal{N}.$$

It is easy to see that $dim\mathcal{M}' = l_1 \leq n$, $dim\mathcal{N}' = l_2 \leq n$ and

$$\forall x \in \mathcal{M}', x = \sum_{i=1}^{n} \alpha_i e_i + z_1, z_1 \in \mathcal{M}, \quad \forall y \in \mathcal{N}', y = \sum_{i=1}^{n} \beta_i e_i + z_2, z_2 \in \mathcal{N}.$$

Thus, for $e_i \in \mathcal{H}_1$, $e_i = x + y$, $x \in \mathcal{M}'$, $y \in \mathcal{N}'$, where

$$x = \sum_{i=1}^{n} \alpha_i e_i + z_1, z_1 \in \mathcal{M}, \quad y = \sum_{i=1}^{n} \beta_i e_i + z_2, z_2 \in \mathcal{N}.$$

Since $\mathcal{M} \perp \mathcal{N}$, we obtain $x = 0$ or $y = 0$. This implies that

$$e_i \in Pker T_2$$

or

$$e_i \in (I - P)ker T_2 \ (i = 1, 2, \cdots, n).$$

Without loss of generality, we assume that

$$e_1, e_2, \cdots, e_k \in P ker T_2, \quad e_{k+1}, e_{k+2}, \cdots, e_n \in (I - P) ker T_2.$$

By a routine proof, P admits a block upper triangular matrix representation with respect to the decomposition $\mathcal{H}_1 \oplus \mathcal{H} \oplus \mathcal{H} \oplus \mathcal{H}$.

Since P is a projection,

$$P = \begin{bmatrix} P_1 & 0 & 0 & 0 \\ 0 & P_2 & 0 & 0 \\ 0 & 0 & P_3 & 0 \\ 0 & 0 & 0 & P_4 \end{bmatrix}.$$

Note that $\overline{A} \in (RI)$, thus $P_2 = P_3 = P_4 = 0$ or I. Without loss of generality, assume that $P_2 = P_3 = P_4 = 0$. Since $PT_2 = T_2 P$, $P_1 E \overline{A} = 0$. But $E \overline{A}$ is surjective, thus $P_1 = 0$ and $P = 0$. So T is similar to an irreducible operator.

The proof of next lemma is similar to the proof of Lemma 6.2.17.

Lemma 6.2.20 *Suppose that* $T \in \mathcal{N}_k(\mathcal{H})$ *and* $T = 0_{\mathcal{H}_1} \oplus A$ *is of the form (6.2.2). Then* T *is similar to an irreducible operator.*

So far we have proved the sufficiency of Theorem 6.2.9. As for the necessity, we need only to show that if $T^2 = 0$ or $ran T$ is finite rank, then $T \notin (RI)$.

(1) If $T^2 = 0$. We may assume that

$$Q = \begin{bmatrix} 0 & A \\ 0 & 0 \end{bmatrix} \begin{matrix} \mathcal{H} \\ \mathcal{H} \end{matrix}$$

and $dim \mathcal{H} = \infty$. By the proof of Lemma 6.2.15,

$$Q \cong \begin{bmatrix} 0 & \overline{A} \\ 0 & 0 \end{bmatrix},$$

where $\overline{A}$ is a positive operator. Suppose that P_1 is a nontrivial projection commuting with $\overline{A}$. Set $P = P_1 \oplus P_2$, then P commutes with Q.

(2) If T is finite rank, then T^* is finite rank. Denote $\mathcal{M} = \bigvee \{ran T, ran T^*\}$, then $\mathcal{M}$ is a reducing subspace of T. Thus the proof of the necessity of theorem 6.2.9 is now complete,

Next, we are going to discuss normal operators.

Theorem 6.2.21 *Suppose that $N \in \mathcal{L}(\mathcal{H})$ is a normal operator. Then N is similar to an irreducible operator if and only if the following conditions are satisfied.*

   (i) $\lambda - N$ is not finite rank for all $\lambda \in \mathbf{C}$;

   (ii) N does not satisfy any quadratic equation $ax^2 + bx + c = 0$, where

$$|a| + |b| + |c| \neq 0.$$

In order to prove Theorem 6.2.21, we need several lemmas.

Lemma 6.2.22 *Suppose that $A \in \mathcal{L}(\mathcal{H})$ and each operator in the similarity orbit $\mathcal{S}(A)$ of A is irreducible, then each operator in $\mathcal{S}(\mathcal{A}''(A))$ is also irreducible, where $\mathcal{S}(\mathcal{A}''(A))$ denotes the similarity orbit of $\mathcal{A}''(A)$.*

Proof Assume that $B \in \mathcal{A}''(A)$ and X is invertible. Since $XAX^{-1} \notin (RI)$, XAX^{-1} commutes with some nontrivial projection P, i.e., $P \in \mathcal{A}'(XAX^{-1})$. Thus $X^{-1}PX \in \mathcal{A}'(A)$ and therefore $BX^{-1}PX = X^{-1}PXB$ or $XBX^{-1}P = PXBX^{-1}$. Namely $XBX^{-1} \notin (RI)$.

Lemma 6.2.23 *Suppose that $A_i \in \mathcal{L}(\mathcal{H})$ $(i = 1, 2, 3)$ and $A = A_1 \oplus A_2 \oplus A_3$ satisfying $ker\tau_{A_i, A_j} = \{0\}$ $(i \neq j)$. Then A is similar to an irreducible operator.*

Proof By Lemma 6.2.22, it is sufficient to show that there is some operator in $\mathcal{A}''(A)$ that is similar to an irreducible operator. Since $ker\tau_{A_i, A_j} = \{0\}$ $(i \neq j)$, calculations indicate that

$$\mathcal{A}'(A) = \{B_1 \oplus B_2 \oplus B_3 : B_i \in \mathcal{L}(\mathcal{H}), i = 1, 2, 3\}.$$

Therefore, it is sufficient to prove that $B = \alpha_1 \oplus \alpha_2 \oplus \alpha_3$ is similar to an irreducible operator, where $\{\alpha_i\}_{i=1}^3$ are pairwise distinct complex numbers.

Define

$$T = \begin{bmatrix} \alpha_1 & 1 & D \\ 0 & \alpha_2 & 1 \\ 0 & 0 & \alpha_3 \end{bmatrix},$$

where D is an irreducible operator. Suppose that $P \in \mathcal{A}'(T)$ is a projection and

$$P = \begin{bmatrix} P_{11} & P_{12} & P_{13} \\ P_{21} & P_{22} & P_{23} \\ P_{31} & P_{32} & P_{33} \end{bmatrix}.$$

Since $\{\alpha_i\}_{i=1}^3$ are pairwise distinct, $P_{ij} = 0$ $(i \neq j)$ and $P_{11} = P_{22} = P_{33}$. It follows from $P_{11}D = DP_{11}$ and $D \in (RI)$ that $P_{11} = P_{22} = P_{33} = 0$ or I. Thus $T \in (RI)$. It is easy to see that $T \sim \alpha_1 \oplus \alpha_2 \oplus \alpha_3$.

Lemma 6.2.24 *Suppose that $A_1 \in \mathcal{L}(\mathbf{C}^m)$ $(m < \infty)$, $A_2, A_3 \in \mathcal{L}(\mathcal{H})$ satisfying $\ker \tau_{A_i, A_j} = \{0\}$ $(i \neq j)$, then $A = A_1 \oplus A_2 \oplus A_3$ is similar to an irreducible operator.*

Proof By Lemma 6.2.22, it is sufficient to prove that $B = \alpha_1 \oplus \alpha_2 \oplus \alpha_3$ is similar to an irreducible operator, where $\{\alpha_i\}_{i=1}^3$ are pairwise distinct complex numbers. Without loss of generality, we assume that $\mathcal{H} = L^2(0,1)$. Define

$$ T = \begin{bmatrix} \alpha_1 & F & 0 \\ 0 & \alpha_2 & M \\ 0 & 0 & \alpha_3 \end{bmatrix}, $$

where M is "multiplication by the idempotent variable", i.e., $(Mf)(t) = tf(t)$, F is a surjective operator from $L^2(0,1)$ to $\mathbf{C}^m$ given by

$$ Ff = \left(\int_0^1 tf(t)dt, \int_0^1 t^2 f(t)dt, \cdots, \int_0^1 t^m f(t)dt \right), \quad f \in L^2(0,1). $$

It is not difficult to see that $T \sim B$. It is sufficient to prove that $T \in (RI)$.

Suppose that $P \in \mathcal{A}'(T)$ is a projection. Since $\alpha_i \neq \alpha_j$ $(i \neq j)$, $P = P_1 \oplus P_2 \oplus P_3$, where each P_i is a projection and $P_2 M = M P_3$, $P_1 F = F P_2$. Thus

$$ P_2 M^2 = M P_3 M = M(M P_3)^* = M(P_2 M)^* = M^2 P_2. $$

Therefore

$$ P_2 M^{2k} = M^{2k} P_2 \text{ for } k \geq 1. $$

Using functional calculus we have $P_2 M = M P_2$ and $P_2 = P_3$. The spectral theory of self-adjoint operators asserts that there exists a Borel subset $E \subset [0,1]$ such that $P_2 f = \chi_E f$ for all $f \in \mathcal{H}$, where χ_E denotes the characteristic function of E. If the Lebesgue measure of E, $\mu(E) = 1$, then $P_2 = P_3 = 1$. Since $P_1 F = F P_2$ and since F is surjective, $P_1 = 1$, i.e., $P = 1$. If $\mu(E) < 1$, let $\lambda_1, \lambda_2, \cdots, \lambda_m$ be nonzero, pairwise distinct Lebesgue points of $F = [0,1] \setminus$ ($\lambda \in F$ is a Lebegue point of F if $\lim \mu(F \cap [\lambda - \varepsilon, \lambda + \varepsilon])/2\varepsilon = 1$).

Since the matrix

$$\begin{bmatrix} \lambda_1 & \lambda_2 & \cdots & \lambda_m \\ \lambda_1^2 & \lambda_2^2 & \cdots & \lambda_m^2 \\ \cdots & \cdots & \cdots & \cdots \\ \lambda_1^m & \lambda_2^m & \cdots & \lambda_m^m \end{bmatrix}$$

is invertible, there is a $\delta > 0$ such that the matrix $(\alpha_{ij})_{m \times m}$ is invertible provided that $|\alpha_{ij} - \lambda_j^i| < \delta$ $(i, j = 1, 2, \cdots, m.$ Choose a sufficiently small $\varepsilon > 0$ such that

$$\left| \frac{1}{2\varepsilon} \int\limits_{[\lambda_i - \varepsilon, \lambda_i + \varepsilon] \cap F} t^i dt - \lambda_j^i \right| < \delta, \quad i, j = 1, 2, \cdots, m.$$

Set $f_j = \frac{1}{2\varepsilon} \chi_{[\lambda_i - \varepsilon, \lambda_i + \varepsilon] \cap F}$. Since $E \cap F = \emptyset$,

$$P_1 F f_j = F P_2 f_j = F \chi_E f_j = 0 \quad (j = 1, 2, \cdots, m).$$

On the other hand, if $\{e_k\}_{k=1}^n$ is an ONB of $\mathbf{C}^m$, we have

$$0 = P_1 F f_j = P_1 \left(\int_0^1 t f_j dt, \int_0^1 t^2 f_j dt, \cdots, \int_0^1 t^m f_j dt \right)$$

$$= P_1 \left[\sum_{k=1}^m \left(\frac{1}{2\varepsilon} \int\limits_{[\lambda_i - \varepsilon, \lambda_i + \varepsilon] \cap F} t^k dt \right) e_k \right]$$

$$= \sum_{k=1}^m \left(\frac{1}{2\varepsilon} \int\limits_{[\lambda_i - \varepsilon, \lambda_i + \varepsilon] \cap F} t^k dt \right) P_1 e_k.$$

Define

$$\alpha_{ij} = \frac{1}{2\varepsilon} \int\limits_{[\lambda_i - \varepsilon, \lambda_i + \varepsilon] \cap F} t^i dt,$$

then $(\alpha_{ij})_{m \times m}$ is invertible. Thus $P_1 e_k = 0$ $(k = 1, 2, \cdots, m)$, i.e., $P_1 = 0$. Since $F P_2 = P_1 F, F P_2 = 0$. Note that $F P_2 e = (\int_0^1 t \chi_E dt, \int_0^1 t^2 \chi_E dt, \cdots, \int_0^1 t^m \chi_E dt)$, where $e \in \mathcal{H}_2, e(t) = 1$. Therefore $\mu(E) = 0$, i.e., $P_2 = P_3 = 0$ and $P = 0$. Thus $T \in (RI)$.

Now we are in a position to prove Theorem 6.2.21.

Proof of Theorem 6.2.21 By the arguments similar to that used in the proof of Theorem 6.2.9 we can prove the necessary condition of the theorem. Now we prove the sufficient condition. If (i) and (ii) of the theorem are satisfied, then there are three possibilities:

(a) $\sigma(N)$ consists of infinitely many points;

(b) $\sigma_e(N)$ consists of at least three points;

(c) $\sigma(N)$ is a finite set, $\sigma_e(N)$ consists of two points and $\sigma(N)\backslash\sigma_e(N) \neq \emptyset$.

In cases (a) or (b), there exist pairwise disjoint Borel sets σ_1, σ_2 and σ_3 such that

$$\sigma(N) = \sigma_1 \cup \sigma_2 \cup \sigma_3$$

and

$$dimE(\sigma_i) = \infty \quad \text{for} \quad i = 1, 2, 3,$$

where $E(\cdot)$ is the spectral measure of N. Set $N_j = N|_{E(\sigma_i)\mathcal{H}}$ $(i = 1, 2, 3)$. Then $ker\tau_{N_i,N_j} = \{0\}$ $(i \neq j)$. By Lemma 6.2.23, N is similar to an irreducible operator.

In Case (c), assume that $\sigma_e(N) = \{\lambda_1, \lambda_2\}$. Thus

$$dimE(\{\lambda_i\})\mathcal{H} = \infty \quad (i = 1, 2)$$

and

$$0 < dimE(\sigma(E)\backslash\sigma_e(N)) < \infty.$$

By Lemma 6.2.24 N is similar to an irreducible operator.

Proposition 6.2.25 *Given* $T \in \mathcal{L}(\mathcal{H})$ *such that* $\mathcal{A}'(T)$ *is abelian,* T *is similar to an irreducible operator.*

Proof (i) Suppose that there exists a projection $P \in \mathcal{A}'(T)$ such that

$$dimP\mathcal{H} = \infty$$

and

$$A = T|_{P\mathcal{H}} \in (RI).$$

Fix a $\lambda \notin \sigma(A)$. Then

$$\lambda \oplus A \in \mathcal{L}((P\mathcal{H})^\perp \oplus P\mathcal{H}) \in \mathcal{A}'(T).$$

Since $\mathcal{A}'(T)$ is abelian, $\mathcal{A}'(T) = \mathcal{A}''(T)$. Thus

$$\lambda \oplus A \in \mathcal{A}''(T).$$

Choose $D \in \mathcal{L}(P\mathcal{H}, (P\mathcal{H})^{\perp})$ such that D is surjective, then

$$\lambda \oplus A \sim B = \begin{bmatrix} \lambda \, D \\ 0 \, A \end{bmatrix} \begin{matrix} (P\mathcal{H})^{\perp} \\ P\mathcal{H} \end{matrix}.$$

By Lemma 6.2.22 it suffices to prove that $B \in (RI)$. Suppose that

$$P = \begin{bmatrix} P_{11} \, P_{12} \\ P_{21} \, P_{22} \end{bmatrix} \in \mathcal{A}'(B)$$

is a projection. Since $\lambda \notin \sigma(A), P_{21} = P_{12} = 0$. Since A is irreducible, $P_{22} = 0$ or I. It follows from $DP_{22} = P_{11}D$ and D is surjective that $P_{22} = P_{11} = 0$ or I and $B \in (RI)$.

(ii) If there is not any projection $P \in \mathcal{A}'(T)$ with infinite rank such that $T|_{P\mathcal{H}} \in (RI)$, then we can find projections Q_1, Q_2 and $Q_3 \in \mathcal{A}'(T)$ such that

$$dim ran Q_i = \infty \ (i = 1, 2, 3).$$

Set $T_i = T|_{ranQ_i} \ (i = 1, 2, 3)$, then $T = T_1 \oplus T_2 \oplus T_3$. Thus

$$F = \begin{bmatrix} \lambda_1 & & 0 \\ & \lambda_2 & \\ 0 & & \lambda_3 \end{bmatrix} \begin{matrix} ranQ_1 \\ ranQ_2 \\ ranQ_3 \end{matrix} \in \mathcal{A}'(T) = \mathcal{A}''(T)$$

where $\{\lambda_i\}_{i=1}^{3}$ are pairwise distinct numbers. By Lemma 6.2.23, F is similar to an irreducible operator and by Lemma 6.2.22, T is similar to an irreducible operator.

Proposition 6.2.26 *Every Cowen-Douglas operator is similar to an irreducible operator.*

In order to prove Proposition 6.2.26, we need some lemmas.

Lemma 6.2.27 *Given $B \in \mathcal{B}_n(\Omega)$ and $\lambda \in \Omega$. Denote*

$$\mathcal{H}_k = ker(B - \lambda)^k \ominus ker(B - \lambda)^{k-1} \ (k = 1, 2, \cdots),$$

then

$$\mathcal{H} = \bigoplus_{k=1}^{\infty} \mathcal{H}_k$$

and

$$B = \begin{bmatrix} \lambda & \overline{B}_{12} & \overline{B}_{13} & \star \\ & \lambda & \overline{B}_{23} & \\ & & \lambda & \ddots \\ 0 & & & \ddots \end{bmatrix},$$

where $\dim \mathcal{H}_k = n$ *and each* $\overline{B}_{k\,k+1}$ *is invertible. Furthermore, there is a unitary operator* U *such that*

$$UBU^* = \begin{bmatrix} \lambda & B_{12} & & \star \\ 0 & \lambda & B_{23} & \\ & & \lambda & \ddots \\ 0 & & & \ddots \end{bmatrix} \begin{matrix} \mathcal{H}_1 \\ \mathcal{H}_2 \\ \mathcal{H}_3 \\ \vdots \end{matrix}, \qquad (6.2.14)$$

where $B_{k\,k+1}$ *is a positive invertible operator for each* i.

Proof Without loss of generality, we assume that $\lambda = 0 \in \Omega$. It is easily seen that $\dim \mathcal{H}_n = n$ and

$$B = \begin{bmatrix} 0 & \overline{B}_{12} & \overline{B}_{13} & \star \\ & 0 & \overline{B}_{23} & \\ & & 0 & \ddots \\ 0 & & & \ddots \end{bmatrix}.$$

Let $k_0 = min\{k : ker\overline{B}_{k\,k+1} \neq \{0\}, k \geq 1\}$. Then there exists a vector $x_{k_0+1} \neq 0$ such that $\overline{B}_{k_0\,k_0+1} x_{k_0+1} = 0$. Note that $\mathcal{M} = \bigoplus_{k=1}^{k_0} \mathcal{H}_k = kerB^{k_0}$ and $dim\mathcal{M} = k_0 n$. But $y = (0, \cdots, 0, x_{k_0+1}, 0, \cdots) \in kerB^{k_0}$ and $y \notin \mathcal{M}$. This contradiction indicates that each $\overline{B}_{k\,k+1}$ is invertible. Thus $\overline{B}_{12} = U_1 B'_{12}$, where U_1 is a unitary operator and B'_{12} is a positive invertible operator. Similarly,

$$U_1 \overline{B}_{23} = U_2 B'_{23}, U_2 \overline{B}_{34} = U_3 B'_{34}, \cdots, U_k \overline{B}_{k+1\,k+2} = U_{k+1} B'_{k+1\,k+2}, \cdots$$

and each U_k is a unitary and each $B'_{k\,k+1}$ is positive and invertible.

Define $U = \bigoplus\limits_{k=0}^{\infty} U_k$, where $U_0 = I$, then U is a unitary operator and

$$UBU^* = \begin{bmatrix} 0 & B_{12} & & \star \\ 0 & 0 & B_{23} & \\ & & 0 & \ddots \\ 0 & & & \ddots \end{bmatrix},$$

where $B_{k\,k+1} = U_k B'_{k\,k+1} U_k^*$ $(k \geq 1)$ is positive and invertible.

Lemma 6.2.28 *Given $B \in \mathcal{B}_n(\Omega)$ with the representation (6.2.14). If $B_{1\,k_0} \in (RI)$ for some $k_0 \geq 2$, then $B \in (RI)$.*

Proof Suppose that $P \in \mathcal{A}'(B)$ is a projection, then P admits the following expression with respect to the decomposition $\mathcal{H} = \bigoplus\limits_{k=1}^{\infty} \mathcal{H}_k$,

$$P = \begin{bmatrix} P_1 & P_{12} & & \star \\ & P_2 & P_{23} & \\ & & P_3 & \ddots \\ 0 & & & \ddots \end{bmatrix}.$$

Since $P^* = P$, $P = P_1 \oplus P_2 \oplus \cdots$. Since $P_k B_{k\,k+1} = B_{k\,k+1} P_{k+1}$ and since $B_{k\,k+1}$ is positive and invertible, $P_k = P_1$ for $k \geq 2$. Thus

$$B_{1k_0} P_{k_0} = P_1 B_{1k_0}.$$

Since $B_{1k_0} \in (RI)$, $P_k = P_1 = 0$ or I $(k = 2, 3, \cdots)$, i.e., $P = 0$ or I. Therefore $B \in (RI)$.

Proof of Proposition 6.2.26 without loss of generality, assume that $0 \in \Omega$. Because of Lemma 6.2.27, we may assume that

$$B = \begin{bmatrix} 0 & B_{12} & & \star \\ & 0 & B_{23} & \\ & & 0 & \ddots \\ 0 & & & \ddots \end{bmatrix} \begin{matrix} \mathcal{H}_1 \\ \mathcal{H}_2 \\ \mathcal{H}_3 \\ \vdots \end{matrix},$$

where each B_{kk+1} is positive and invertible. Set

$$B_1 = (B_{12}, B_{13}, \cdots) \in \mathcal{L}(\mathcal{H}_1^{\perp}, \mathcal{H}_1)$$

and

$$B_2 = \begin{bmatrix} 0 & B_{23} & & \star \\ & 0 & B_{34} & \\ & & 0 & \ddots \\ 0 & & & \ddots \end{bmatrix} \in \mathcal{L}(\mathcal{H}_1^\perp, \mathcal{H}_1^\perp).$$

Let J_n be the $n \times n$ Jordan nilpotent. Then $J_n \in \mathcal{L}(\mathbf{C}^n) \cap (SI)$. Set

$$E = B_{12}^{-1}(B_{13} - J_n)$$

and

$$X_1 = \begin{bmatrix} 1 & E & 0 & \star \\ & 1 & 0 & \\ & & 1 & \ddots \\ 0 & & & \ddots \end{bmatrix} \in \mathcal{L}(\mathcal{H}_1^\perp, \mathcal{H}_1^\perp).$$

Set

$$X = \begin{bmatrix} 1 & 0 \\ 0 & X_1 \end{bmatrix} \in \mathcal{L}(\mathcal{H}).$$

Then X is invertible and

$$XBX^{-1} = \begin{bmatrix} 1 & 0 \\ 0 & X_1 \end{bmatrix} \begin{bmatrix} 0 & B_1 \\ 0 & B_2 \end{bmatrix} \begin{bmatrix} 1 & 0 \\ 0 & X_1^{-1} \end{bmatrix}$$

$$= \begin{bmatrix} 0 & B_1 X_1^{-1} \\ 0 & X_1 B_2 X_1^{-1} \end{bmatrix}$$

$$= \begin{bmatrix} 0 & B_{12} & J_n & \star \\ & 0 & B_{23} & \\ & & 0 & \ddots \\ 0 & & & \ddots \end{bmatrix}.$$

By Lemma 6.2.28, $XBX^{-1} \in (RI)$.

6.3 Application of Operator Structure Theorem

Theorem 6.3.1 *Let Ω be a connected open subset of $\mathbf{C}$ and $(\overline{\Omega})^0$ be finitely connected. Assume that $\{P_k(z)\}_{k=1}^m, z\in\Omega$ is a class of $M_n(\mathbf{C})$-valued holomorphic functions such that:*

   (i) $P_k(z)\in M_n(H^\infty(\Omega))$, $1\leq k\leq m$;

   (ii) $P_i(z)P_j(z) = \delta_{ij}P_j(z), z\in\Omega, 1\leq i,j\leq m$;

   (iii) $\sum_{k=1}^m P_k(z) = I_n$.

Then for fixed $z_0\in\Omega$, there exists an $M_n(\mathbf{C})$-valued holomorphic invertible function $X(z)\in M_n(H^\infty(\Omega))$ such that

$$X(z)P_k(z)X^{-1}(z) = P_k(z_0),$$

where $1\leq k\leq m$ and $X(z_0) = I_n$.

Proof Let $\mathcal{H} = L_a^2(\Omega^*)$ denote the Bergman space on Ω^* and B_z be the "multiplication by z" on $L_a^2(\Omega^*)$. Then $B_z^*\in\mathcal{B}_1(\Omega)$ and $\mathcal{A}'(B_z)\cong\mathcal{A}'(B_z^*)\cong H^\infty(\Omega)$. Set $A = B_z^*$ and $T = A^{(n)}$, then $\mathcal{A}'(T)$ is isometrically isomorphic to $M_n(H^\infty(\Omega))$. Thus there exist $P_k\in\mathcal{A}'(T)$ such that $P_k|_{ker(T-z)} = P_k(z)$ for $k : 1\leq k\leq m$,

$P_iP_j = \delta_{ij}P_i, 1\leq i,j\leq m$ and $\sum_{k=1}^m P_k = I_{\mathcal{H}^{(n)}}$.

Denote $\mathcal{H}_k = P_k\mathcal{H}^{(n)}$ $(1\leq k\leq m)$, then

$$\mathcal{H}^{(n)} = \mathcal{H}_1\dotplus\cdots\dotplus\mathcal{H}_n.$$

Denote $T_k = T|_{\mathcal{H}_k}$. Since $T^{(n)}$ has a finite (SI) decomposition up to similarity (Theorem 5.5.11), there exist invertible X_i such that

$$X_iT_iX_i^{-1} = A^{(n_i)}.$$

Set $X = X_1\dotplus\cdots\dotplus X_m$, then $X\in\mathcal{A}'(T)$ and

$$XP_kX^{-1} = 0_{\mathcal{H}_1}\oplus\cdots\oplus 0_{\mathcal{H}_{k-1}}\oplus I_{\mathcal{H}_k}\oplus 0_{\mathcal{H}_{k+1}}\oplus\cdots\oplus 0_{\mathcal{H}_m}.$$

Denote $Y(z) = X|_{ker(T-z)}$, then $Y(z)\in M_n(H^\infty(\Omega))$ and

$$Y(z)P_k(z)Y(z)^{-1} = 0_{ker(T_1-z)}\oplus\cdots\oplus I_{ker(T_k-z)}\oplus 0_{ker(T_{k+1}-z)}\oplus\cdots\oplus 0_{ker(T_m-z)}.$$

Set $X(z) = Y^{-1}(z_0)Y(z)$, then $X(z)\in M_n(H^\infty(\Omega))$ and

$$X(z)P_k(z)X^{-1}(z) = P_k(z_0)$$

and $X(z_0) = I_n, 1\leq k\leq m$.

In the following, we will discuss the winding number of some analytic functions.

Let D be the unit disk and f be an analytic function on $\overline{D}$. For $\alpha \in D$, the winding number of $f(e^{it})$ with respect to $f(\alpha)$ is given by

$$W(f, f(\alpha)) = \frac{1}{2\pi i} \int_0^1 \frac{f'(e^{it})}{f(e^{it}) - f(\alpha)} dt.$$

Lemma 6.3.2 *Given $f \in H^\infty$. If there is an $\alpha \in D$ such that the inner function $inn(f - f(\alpha))$ of $f - f(\alpha)$ is a finite Blachke product, then there exists a natural number n such that $T_f \sim T_g^{(n)}$, where $g \in H^\infty$ and $T_g \in (SI)$.*
Proof Denote $h = inn(f - f(\alpha))$, which is a finite Blachke product. By Corollary 2.1 of [Cowen, C.C. (1978)], there exists a finite Blachke product φ such that

$$\mathcal{A}'(T_f) = \mathcal{A}'(T_\varphi).$$

Note that T_φ is an isomorphic operator with $codimran T_\varphi = n < \infty$, thus

$$\mathcal{A}'(T_\varphi) \cong \mathcal{A}'(T_{z^n}) \cong M_n(H^\infty).$$

By Theorem 5.5.11, $\bigvee(\mathcal{A}'(T_f)) \cong \mathbf{N}$. If

$$codimran T_\varphi = 1,$$

then

$$\mathcal{A}'(T_\varphi) \cong \mathcal{A}'(T_z) \cong H^\infty.$$

Thus

$$T_f \in (SI).$$

If $codimran T_\varphi > 1$, by von-Neumann-Wold theorem, there exists a unitary operator U such that $UT_\varphi U^* = T_{z^n}$ and $\mathcal{A}'(UT_\varphi U^*) = \mathcal{A}'(T_z^{(n)}) \cong M_n(H^\infty)$. Let $T_z^{(n)}$ act on $(H^2)^{(n)}$ and P_i be the projection from $(H^2)^{(n)}$ to the i-th copy of subspace H^2, then $T_z^{(n)}|_{P_i(H^2)^{(n)}} = T_z$. Let $T_i = UT_f U^*|_{P_i(H^2)^{(n)}}$, then

$$UT_f U^* = \bigoplus_{i=1}^n T_i \quad \text{and} \quad T_i \in (SI).$$

Clearly, $T_i T_z = T_z T_i$ $(i = 1, 2, \cdots, n)$. Thus there exist $f_1, f_2, \cdots, f_n \in H^\infty$ such that $T_i = T_{f_i}$. This implies that

$$UT_f U^* = \bigoplus_{i=1}^{n} T_{f_i}, \quad T_i \in (SI).$$

Note that $\bigvee (\mathcal{A}'(T_f)) \cong \mathbf{N}$. By Theorem 4.2.1, $T_{f_i} \sim T_{f_1}$ $(i = 2, 3, \cdots, n)$. Thus $T_f \sim T_{f_1}^{(n)}$. Denote $f_1 = g$, the proof of the theorem is complete.

Theorem 6.3.3 *Let f be an analytic function on $\overline{D}$, if $T_f \notin (SI)$ and $W(f, f(\alpha_0))$*
$= p$, a prime number, for some $\alpha_0 \in D$, then for each $\alpha \in D$, $W(f, f(\alpha)) = kp$, where k is a natural number.
Proof Since $W(f, f(\alpha_0)) = p$ is a prime number, $inn(f - f(\alpha_0))$ is a finite Blachke product. Since $T_f \notin (SI)$, it follows from Lemma 6.3.2 that $T_f \sim T_g^{(n)}$ for some $g \in H^\infty$ and $T_g \in (SI)$. Thus

$$codim\, ran(T_f - f(\alpha_0)) = n\, codim\, ran(T_g - g(\alpha_0)) = p.$$

But p is a prime number, thus $codim\, ran(T_g - g(\alpha_0)) = 1$ and $n = p$. Therefore $W(f, f(\alpha)) = pW(g, g(\alpha)) = kp$ for all $\alpha \in D$.

6.4 Remark

Lemma 6.1.3 is given by [Fang, J.S.(2003)]. Theorem 6.1.1 is due to [Fang, J.S.(2003)], [Jiang, C.L. (1991)]. Section 6.2 is the work of [Jiang, C.L. and He, H. (2004)], [Jiang, C.L., Guo, X.Z. and Ji, K.]. Example 6.2.6 is given by [Fong, C.K. and Jiang, C.L. (1993)]. All the contents of Section 6.3 are given by [Fang, J.S.(2003)], [Jiang, C.L. (1994)].

6.5 Open Problems

1. Let $A \in \mathcal{L}(\mathcal{H})$. If A^2 is irreducible, does A have non-trivial invariant subspaces?
2. Give the necessary and sufficient conditions for $A \sim B$ if $A^{(2)} \sim B^{(2)}$, where $A, B \in \mathcal{L}(\mathcal{H})$.
3. Conjecture: Given $A \in \mathcal{L}(\mathcal{H})$, if $\sigma(A)$ is uncountable, then A is quasisimilar to an irreducible operator.

Bibliography

Admas, R.A.(1988). Sobolev space, Academic press, New York-San Francisco-London.

Aleman, A., Richter, S. and Sundberg, C.(1996). Bergling's Theorem for the Bergman space, *Acta Math. 177*, pp. 275-310.

Antonevich, A. and Krupmk, N. (2000). On trivial and non-trivial n-homogeneous C^*-algebras, *Integr. Equ. Oper. Theory, 38*, pp. 172-189.

Apostal, C., Bercobici, H., Foias, C. and Pearcy, C. (1985). Invariant subspaces, dilation theory, and the structure of the predual of a dual algebra, *J. Funct. Anal, 63*, pp. 369-404.

Apostal, C., Fialkow, L.A. Herrero, D.A. and Voiculescu, D. (1984). Approximation of Hilbert space space operator II, Research Notes in Math. 102, Longman, Harlow, Essex.

Apostal, C., Fioas, C. and Pearcy, C.M. (1979). That quasinilpotent operators are norm-limits of nilpotent operators revisited, *Proc. Amer. Math. Soc. 73*, pp. 61-64.

Apostal, C. and Voiculescu, D. (1974). On a problem of Halmos, *Rev. Roum. Math. Pures Apple. 19*, pp. 283-284.

Arveson, W.B. (1976). An invitation to C^*-algebras. Graduate Texts in Mathematics, Series 39, Springer-Verlag.

Aupetit, B. (1991). A primer on spectral theory, Springer-Verlag, Berlin.

Azoff, E.A., Fong, C.K. and Gilfeather., F. (1976). A reduction theory for non-self-adjoint operator algebras, *Trans. Amer. Math. Soc. 224*, pp. 351-366.

Baker, I.N., Deddens, J.A. and Uliman, J.L. (1974). A theorem on intire functions with application to Toeplitz operators, *Duke. Math. J.41*, pp. 736-745.

Ball, J.A. (1975). Hardy space expectation oprators and reducing subspaces, *Proc. Amer. Math. Soc. 47*, pp. 351-357.

Beauzamy, B. (1988). Introduction to operator theory and invariant subspaces, Elsevier Science Publications, North-Holland.

Bercovici, H. (1987). Three test problems for quasisimilarity, *Canad. J. Math. 39*, pp. 880-892.

Bercovici, H. (1988). Operator theory and Arithmetic in H^∞, Amer. Math. Soc. Prov. RI.

Blanckdar, B. (1986). K-theory for operator algebras, Springer-Verlag, Heidelberg.

Bonsall, F.F. and Duncan, J. (1973). Complex normal algebra, Springer-Verlag, Heidelberg.

Brown, L.G., Douglas, R.G and Fillmore, P.A. (1977). Unitary equivalence modulo the compact operators and extensions of C^*-algebras, *Ann. Math. 105*, pp. 265-324.

Antonevich, A. and Krupmk, N. (2000). On trivial and non-trivial n-homogeneous C^*-algebras, *Integr. Equ. Oper. Theory, 38*, pp. 172-189.

Cao, Y., Fang, J.S. and Jiang, C.L.(2002). K-groups of Banach algebras and Decomposition of strongly irreducible operators, *J. Oper. Theory 48*, pp. 235-253.

Choi, M.D., Lausee, C. and Radjavi, H. (1981). On commutators and invariant subspaces, *Linear Multilinear algebra 9*, pp. 329-340.

Conway, J.B. (1978). Functions of one complex variable, Grad. Texts in math. 11, New York-Heidelberg-Berlin:Springer-Verlag.

Conway, J.B. (1990). A course in functional analysis, Grad. Texts in Math. 95, New York-Heidelberg-Berlin:Springer-Verlag.

Conway, J.B. (1995). Functions of one complex variable II, Grad. Texts in Math. 95, New York-Heidelberg-Berlin:Springer-Verlag.

Conway, J.B. and Gillespie, T.A. (1985). Is a self-adjoint operator determined by its invariant subspace lattice? *J. Funct. Anal. 64*, pp. 178-189.

Conway, J.B., Herrero, D.A. and Morrel, B.B. (1989). Completing the Riesz-Dundord functional calculus, Memories of AMS 82.

Cowen, C.C. (1978). The commutant of an analytic Toeplitz operator, *Trans. Amer. Math. Soc. 239*, pp. 1-31.

Cowen, M.J. and Douglas, R. (1977). Complex geometry and operator theory, *Bull. Amer. Math. Soc. 83*, pp. 131-133.

Cowen, M.J. and Douglas, R. (1978). Complex geometry and operator theory, *Acta. Math. 141*, pp. 187-261.

Cowen, M.J. and Douglas, R. (1980). Operators possessing an open set of eigenvalues, Functions, Series, Operators, *Colloquia Mathematical Societies Janos Bolyai, 35*, pp. 323-341.

Cuckovic, Z. (1994). Commutant of Toeplitz operators on the Bergman spaces, *Pacifics J. Math. 162*, pp. 277-285.

Curto, R.E. and Fialkow, L.A. (1989). Similarity, quasisimilarity and operator factorizations, *Trans. Amer. Math. Soc., 314(1)*, pp. 225-254.

Dadarlat, M. and Gong, G. (1997). A classification theorem of AH algebras of real rank zero, *Geometric and functional analysis, 7, No.4*, pp. 646-711.

Davidson, K.R. (1988). Nest algebras, Research Notes in Mathematics 191, Longman-Harlow-Essex.

Davidson, K.R. and Herrero, D.A. (1990). The Jordan form of bitriangular operator, *J. Funct. Anal. 94*, pp. 27-73.

Deddens, J.A. and Wong, T.K. (1973). The commutant of analytic of Toeplitz operators, *Trans. Amer. Math. Soc. 184*, pp. 261-273.

Dixmier, J. and Foias, C. (1970). Surle spectre ponctuel d'nn operator, Colloq. Math. J. Bolyai. 5 Hhilbert space operators and operator algebras, Tihang(Hungary).

Douglas, R.G. (1972). Banach algebras techniques in operator theory, New York and London: Academic press.

Duren, P.L. (1970). Theory of H^p space, Academic press, New York.

Elliott, G. and Gong, G. (1996). On the classification of C^*-algebras of real rank Zero II., *Annals of Mathematics, 144*, pp. 497-610.

Elliott, G., Gong, G. and Li, L. On the classification of simple inductive limit C^*-algebras., II: The Isomorphism Theorem., Preprint.

Fang, J.S.(2003). K-theory and operator theory, Doctorate dissertation.

Fang, J.S. and Jiang, C.L. (1999). Strongly irreducible operators on Hilbert space, *Annal. Math. (Chinese) 20A:6*, pp. 707-714.

Fang, J.S., Jiang, C.L. and Wu, P.Y. (2003). Direct Sum of irreducible operators., *Studia Math., 155(1)*, pp. 37-49.

Feldman, N.S. (1999). Pointwise multipliers from the Hardy space to the Bergman space, *Illinois J. Math. 43*, pp. 211-220.

Fialkow, L.A. (1981). A note on the range of the operator $X \mapsto AX - XB$, *Illinois J. Math. 25*, pp. 112-124.

Fisher, S.D. (1983). Functional theory on planar domains, Wiley-interscience Series, John Wiley and Sons, New York.

Fong, C.K. and Jiang, C.L. (1993). Approximation by Jordan type operators., *Houston. J. Math., 19*, pp. 51-62.

Fong, C.K. and Jiang, C.L. (1994). Normal Operators similar to irreducible operators., *Acta Math. Sinica. New series 10*, pp. 192-235.

Gardner, B.J. (1989). Radical theory, Pitman Research Notes in Mathematics Series 198, New York.

Gardner, B.J. and Wiegandt, R. (2004). Radical theory of rings, Pure and Applied Mathematics, Marcel Dekker 261. New York, xii, 387.

Garnett, J.B. (1981). Bounded analytic functions, Academic press, New York.

Gilfeather, F. (1972). Strong reducibility of operators, *Indiana Univ. Math. J. 22*, pp. 393-397.

Gray, L.J. (1977). Jordan representation for a class of nilpotent operators, *Indiana Univ. Math. J. 26*, pp. 57-64.

Griffiths, P. (1985). Algebraic curve, Peking University Press, Beijing.

Halmos, P.R. (1968). Irreducible operators, *Mich. Math. J. 15*, pp. 215-223.

He, H. and Ji, K. K-theory and similarity classification of operators II, J. Math. Illinous, to appear.

Hedenmalm, H., Korenblum, B. and Zhu, K.H. (2000). Theory of Bergman spaces, Springer -Verlag, New York.

Herrero, D.A. (1974). Normal limits of nilpotent operators, *Indiana Univ. Math. J.23*, pp. 1097-1108.

Herrero, D.A. (1979). Quasisimilar operators with different spectra, *Acta Sci. Math.(Szeged) 41*, pp. 101-118.

Herrero, D.A. (1984). Compact perturbation of nest algebras, index obstructions and a problem of Arveson, *J. Funct. Anal. 55*, pp. 78-119.

Herrero, D.A. (1987). Spectral pictures of operators in the Cowen-Douglas class

$\mathcal{B}_n(\Omega)$ and its closure, *J. Operator Theory 10*, pp. 213-222.

Herrero, D.A. (1990). Approximation of Hilbert space operators I, 2nd ed., Research Notes in Math. 224, Longman, Harlow, Essex.

Herrero, D.A. and Jiang, C.L. (1990). Limits of strongly irreducible operators and the Riesz decomposition theorem, *Mich. Math. J.37*, pp. 283-291.

Herrero, D.A., Taylor, T.J. and Wang, Z.Y. (1988). Variation of the point spectrum under compact perturbations, Operator theory: Advances and Applications, 32.

Hoover, T.B. (1970). Quasisimilarity and hyperinvariant subspaces, Dissertation, Univ. of Michigan, Ann Arbor, Mich.

Ji, Y.Q. and Jiang, C.L. (2002). Small compact perturbation of strongly irreducible operators, *Integr. Equ. Oper. Theory 43*, pp. 417-449.

Ji, Y.Q., Jiang, C.L. and Wang, Z.Y. (1997). The unitary orbit of strongly irreducible operators in the nest algebra with well-ordered set., *Mich, Math. J.44*, pp. 85-98.

Ji, Y.Q., Jiang, C.L. and Wang, Z.Y. (1996). Essentially normal+small compact=strongly irreducible, *Chinese Ann. Math. 24*, pp. 41-72.

Ji, Y.Q., Li, J.X. and Sun, S.L. (2003). The essential spectrum and Banach reducibility of operator weighted shifts, *Acta Math. Sinca, Vol 46, No. 3*, pp. 413-424.

Ji, Y.Q. and Yang, Y.H. (2003). A class of operators of strongly irreducible decomposition unique with respect to similarity, *Journal of Jilin Univ., No. 3*, pp. 193-201.

Jiang, C.L. (1991). Strongly irreducible operator and Cowen-Douglas operators, *Northeast Math. J. 7(1)*, pp. 1-3.

Jiang, C.L. (1994). Similarity, reducibility and approximation of the Cowen-Douglas operators, *J. Operator Theory, 32(1)*, pp. 77-89.

Jiang, C.L. (2004). Similarity classification of Cowen-Douglas operators, *Cann. J. Math., 56*, pp. 742-775.

Jiang, C.L. The commutant of strongly irreducible operators, preprint.

Jiang, C.L., Guo, X.Z. and Ji, K. K-group and similarity classification of operators I, J. Funct. Anal., to appear.

Jiang, C.L., Guo, X.Z. and Yang, Y.F. (2001). Nilpotent operator similar to irreducible operators, *Science in China Ser. A vol 31*, pp. 673-686.

Jiang, C.L. and He, H. (2004). Quasisimilarity of Cowen-Douglas operators, *Science in China Ser. A vol47, No. 2*, pp. 297-310.

Jiang, C.L., Jin, Y.F. and Wang, Z.Y. About operator weighted shift, Acta Math. Sinca, to appear.

Jiang, C.L. and Li, J.X. (2000). The irreducible decomposition of Cowen-Douglas operators and operator weighted shifts, *Acta Sci. Math. (Szeged)66*, pp. 679-695.

Jiang, C.L., Sun, S.L. and Wang, Z.Y. (1997). Essentially normal+compact operator=strongly irreducible operator, *Tran. Amer. Math. Soc. 349*, pp. 217-233.

Jiang, C.L. and Wang, Z.Y. (1996a). A class of strongly irreducible operators with nice property, *J. Operator Theory, 36*, pp. 3-19.

Jiang, C.L. and Wang, Z.Y. (1996b). The spectral picture and the closure of similarity orbit of strongly irreducible operators., *Integral Equation and Operator Theory, 24*, pp. 81-105.

Jiang, C.L. and Wang, Z.Y. (1998). Strongly irreducible operators on Hilbert space., π-Pitman Research Notes in Mathematics Series, 389, Longman, Harlow.

Jiang, C.L. and Wu, P.Y. (1998). Sums of strongly irreducible operators., *Houston Journal of Mathematics, 24*, pp. 467-481.

Jiang, Z.J. (1979). Topics in operator Theory, Seminar Reports in Functional Analysis., Jilin University(Chinese).

Jiang, Z.J. (1981). On the structure of linear operators(Chinese), The report on the national symposium on operator theory, Jiujiang.

Jiang, Z.J. and Sun, S.L. (1992). On completely irreducible operators(Chinese), *Acta. Sci. Natur. Univ. Jilin (4)*, pp. 20-29.

Jin, Y.F. and Wang, Z.Y.(1). About the commutant of Cowen-Douglas operators, Acta Math. Sinica, to appear.

Jin, Y.F. and Wang, Z.Y.(2). The unilateral weighted shift with zero weights, to appear.

Kadison, R.V. and Singer, I.M. (1957). Three test problem in operator theory, *Pacific J. Math. 7*, pp. 1101-1106.

Kato, T. (1984). Perturbation theory for linear operator, Berlin, Heidelberg, New York, Tokyo: Springer-Verlag.

Lambert, A (1971). Strictly cyclic operator algebra, *Pacific J. Math. 38*, pp. 717-766.

Liu, Y.Q. and Wang, Z.Y. (2004). The commutant of multiplication operators on Sobolev disk algebra, *Journal of Analysis and Applications, 2*, pp. 65-86.

Liu, Y.Q. and Wang, Z.Y. Invariant subspaces of Soblov disk algebra, to appear.

Lomonosov, V.I. (1973). Invariant subspace for family of operators which commute with a completely continuous operators, *Functional Anal. Appl. 7*, pp. 213-214.

Putnam, I. (1989). The C^*-algebras associated with minimal homeomorphisms of Cantor set, *Pacific J. Math. 146*, pp. 329-353.

Ong, S.C. (1987). What kind of operators have few invariant subspaces? *Linear Algebra Appl. 95*, pp. 181-185.

Pearcy, C. and Petrovic, S. (1994). On polynomially bounded weighted shifts. *Houston Journal of Mathematics, 20*, pp. 27-45.

Radjaval, H. and Rosenthal, P. (1973). Invariant subspaces, Springer-Verlag.

Richter, S. (1987). Invariant subspaces in Banach spaces of analytic functions, *Trans. Amer. Math. Soc. 304*, pp. 585-618.

Robati, B.K. and Vaezpour, S.M. (2001). On the commutant of operators of multiplication by univalent functions, *Proc. Amer. Math. Soc. 129*, pp. 2379-2383.

Rosenblum, M. and Rovnyak, J. (1985). Hardy classes and operator theory, Oxford University Press.

Rudin, W. (1974). Real and complex analysis, McGraw-Hill Book Company, New York.

Seddighi, K. (1983). Essential spectra of operators in the class $\mathcal{B}_n(\Omega)$, Proc. Amer. Math. Soc.

Shields, A.L. (1974). Weighted shift operators and analytic function theory, *Math. Surveys, Vol. 13, Amer. Math. Soc. Prov. RI*, pp. 49-128.

Shields, A.L. and Wallen, L.J. (1971). The commutant of certain Hilbert space operators, *Indiana Univ. Math. J. 20*,pp. 777-788.

Stessin, M. and Zhu, K. (2003). Generalized Factorization in Hardy spaces and the commutant of Toeplitz operators, *Canada J. Math. 55*, pp. 379-400.

Suarez, D. (1996). Backward shift invariant spaces in H^2, *Ind. Univ. Math. J. 46*, pp. 593-618.

Sunder, V.S. (1998). Functional analysis:Spectral theory, Basel.

Takesaki, M. (1979). Theory of operator algebras I, Springer, New York.

Taylor, J. (1975). Banach algebras and topology, Algebras in Analysis, ed. J.H. Williamson, Academic Press.

Thomson, J.E. (1993). Commutants of analytic Toeplitz operators, *Proc. Amer. Math. Soc. 117*, pp. 1023-1030.

Tolokommokov, V. (1993). Stable rank of H^∞ in multiply connected domains, *Proc. Amer. Math. Soc. 117*, pp. 1023-1030.

Turner, T.R. (1971). Double commutant of singly generated operator algebras thesis, University of Michigan.

Voiculescu, D. (1976). A non-commutative Weyl-von Neumann theorem, *Rev. Roum. Math. Pure et appl. 21*, pp. 97-113.

Voiculescu, D. (1978). Some results on norm-ideal perturbations of Hilbert space operators, Institutul de Mathematics, Bucharest.

Wang, Z.Y. (1993). The multiplication operators on Sobolev space, *the Dissertation in the first Chinese postdoctoral science symposium*, pp. 1167-1170.

Wang, Z.Y. and Liu, Y.Q. Multiplication operators on Sobolev disk algebra, to appear.

Wang, Z.Y. and Xue, Y.F. (2000). On the unique (SI) decomposition of the Jordan operator $\bigoplus_{i=1}^{n} S(\theta)$ for certain inner function θ, *Integr. Equa. Oper. theory, vol 36, No. 3*, pp. 370-377.

Wegge-Olsen, N.E. (1993). K-theory and C^*-algebras, Oxford Univ. Press, Oxford.

Xu, X.M. (1999). Composite operator theory, Science Press, Beijing.

Xue, Y.F. (1989). Some properties of the C^*-algebra generated by the $\mathcal{B}_n(\Omega)$-class operator(Chinese), *Chinese Anna. Math. Ser. A, 10(4)*, pp. 441-446.

Yan, C.Q. (1993). Irreducible decomposition of Cowen-Douglas operator, *Northeast Math. J. 42*, pp. 261-267.

Yu, C.L. (1987). A completely unitary invariant and reducibility of a class of operators in $\mathcal{B}_2(\Omega)$, *Northeast Math. J.3*, pp. 410-417.

Yue, H. (2002). Every operator can be written as the sums of two strongly irreducible operators, *Mathematics in practice and theory, 24*, pp. 20-26.

Index